GRAPHENE - NEW TRENDS AND DEVELOPMENTS

Edited by **Farzad Ebrahimi**

Graphene - New Trends and Developments
http://dx.doi.org/10.5772/59823
Edited by Farzad Ebrahimi

Contributors

Mineo Hiramatsu, Hiroki Kondo, Masaru Hori, Miroslav Pardy, Pavel Vyacheslavovich Ratnikov, Andrei Silin, Hao-Li Zhang, Hang-Xing Wang, Farzad Ebrahimi, Ashok K Sood

Published by InTech

Janeza Trdine 9, 51000 Rijeka, Croatia

Publishing Process Manager Iva Lipovic

Technical Editor InTech DTP team

Cover InTech Design team

First published November, 2015

Printed in Croatia

Additional hard copies can be obtained from orders@intechopen.com

Graphene - New Trends and Developments, Edited by Farzad Ebrahimi

p. cm.

ISBN 978-953-51-2220-3

For EU product safety concerns:
IN TECH d.o.o., Prolaz Marije Krucifikse Kozulić 3, 51000 Rijeka, Croatia,
info@intechopen.com or visit our website at intechopen.com.

Meet the editor

Dr. Farzad Ebrahimi was born in Qazvin, Iran, in 1979. He graduated in mechanical engineering from the University of Tehran, Iran, in 2002. He received his MSc and PhD in mechanical engineering, with a specialization in applied design, from the University of Tehran, Iran, in 2009. Since 2002, he has been working at the "Smart Materials and Structures Lab" Research Center of the Faculty of Mechanical Engineering at the University of Tehran, where he is a researcher of smart functionally graded materials and structures. He joined the department of mechanical engineering of Imam Khomeini International University as an assistant professor in 2010.Dr. Ebrahimi is involved in several international journals as editor and reviewer and serves on the editorial board of the SAGE publication *Journal of Advances in Mechanical Engineering*. He is the author of the books *Smart Functionally Graded Plates* and *Progress in Analysis of Functionally Graded Structures,* published by Nova Science Publishers, NY. He also served as the editor of the book *Nanocomposites - New Trends and Developments*, published by In-Tech - Open Access Publisher. His research interests focus on the areas of smart materials and structures, nanostructures, vibration nanocomposites, composite materials and structures, and he has published several researches in these fields. His research in these areas has been presented at international conferences and appeared in academic journals such as *Composite Structures, Composites Part B, Journal of Mechanical Science and Technology, Smart Materials and Structures, European Journal of Mechanics*, and *Archive of Applied Mechanics*. Dr. Ebrahimi also has strong collaboration with Iranian industries on gas and oil projects, and he serves as an ad hoc referee in several top academic journals.

Contents

Preface

Graphene is one of the most promising materials in the world and has attracted much attention on the global scale. This Nobel Prize-winning material combines multiple outstanding physical and chemical characteristics such as extremely high mechanical strength, hardness, and adjustable thermal and electrical conductivity, as well as excellent surface and optical features through chemical marking. Besides, graphene is the strongest and the most stretchable material ever known and can be an efficient substitute for silicon. As a result, graphene has exhibited appealing application potentials in various areas that cover a range of different fields in human life, and its extraordinary properties have received a great deal of attention by many researchers.

This book is a result of contributions of experts from the international scientific community working on different aspects of graphene science and applications and reports on the state-of-the-art research and development findings on graphene through original and innovative research studies. Through its seven chapters, the reader will have access to works related to the theory and characterization of various planar heterostructures and nanoplatforms based on graphene and also the Compton effect in graphene and in the graphene-like dielectric medium, while it introduces photoactive graphene from functionalization to applications and also the modeling and control of a smart single-layer graphene sheet. Besides, it presents reviews on the modeling, synthesis, and properties of graphene and graphene technology and its applications in electronic devices.

The authors of each chapter give a unique insight about the specific intense research area of graphene. The book is addressed not only to researchers but also to professional engineers, students, and other experts in a variety of disciplines, both academic and industrial, seeking to gain a better understanding of what has been done in the field recently and what kind of open problems exist in this area.

Dr. Farzad Ebrahimi
Department of Mechanical Engineering
Faculty of Engineering Imam Khomeini International University,
Qazvin, I.R. Iran

Chapter 1

A Review on Modeling, Synthesis, and Properties of Graphene

Farzad Ebrahimi and Ebrahim Heidari

Additional information is available at the end of the chapter

http://dx.doi.org/10.5772/61564

Abstract

Every few years, a new material with unique properties emerges and fascinates the scientific community, typical recent examples being high-temperature superconductors and carbon nanotubes. Graphene is the latest sensation with unusual properties, such as half-integer quantum Hall effect and ballistic electron transport. In this work, we introduce graphene and its property such as mechanical, thermal, and electrical properties. Furthermore, we briefly mention some of processes that are used for production of graphene. We will continue our discussion to review the approaches that have arose to analyze graphene and other nanoplates.

Keywords: Graphene, SLGSs, Mechanical, thermal and electrical properties

1. Introduction

Graphene is the name given to a two-dimensional sheet of sp^2-hybridized carbon. Its extended honeycomb network is the basic building block of other important allotropes; it can be stacked to form 3D graphite, rolled to form 1D nanotubes, and wrapped to form 0D fullerenes. Graphene is the name given to a two-dimensional sheet of sp^2-hybridized carbon [1]. Graphene is a wonder material with many superlatives to its name. It is the thinnest known material in the universe and the strongest ever measured. Its charge carriers' exhibit giant intrinsic mobility, have zero effective mass, and can travel for micrometers without scattering at room temperature [2]. Synthesis and characterization of graphenes pose challenges, but there has been considerable progress in the last year or so [3]. A great deal of research has been conducted to explore the promising properties of the single-layered graphene sheets (SLGSs) after the appearance of the new method of graphene sheet preparation. Stankovich et al. have proposed their findings related to the synthesis and exfoliation of isocyanate-treated graphene oxide nanoplatelets [4]. Implementing the chemical reduction, they have also been able to produce the graphene-based nanosheets [5]. In addition, Ferrari has reported the Raman spectroscopy of the SLGS [6]. Furthermore, Katsnelson and Novoselov have explored the unique electronic properties of the SLGSs [7]. They have stated that the graphene sheet is an unexpected bridge between condensed matter physics and quantum electrodynamics. On the other hand, Bunch et al. have reported the experimental results of using electromechanical resonators made from suspended single- and multi-layered graphene sheets [8]. The superior mechanical, chemical, and electronic properties of nanostructures make them favorable for nano engineering applications [9]. Graphene sheets are one of the most important nano-sized structural elements which are commonly used as components in micro-electro-mechanical systems (MEMS) and nano-electro-mechanical systems (NEMS) [8, 10]. Furthermore, it has been revealed that adding graphene sheets to polymer matrix could greatly improve the mechanical properties of the host polymer [11]. In addition, nanostructures such as armchair carbon nanotubes and nanoplates have shown significant potential applications in the field of environmental technologies [12]. Nano-mechanical resonators are one of most important NEMS devices which have received increasing attention from the scientific community in recent years [13-16]. The nano mechanical resonators may operate at very high frequencies up to gigahertz range [17]. The potential applications of the SLGSs as mass sensors and atomistic dust detectors have further been investigated [10]. Also, the promising usage of the SLGS as strain sensor has been examined [18].

Since graphene has a prominent application in human's life, the necessity of mechanic analytical approach for graphene is drastically felt. There are many approaches to analyze a graphene and other nonoplates mechanically, however, they can be divided into two bunches: first, the methods that consider graphene or other nonoplates downright, and second, the methods that consider interactions between graphene and other nonoplates with their surrounding environment.

The first cluster of approaches is often the Molecular Dynamics (MD) method. It is very powerful method has furthered scientists in their case studies. Sheehan et al. utilized the molecular dynamics methodology for analyzing the effect of solvents on reaction kinetics and post reaction separation is presented [19]. Kresse et al. used ab initio molecular dynamics to predict the wave functions for new ionic positions using sub-space alignment [20].With the ability to examine atomic-scale dynamics in great detail, researchers have used MD to gain new insight into problems that have been resistant to theoretical solution, such as solid fracture [21], surface friction [22], and plasticity [23]. For example, a 10-nm cubic domain of a metal can be simulated only for times less than around 10^{-10} s, even on very large parallel machines [22]. Increases in this simulation time require a proportional reduction in the number of atoms simulated. The results of such a simulation therefore can rarely be compared directly to experiments, since laboratory observations of these sorts of mechanical phenomena are usually made on much larger length and time scales. One possible approach that can be applied to many problems is to use MD only in localized regions in which the atomic-scale dynamics are important and a continuum simulation method (such as finite elements) everywhere else. This general approach has been taken by several different groups of researchers. Abraham et al. [24,

25] have developed a coupled finite element, MD, tight-binding (FE/MD/TB) method in which the three methods are used concurrently in different regions of the computational domain. Another method developed recently is the quasi continuum method [26-31], in which atomic degrees of freedom are selectively removed from the problem by interpolating from a subset of representative atoms, similar to finite element interpolation in which atomic degrees of freedom are selectively removed from the problem by interpolating from a subset of representative atoms, similar to finite element interpolation. Finally Wagner et al. [32] submitted a professional coupling of atomistic and continuum simulation method that is called bridging-scale method. In this review, we are going to represent some properties of graphene and study briefly the mechanic analytical approaches that we mentioned before.

2. Structure, synthesis, and properties

Graphene has a honeycomb network that could have ripple in the surface. Ripples can induce the local electrical and optical properties of graphene. Three different types of graphenes can be defined: single-layer graphene (SG), bilayer graphene (BG), and few-layer graphene (FG, number of layers).

Typically, the important properties of graphene are a quantum Hall effect at room temperature, an ambipolar electric field effect along with ballistic conduction of charge carriers, tunable band gap, and high elasticity. Although graphene is expected to be perfectly flat, ripples occur because of thermal fluctuations. Ideally, graphene is a single-layer material, but graphene samples with two or more layers are being investigated with equal interest.

There are now four primary ways to produce 'pristine' graphene:

a. Epitaxial graphene: This method involves chemical vapor deposition (CVD growth) on epitaxially matched metal surfaces

b. Micromechanical Exfoliation or micromechanical cleavage: in which highly oriented pyrolitic graphite (HOPG) is pealed using Scotch-tape and deposited on to a silicon substrate.

c. Exfoliation of graphite in solvents: Gram quantities of single-layer graphene have been prepared by employing a solvothermal procedure and subsequent by sonication. In this process, the solvothermal product of sodium and ethanol is subjected to low-temperature flash pyrolysis yielding a fused array of graphene sheets, which are dispersed by mild sonication. A single-layer graphene can be produced in good yields by the solution-phase exfoliation of graphite in an organic solvent, such as N-methylpyrrolidone (NMP). This process works because the energy required to exfoliate graphene is balanced by the solvent–graphene interaction.

d. "Other methods,": such as

- Substrate-free gas-phase synthesis of graphene platelets in a microwave plasma reactor
- Arc discharge synthesis of multi-layered graphene

- Graphene can be grown on metal surfaces by surface segregation of carbon or by decomposition of hydrocarbons.
- etc.

2.1. Mechanical properties

Pristine graphene structures are found in 2D plane sheets. It has a hexagonal crystal lattice which resulted in covalent bonds between carbon atoms. In the environment, graphene are discovered in tow forms: "monolayer" and "free- standing." With the first form, we find graphene parts as a cover over a substrate material such as SiC. However, we are able to find graphene individually and independent from other materials in the environment which corresponds with the second form, "free-standing graphene" [33].

Mechanical properties for any crystal material are affected by pristine Lattice and defects are comprised of dislocations and grain boundaries [34, 35]. For example, we can mention elastic properties of materials that are affected by atoms interactions and lattice geometry, whereas strength and plastic flow stress as another properties of materials are affected by characteristics of defects. Indeed caused defects in the material severely decrease strength of it in comparison of ideal material. Anyway, we are always not able to impede the existence of defect and its effect in the materials. However, there is one exception; nano-materials can be discovered defect-free initially, and this is the main reason of superiority of strength for these materials [36]. Graphene as a nano-material is not excepted in this issue.

Lee and his co-workers performed the pioneer empirical analysis of elastic properties and strength of pristine graphene [37]. A deposited graphene membrane onto a substrate material that possesses some cavities on the surface is loaded by the tip of the atomic force microscope (Fig. 1) [37], and it was discovered that graphene brings out both nonlinear elastic behavior and brittle fracture. Thus, for nonlinear elastic behavior, we can write: $\sigma = E\varepsilon + D\varepsilon^2$, where σ is the applied stress (the symmetric second Piola Kirchhoff stress), E is the Young modulus, ε is the elastic strain (uniaxial Lagrangian strain), and D is the third-order elastic stiffness. This experiment is convoyed by this result as follows: Young modulus of $E = 1.0$ TPa, and the third-order elastic stiffness of $D = -2.0$ TPa. The Young modulus they found is very close to the Young modulus of nanotubes. They also found that brittle fracture happens at an intrinsic stress as much as $\sigma_{int} = 130GPA$, which is very huge and magnitude.

Simulating by computer [38] shows $E = 1.05$ TPa and $\sigma_{int} = 110\ GPa$ for Young modulus and brittle fracture of graphene, which is compatible with explorations of Lee and his co-workers. All of these explorations prove that graphene can be very useful for structural applications and for the cases that we are dependent on high strength. Furthermore, graphene is flexible and can be bent easily, which make it more desirable and attractive.

The propagation of crack in monolayer graphene has been studied empirically and also analytically (molecular dynamics) by considering Crystallographic characteristics [39]. It has been evident that the sources of cracks in monolayer graphene membranes are unavoidable, mechanically applied stresses that are exerted during their processing. Cracks or tears

http://dx.doi.org/10.5772/61564

propagate along the sides of the hexagonal crystal lattice and defray an occasional direction change as much as 30° in vertices of hexagonal.

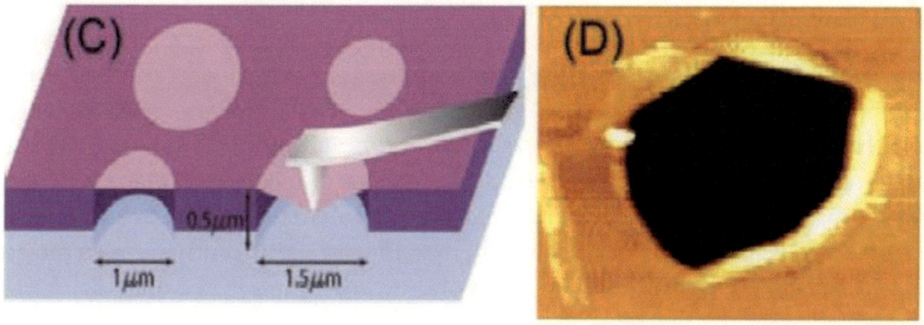

Figure 1. Mechanical testing of graphene. Schematic of nanoindentation on suspended graphene membrane (left figure). Atomic force microscope image of a fractured graphene membrane (right figure).

It is sometimes seen that propagation can go under the TEM electron beam [40]. Kim et al. [39] used this assumption that the propagation of crack is motivated by incorporating the effects of stress concentrations at the tip of the crack and the ionization influence of electron beam. Under the simultaneous effect of these problems, atomic bonds break in the vicinities of a crack tip, and propagation takes place.

Novoselov et al. [40] performed a computer simulation on the exfoliation of graphene sheets from adhesive substrate to examine crystallographic features of cracks which happens. The idea of this simulation developed from this fact that graphene ribbons became tapered as they were produced by exfoliation process (Fig. 2). They found that tear angle is affected by adhesive strength. Their simulation showed that, with the low strength adhesive, tearing takes place in the conqueror armchair direction of the hexagonal crystal lattice of graphene, and meanwhile, occasional change in direction is observed rarely. They also concluded that any increase in adhesive strength results in more tear angel; hence, in pretty high adhesive strength a change of 60° in the direction of tearing will present.

As a material, graphene is not excluded from defects. The defects that graphene may obsess are: vacancies [41], Stone Wales defects [41], dislocations [42], and grand boundaries of GBs. Among these defects, dislocations and GBs are very common and play a prominent role in mechanical properties of graphene. For instance, dislocations can cause plastic flow in graphene, whereas GBs decrease its strength characteristics [44, 45]. Dislocations can also violate translational symmetry of graphene [46].

There is no getting around GBs in graphene specimens when it is produced in a large area; thus, their effects on mechanical properties have been always noticeable from both fundamental and applied sides. Empirical data [44] have proved that free-standing polycrystalline graphene under concentrated load has severe low failure strength when found in polycrystalline state than when produced as a single crystal. In this experiment, they used a tip of atomic force microscopy (AFM) to exert a concentrated load on a polycrystalline graphene membrane, and it was observed that the force needed to cause a tear in the graphene along GBs is an amount of 100 nN [44], this is while the force for tearing a single-crystal exfoliated graphene is not more than 1.7 mN [37].

Vargas et al. [45] did similar test using atomic force microscopy and molecular dynamics simulations to study the mechanical characteristics of a graphene with polycrystalline structure that is obtained by chemical vapor deposition. They used nanoindentation measurements and found that out-of-plane ripples effectively decrease in-plane stiffness in the mentioned graphene. They also found that GBs effectively decrease the breaking strength of the graphene. Molecular dynamics showed them voids can significantly weaken the graphene membranes. In fact, GB is a place where amorphous carbon and iron oxide nanoparticles are absorbed [45].

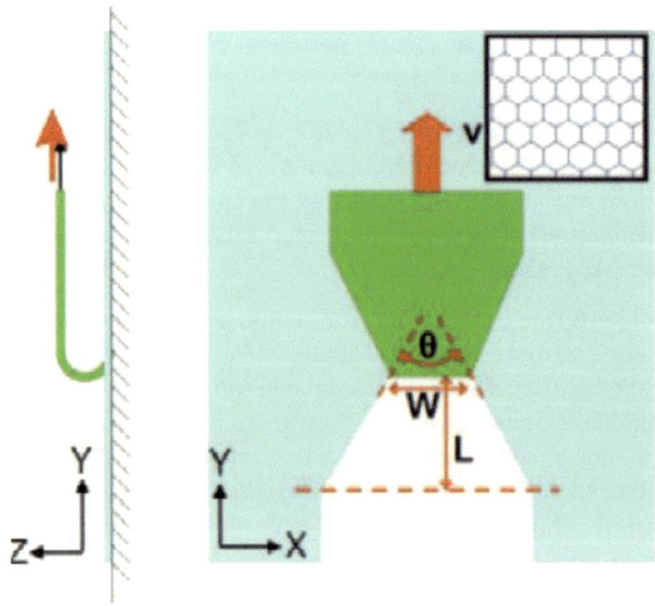

Reprinted with permission from D. Sen, K.S. Novoselov, P. Reis, M.J. Buehler, Small, Volume 6, 1108,2010.

Figure 2. Schematic diagram of the setup for the tearing studies of graphene: side and top views. The inset shows the sheet orientation. An initial flap of 8 nm in width is cut in the sheet, folded back, and moved at a constant speed.

2.2. Thermal Properties

The heat flow direction in a two dimensional graphene can be divided into in-plane and out-of-plane directions. In-plane heat flow is greater than out-of-plane one and is developed due to covalent sp^2 bonding between carbon atoms, while the latter is emanated from weak van der Waals coupling.

Graphene transistors and interconnectors are beneficiaries of in-plane heat flow depending on a certain channel length. Although thermal coupling with substrate materials is miserly, it is a prominent reason for the dissipation of heat flow. We can regulate the heat flow by phonon scattering, edges, or interfaces. Ultimately, the unusual thermal properties of graphene stem from its 2D nature, forming a rich playground for new discoveries of heat-flow physics and potentially leading to novel thermal management applications.

By reviewing thermal properties, with no respect to material, one that should be considered is the specific heat. This is a quantity that implies two things: first, the thermal energy that a body is capable to store, and second, the rate of cooling and heating that a body will experiment. The later can be modeled by the thermal time constant $\tau \approx$ RCV, where τ is the thermal time constant, R is the thermal resistance for heat dissipation (the inverse of conductance, $R=1/G$) and V is the volume of the body. Thermal time constants can be varied from 0.1 ns for a single graphene sheet or carbon nanotube (CNT) and 10 ns for nanoscale transistors to 1 ps for the relaxation of individual phonon modes [47 -49].

The specific heat of graphene is divided into phonons (lattice vibrations) and free electrons contributions, $C=C_p+C_e$, Knowing that phonon contributions are dominant [50 -52]. Phonon specific heat as a dominant coefficient becomes constant at very high temperature near in-plane Debye temperature $\Theta_D \approx 2100$ K, At this time, we have $C_p=3N_Ak_B \approx 25$ J mol^{-1}K^{-1} ≈ 2.1 J g^{-1}K^{-1}, also known as the Dulong–Petit limit, where N_A is the Avogadro' number and k_B is the Boltzmann constant. This property belongs in a classical solid behavior when six atomic degrees of motion are entirely exited and each carries 1/2 k_BT energy.

In heat transport exploration, it is assumed that the thickness of a graphene monolayer is about the graphite interlayer spacing $h \approx 3.35\ \mathring{A}$. Graphene has one of the highest in-plane thermal conductivities at room temperature, about 2000–4000 W m^{-1} K^{-1}, when it was found in freely suspended samples [53, 54]. This range corresponds with values between isotopically purified samples (0.01% C instead of 1.1% natural abundance) as a right end with large grains [55] and isotopically mixed samples or those with smaller grain sizes as a left end.

Disorders with no respect of their source introduce more phonon scattering, and this results in a descendant of conductivity lower than the mentioned range [56]. Figure 3a compares the thermal conductivity of natural diamond (about 2200 W m^{-1} K^{-1}) with those of other related materials at room temperature [57, 58]. Figure 3b exhibits the thermal conductivity of materials in Figure 3a with respect to lack of disorders.

Heat flow is strongly limited by weak van der Waals interaction in both of directions: cross-plane (along the c axis) and perpendicular to a graphene sheet, knowing that the van der Waals interaction in the perpendicular direction is between graphene and adjacent substrates, such as Sio_2. As we can see in Figure 3a, the thermal conductivity along the c axis of pyrolytic graphite is just ~6 W m^{-1} K^{-1} at room temperature. The relevant metric for heat flow across such interfaces is the thermal conductance per unit area, $G'' = Q'' / \Delta T \approx 50$ MW m^{-2} K^{-1} at room temperature [60 -62]. This is approximately equivalent to the thermal resistance of an ~ 25-nm layer of SiO_2 [59] and could become a limiting dissipation bottleneck in highly scaled graphene devices and interconnects [63], as discussed in a later section. When we have a few layers of graphene (from 1 to 10 layers), it can be expertized that interlayer resistance, 1/ G'', remains almost constant and pretty smaller than the resistance between graphene and its nongraphene environment [61]. Indeed, the interlayer thermal conductance of bulk graphite is ~ 24 GW m $^{-2}$ K $^{-1}$ if the typical 3.35- $\mathring{A}$ spacing and the c axis thermal conductivity are assumed.

It must be remarked that surface effects are able to decrease the thermal conductivity of graphene because of the sensitivity of phonon propagation to surface or edge perturbation, and as a result of this, the in-plane thermal conductivity of freely suspended graphene is drastically lower than a graphene nanoribbon or a graphene contacted with a substrate.

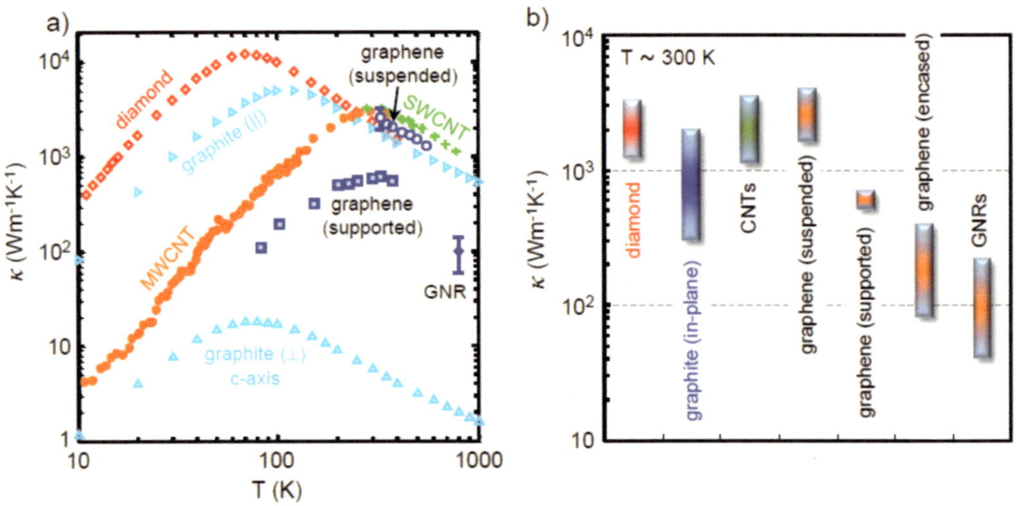

Figure 3. (a) Thermal conductivity κ as a function of temperature for representative data of suspended graphene [55], SiO_2-supported graphene [64], ~20-nm-wide graphene nanoribbons (GNRs)[63], suspended single-walled CNTs (SWCNTs)[66], multi-walled CNTs (MWCNTs)[67], type IIa diamond, graphite in-plane and out-of-plane. Additional data for graphene and related materials are also summarized in Ref.[54]. (b) Room temperature ranges of thermal conductivity data κ for diamond [57], graphite (in-plane) [54], carbon nanotubes (CNTs) [54], suspended graphene [54, 55], SiO_2-supported graphene [64], SiO_2-encased graphene [65], and GNRs [63].

It has been seen that the in-plane thermal conductance G of graphene can reach a significant fraction of the theoretical ballistic limit in sub-micrometer samples, owing to the large phonon mean free path ($\lambda \approx 100$ to 600 nm in supported and suspended samples, respectively). However, thermal properties of graphene could be highly tunable, so that makes it useful for heat- sinking applications when we regulate it in ultra-high thermal conductivity, and it is useful for thermoelectric applications when it is regulated for ultra-low thermal conductivity.

2.3. Electrical Properties of Graphene

The electronic properties of graphene are one of the prominent cases that relating experimental researches have dealt with. The controllable continuous transformation of charge carriers from holes to electrons was one of the most notable features in pioneering researches.

An example of the gate (or gate electrode is a region at the top of the transistor whose electrical state determines whether the transistor is on or off) dependence in single-layer graphene is shown in Fig. 4a. This dependency is much weaker in multiple layers of graphene because electric field is screened by the other layers.

The high electronic mobility of graphene permits the development of quantum hall effect (an effect that is visible in conductor and semiconductors materials; when there are both electrical and magnetic field at the same time in a conductor or semiconductor material, it can arise an electric potential perpendicular to the magnetic field that causes electric current perpendicular to the magnetic field) at low temperatures and high magnetic fields, for both electrons and holes (Fig. 4b) (Novoselov, et al., 2005; Zhang, et al., 2005). As seen in Fig. 4b, at room temperature, the Hall conductivity σ_{xy} reveals clear plateaus at $2e^2/h$ for both electrons and holes, while the longitudinal conductivity ρ_{xx} approaches zero. (Quantum Hall effect is measured by $\sigma = \upsilon \frac{e^2}{h}$, where "$e$" is the elementary charge, h is the Planck' constant, and υ is the "filling factor." If υ is an integer, it will be an "integer quantum hall effect," and if υ is a fraction, it will be a "fractional quantum Hall effect." Here, at room temperature, the filling factor of graphene is υ=2.)

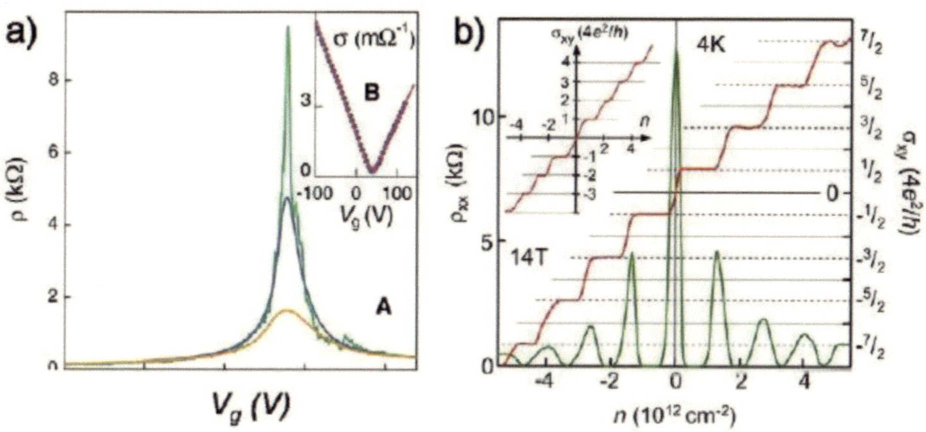

Figure 4. (a) The resistivity of a single layer of graphene vs. gate voltage. (b) The quantum Hall effect in single-layer graphene. Figures taken from (Novoselov, et al. (2005).

For sensing or transistor application, we should utilize the strong gate dependence of graphene. To do this, we should cut graphene into narrow ribbons because graphene does not have a band gap, and thus resistivity variation is small. However, graphene in narrow ribbons, makes an opening in the band gap that is proportional to the width of the ribbon by increasing the momentum of charge carriers in the traverse direction when shrinking them. This band gap in carbon nanotubes is proportional to its diameter. The opening of a band gap in graphene ribbons has recently been observed in wide ribbon devices lithographically patterned from large graphene flakes (Han, et al., 2007) and in narrow chemically synthesized graphene ribbons (Li, et al., 2008).

3. Molecular Dynamics (MD) modeling

Molecular dynamics is nothing but classical dynamics. Indeed classical dynamic equations of motion are valid for slow and heavy particles, with typical velocities $v \ll c$, (where c is the speed of light) and masses $m \gg m_e$, (where m_e is the electron mass). Therefore, we can use them for atoms, ions, and molecular mass only in slow motion (slower than thermal vibration).

This technique is based on computing potential energy that is typically considered only as a function of the system spatial configuration and is described by means of interatomic potentials. These potentials are considered as known input information; they are either found experimentally or are computed by averaging over the motion of the valence electrons in the ion's Coulomb field by means of quantum ab initio methods.

The main equation that we utilize in this technique is the Lagrange equations of motion. For a system of N interacting monoatomic molecules, the Lagrange equation turns Newtonian equations divided into "Dissipative equations" and "Generalized Langevin equations."

Integrals of motion are more functions that are useful for modeling in MD technique. Their notable property is depending only on the initial conditions and staying constant in time. Some of these equations are as follows:

$$\frac{d\boldsymbol{E}}{dt} = 0 \tag{1}$$

$$\frac{d\boldsymbol{P}}{dt} = 0 \tag{2}$$

$$\frac{d\boldsymbol{M}}{dt} = 0 \tag{3}$$

where E is the total energy, P is the total momentum, and M is the total angular momentum in a system with only conservative forces.

4. Lattice Mechanics

Lattice mechanics is an approach to utilize natural symmetries for classical particle mechanics in materials that repetitive or regular atomic structure such as polymers or crystalline materials.

This approach is based on two principles:

Principle 1: Mathematical models such as functions, matrices, operators, and so on are invariant with respect to the symmetry of lattice.

Principle 2: symmetry assumption causes loading is symmetry, and thus we will have symmetry effect.

In this approach, we use the concept of "Unit Cell" instead of particles itself. Unit Cell is an arbitrary part of a lattice, in which we could gain the whole lattice by repeating that in a fixed direction and certain distance.

Hence, we can define displacement vector in a lattice in terms of unit cell as follows:

$$\boldsymbol{u}_n(t) = \boldsymbol{r}_n(t) - \boldsymbol{n} \tag{4}$$

where n is the primary position of the unit cell:

$r_n(t)$, is the position of the unit cell at time t, which is composed of the position of all particles in the unit cell at time t; $\boldsymbol{u}_n(t)$, is the displacement vector for the unit cell. By calculating the total lattice kinetic energy, we are able to get lattice Lagrangian and consequently the equation of motion as follows:

$$\boldsymbol{M}\ddot{\boldsymbol{u}}_n + \frac{\partial U}{\partial \boldsymbol{u}_n} = f_n^{ext} \tag{5}$$

where U is the total Lattice Potential energy and f_n^{ext} is the external load on the Unit Cell n. We absolutely insist that this equation is for each unit cell, that is, with changing n we will have a system of motion differential equations.

Using Taylor series for U results in the following:

$$\boldsymbol{M}\ddot{\boldsymbol{u}}_n(t) - \sum_{n} \boldsymbol{K}_{n-n'} \boldsymbol{u}_{n'}(t) = f_n^{ext}(t) \tag{6}$$

where

$$\boldsymbol{K}_{n-n'} = -\frac{\partial^2 U(\boldsymbol{u})}{\partial \boldsymbol{u}_n \partial \boldsymbol{u}_{n'}}\Big|_{\boldsymbol{u}=0} \tag{7}$$

and n' is indices for neighboring cell.

Since we need neighboring cell in forming equations, we must define another concept as "associate cell". Associate cell is the smallest part of a lattice that represents its mechanical properties perfectly. In lattice mechanic we restrict our studies to associate cell which can comprise several unit cells; this is the consequence of principle I discussed above.

To solve the above-mentioned equation of motion, we should use mathematics like the Fourier transform or the Laplace transform.

In terms of f_{ext}, we have three types of problems:

a. $f_{ext}=0$ or free lattice: the solution is gained by superposing of normal modes called standing waves, because the lattice oscillate around its equilibrium position.

b. $f_{ext}\neq 0$: to solve this type of problems we use Green's function method that requires use of Unit pulse convolution.

c. Quasi-static problem or approximation: this is a name for time-independent problems where any noticeable change of the external forcing occurs during a period of time that is much longer than the characteristic time of atomic vibrations. This leads to eliminate the first term in motion equation, so we have the following:

$$\sum_{n'} \boldsymbol{K}_{n-n'} \boldsymbol{u}_{n'}(t) = f_n^{ext}(t) \tag{8}$$

To solve this equation, we can use green's functions method. One of issue problems in quasi-static approximation is multiscale boundary conditions. We discuss this in a separate clause.

5. Multi scale boundary condition

Each method has its own limitations to use. These limitations are only in time and size scale. The multi scale method tries by dividing the model in different areas in terms of size and time scale (coarse-fine grain) and relating them together generates an assimilated method. In other words, we may be able to solve one part of problem with atomistic modeling and the other part with continuum; the multi scale method uses both of them concurrently and then couples them together.

To couple the methods, we define a region of pseudo-atoms called handshake or pad. The position of pseudo-atoms is determined by the finite element method. The handshake has a duty to absorb the fine grain excitation and transfer the effects of coarse grain surrounding boundary.

If u_1 is the displacement of the pseudo-atom, $\boldsymbol{u}_0$ is the fine grain displacement, and $\boldsymbol{u}_a$ is the coarse grain displacement, then we can say:

$$\boldsymbol{u}_1 = \Theta\{\boldsymbol{u}_0\} + \Xi\{\boldsymbol{u}_a\} \tag{9}$$

where $\boldsymbol{\Theta}$ and $\boldsymbol{\Xi}$ are unknown operators.

6. Multi scale Modeling

The philosophy of arising multiple- scale methods is that, in actuality, nano- materials are always used along with large- scale materials and we are compelled to create a method for modeling them. Atomistic methods such as MD and ab initio are not perfect to model the entire configuration because these methods are limited to time and length scales; thus, they are validated only through fine- scale parts of configuration and not for the other part. There is the same situation for continuum methods because they are validated for large time and length scale, and they are not useable for fine- scale parts of these configurations. Here is the point that the role of multiple- scale methods becomes prominent. Multiple scale methods try to blend the methods that are validated on their own scales of time and length separately.

The base of all these methods is that each scale is modeled by its special method, and their output becomes boundary condition for the other; indeed, there will be exchanges between these methods to model entire of the configuration.

Whatever method we take, it must be free of two issue problems:

- The wavelengths emitted by the MD region are drastically smaller than that can be absorbed by continuum region. It means we have differential energy for these two areas. If our approach is not capable to put in any experience for this redundant energy, it leads to this result that some of the wave will reflect back into the MD domain. This means that we will have spurious heat generation in the MD region and so a mistaken simulation especially in instances of plasticity.
- Furthermore, the transition from the MD region to the continuum region was followed with extension in timescale, which affects the validity of our relations. Hence, we must take such a method that solves this issue.

6.1. MAAD

MAAD or "macroscopic, atomistic, ab initio dynamics" is a multiple- scale approach that incorporates tight binding (TB), molecular dynamics (MD), and finite elements (FE) concurrently to model a configuration of nano- and large- scale materials. In this method, the FE Mesh will be done until we approach the same size as much as atomic spacing; from here, the MD method is entered until we arrived in a physical phenomenon like a crack tip. At this point, we will use the TB approach. Thus, we will have two overlapping regions: FE/MD and MD/TB. Not only here but also in all multiple- scale methods, such overlapping areas are termed "handshake" regions. In this method, each handshake region provides a contribution to the total energy of the system. This contribution is done by linear law in each handshake region. Thus, the total energy that will be used to find equation of motion is as follows:

$$\boldsymbol{H}_{FE} = \boldsymbol{H}_{FE} + \boldsymbol{H}_{FE/MD} + \boldsymbol{H}_{MD} + \boldsymbol{H}_{MD/TB} + \boldsymbol{H}_{TB} \quad (10)$$

There are two problems in this method. First, having finite elements in the scale of atomic space prolongs the process of solving by increasing time steps. Second, that it seems unphysical that continuum relations can be evolved at the same timescales as the atomistic variables.

6.2. Coarse-Grained Molecular Dynamics (CGMD)

This approach couples FE and MD methods together. In this method to derive the governing equations of motion, we use an approximation of coarse-grained energy that converges to the exact atomic energy like the following:

$$E\left(\boldsymbol{u}_k,\dot{\boldsymbol{u}}_k\right)=U_{int}+\frac{1}{2}\sum_{j,k}\left(\boldsymbol{M}_{jk}\dot{\boldsymbol{u}}_j.\dot{\boldsymbol{u}}_k+\boldsymbol{u}_j.\boldsymbol{K}_{jk}.\boldsymbol{u}_k\right) \tag{11}$$

where

$U_{int}=3(N-N_{node})kT$	Internal energy
$\boldsymbol{M}_{jk}\dot{\boldsymbol{u}}_j.\dot{\boldsymbol{u}}_k$	Kinetic energy
$\boldsymbol{u}_j.\boldsymbol{K}_{jk}.\boldsymbol{u}_k$	Potential energy
$\boldsymbol{u}$, $\dot{\boldsymbol{u}}$	Displacement degree of freedom and the velocity

U_{int} represents the thermal energy of those degrees of freedom of coarse grained that have not been included in FE considered nodes. Obviously, when the number of nodes approaches the number of atoms, this term vanished and the full atomistic energy is recovered.

Finally the equation of motion is then given to be as follows:

$$M_{ij}\ddot{u}_j=-G_{ik}^{-1}u_k+\int_{-\infty}^{t}\eta_{ik}\left(t-\tau\right)\dot{u}_k\left(\tau\right)d\tau+F_i\left(t\right) \tag{12}$$

where

M_{ij}	Mass matrix
G_{ij}	Stiffness like quantity
η_{ij}	Time history of memory function
$F_i(t)$	Random force

6.3. Quasi-Continuum Method

This approach is an adaptive FE method. The link between atomistic and continuum is obtained by the use of the Cauchy Born rule. The Cauchy Born rule assumes that the continuum energy density W can be computed using an atomistic potential.

The first Piola-Kirchoff stress tensor in the Cauchy Born rule is

$$\boldsymbol{P} = \frac{\partial W}{\partial \boldsymbol{F}^T} \tag{13}$$

and the Lagrangian tangent stiffness is:

$$\boldsymbol{C} = \frac{\partial^2 W}{\partial \boldsymbol{F}^T \partial \boldsymbol{F}^T} \tag{14}$$

where F is deformation gradient.

The major restriction as well as implication of the Cauchy Born rule is that the deformation of the lattice underlying a continuum point must be homogeneous. This results from the fact that the underlying atomistic system is forced to deform according to the continuum deformation gradient F.

6.4. Coupled Atomistics And Discrete Dislocation (CADD)

This method sets a quasi-static coupling between molecular statics and discrete dislocation plasticity. One of the best assumptions in this approach is that defects within atomistic region are allowed to pass through the atomistic/continuum border into the continuum which is modeled via discrete dislocation mechanics.

Quantities such as stresses, strains, and displacements can be written in the contribution form:

$$A = \tilde{A} + \hat{A} \tag{15}$$

where A is a typical quantity, $\tilde{A}$ is the contribution from the discrete dislocation, and $\hat{A}$ is a correction term we introduce because of the fact that the discrete dislocation solution is for an infinite medium. The continuum energy can be written as

$$E^c = \frac{1}{2} \int_{\Omega_c} (\hat{\boldsymbol{u}} + \tilde{\boldsymbol{u}}) : (\hat{\varepsilon} + \tilde{\varepsilon}) dV - \int_{d\Omega_T} \boldsymbol{T}_0 (\hat{\boldsymbol{u}} + \tilde{\boldsymbol{u}}) dA \tag{16}$$

where T_0 is the traction on the continuum boundary $d\Omega_T$.

The equilibrium condition is

$$\frac{\partial E^c}{\partial \tilde{\boldsymbol{u}}_c} = 0 \tag{17}$$

Where $\tilde{\boldsymbol{u}}$ is displacement field. Using this equation, we could find displacement fields. By the use of displacement field we are able to find the forces exerted on the discrete dislocation as

$$\boldsymbol{P}^i = -\frac{\partial E^c}{\partial \boldsymbol{d}^i} \tag{18}$$

At this point, an iterative procedure involving the discrete dislocation positions, FE positions and atomic positions is solved until all degrees of freedom are at equilibrium.

However, we have some challenge in this method as follows:

- extending this approach to dynamic problems
- simulating the passage of defect from the atomistic to continuum in three dimensions
- annular dislocations that reside in both atomistic and continuum regions at the same time

6.5. Bridging Domain

Bridging Domain is a dynamic multiple scale method that couples MD with continuum. In this method the boundary that overlays the MD region and surrounding continuum region has varying size, termed the bridging domain.

Within overlaying bridging domain, the Hamiltonian is defined as a linear combination of the molecular and continuum Hamiltonians:

$$H = (1-\alpha)\mathrm{H}^{\mathrm{M}} + \alpha H^{C} \tag{19}$$

where α acts to scale the contribution of each domain to the total Hamiltonian.

To make compatibility between the molecular and the continuum regions, we involve Lagrange multipliers in the overlap region as

$$\boldsymbol{g}_I = \left(\sum_J N_J(\boldsymbol{X}_J) u_{iJ} - d_{iI} \right) = 0 \tag{20}$$

Where g_I are the Lagrange Multipliers, u_{iJ} are the FE nodal displacements, and d_{iJ} are the MD displacements.

The coupled equations of motion will be as follows:

$$\bar{M}_I \ddot{\boldsymbol{u}}_I = \boldsymbol{F}_I^{ext} - \boldsymbol{F}_I^{int} - \boldsymbol{F}_I^{L} \tag{21}$$

$$\bar{m}_I \ddot{d}_I = f_I^{ext} - f_I^{int} - f_I^L \tag{22}$$

Where the standard equations are augmented by the Lagrange multiplier-based constraint forces F_I^L and f_I^L . The bar symbols overlaying the FE and MD mass matrices indicate that they need to be modified within the overlapping region.

7. Bridging- Scale Modeling

The bridging scale method that we want to introduce is an incorporation of MD and continuum. The total displacement as our solution is decomposed to fine and coarse- scale as follows:

$$u(x) = \bar{u}(x) + u'(x) \tag{23}$$

where $\bar{u}$ is a coarse- scale solution that will be state in terms of finite element shape functions and $u^{'}$ is a fine- scale solution. These two solutions are orthogonal, that is, one projection onto another is zero.

If $x=x_\alpha$ (we use Greek indices for atoms and Roman indices for coarse- scale nodes) was the initial position for typical atoms of interesting body we can rewrite (23) as:

$$\mathrm{u}(\mathrm{X}_\alpha) = \bar{\mathrm{u}}(\mathrm{X}_\alpha) + \mathrm{u}'(\mathrm{X}_\alpha) \tag{24}$$

$\bar{\mathrm{u}}(\mathrm{X}_\alpha)$ is an average behavior of body since it's interpolation between atoms so we anticipate it is continues function. We can write:

$$\bar{u}(X_\alpha) = \sum_I \mathrm{N}_I^\alpha d_I \tag{25}$$

where, $\mathrm{N}_\mathrm{I}^\alpha = \mathrm{N}_\mathrm{I}(\mathrm{X}_\alpha)$ is a shape function of node I at initial atomic position X_α and d_I is the displacement degree of freedom at node I.

As we said, we want to use MD method in this approach, to do this we have to utilize q_α. q_α is a displacement solution we derive from MD simulation (q_α has the same role as $u(X_\alpha)$ in other words, we are simulating $u(X_\alpha)$ by the use of q_α. Undoubtedly, q_α will have projection onto $\bar{\mathrm{u}}$ (coarse- scale solution) and u (fine- scale solution), and with respect to their orthogonality, each projection will not represent the other one, so we can easily find the fine- scale solution, by subtracting the projection of q_α onto coarse- scale solution, from q_α, that is,

$$u^{'} = q - NW \tag{26}$$

or with indices representation:

$$u^{'}(X_{\alpha}) = q_{\alpha} - \sum_{I} N_{I}^{\alpha} W_{I} \tag{27}$$

$\sum_{I} N_{I}^{\alpha} W_{I}$ is a coarse- scale projection. In other words, to find coarse- scale projection, we have projected q_{α} onto shape functions N_I. However, there is one burning question: how can we find W_I? The answer is laid in least squares weighted residual method. First, we define J function as follows:

$$\boldsymbol{J} = \sum_{\alpha} m_{\alpha} \left(\boldsymbol{u}_{\alpha} - \sum_{I} N_{I}^{\alpha} \boldsymbol{w}_{I} \right)^{2} \tag{28}$$

where m_{α} is the atomic mass. The choice of J (called the mass-weighted square of the fine scale) allows the coarse and fine- scale kinetic energies to be decoupled. Minimizing of J will result in W_I. If we gather all masses in a matrix as follows:

$$\boldsymbol{M}_{A} = \begin{bmatrix} m_{1}\boldsymbol{I}_{3\times 3} & & \\ & m_{2}\boldsymbol{I}_{3\times 3} & \\ & & \ddots \end{bmatrix} \tag{29}$$

J can be writhed as:

$$\boldsymbol{J} = (\boldsymbol{q} - \boldsymbol{N}\boldsymbol{w})^{T} \boldsymbol{M}_{A} (\boldsymbol{q} - \boldsymbol{N}\boldsymbol{w}) \tag{30}$$

Solving for w to minimize this quantity gives

$$\boldsymbol{w} = \boldsymbol{M}^{-1} \boldsymbol{N}^{T} \boldsymbol{M}_{A} \boldsymbol{q} \tag{31}$$

where the coarse- scale mass matrix M is

$$\boldsymbol{M} = \boldsymbol{N}^{T} \boldsymbol{M}_{A} \boldsymbol{N} \tag{32}$$

Substituting W in Equation 26 with Equation 31, we will have

$$u' = q - Pq \tag{33}$$

where the matrix P has been defined as

$$P = NM^{-1}N^{T}M_{A} \tag{34}$$

So we can write the total displacement u_α (with respect to Equation 23-25 and 33) as

$$u = Nd + q - Pd \tag{35}$$

The last term in the above expression is called the "bridging scale"; it is that part of the molecular dynamics calculation that must be subtracted from the total in order to achieve a complete separation of scales. Our sake about the "scales" is fine and coarse scales, and our sake about "complete separation" is the orthogonality of these scales.

Indeed, P was the operator for the projection onto the coarse scale; with respect to Equation 35, we could represent another more perfect operator for this projection as

$$Q = I - P \tag{36}$$

or

$$Q = I - NM^{-1}N^{T}M_{A} \tag{37}$$

Thus, we can rewrite 35 as

$$u = Nd + Qq \tag{38}$$

7.1. Equations of motion

In this step, we will derive the coupled MD and FE equations of motions. First, we adopt the Lagrangian equation definition we submitted in MD discussion for multiscale condition as "multiscale Lagrangian."

7.1.1. Multiscale Lagrangian

$$L(u,\dot{u}) = K(\dot{u}) - U(u) - f_{ext}^{T}u \tag{39}$$

$$K(\dot{u}) = \frac{1}{2}\dot{u}^{T}M_{A}\dot{u} \tag{40}$$

$$\dot{\boldsymbol{u}} = \boldsymbol{N}\dot{\boldsymbol{d}} + \boldsymbol{Q}\dot{\boldsymbol{q}} \tag{41}$$

Substituting $\dot{u}$ in Equation 40 with Equation 41 we will have

$$\boldsymbol{K}(\dot{\boldsymbol{u}}) = \frac{1}{2}\dot{\boldsymbol{u}}^T \boldsymbol{M}_A \dot{\boldsymbol{u}} = \frac{1}{2}\dot{\boldsymbol{d}}^T \boldsymbol{N}^T \boldsymbol{M}_A \boldsymbol{N}\dot{\boldsymbol{d}} + \dot{\boldsymbol{d}}^T \boldsymbol{N}^T \boldsymbol{M}_A \boldsymbol{Q}\dot{\boldsymbol{q}} + \frac{1}{2}\dot{\boldsymbol{q}}^T \boldsymbol{Q}^T \boldsymbol{M}_A \boldsymbol{Q}\dot{\boldsymbol{q}} = \frac{1}{2}\dot{\boldsymbol{d}}^T \boldsymbol{M}\dot{\boldsymbol{d}} + \frac{1}{2}\dot{\boldsymbol{q}}^T \mathrm{M}\dot{\boldsymbol{q}} \tag{42}$$

The second term in the right- hand side of abvoe equation is zero because

$$\boldsymbol{N}^T \boldsymbol{M}_A \boldsymbol{Q} = \boldsymbol{N}^T \boldsymbol{M}_A \left(\boldsymbol{I} - \boldsymbol{N}\boldsymbol{M}^{-1}\boldsymbol{N}^T \boldsymbol{M}_A\right) = \boldsymbol{N}^T \boldsymbol{M}_A - \boldsymbol{M}\boldsymbol{M}^{-1}\boldsymbol{N}^T \boldsymbol{M}_A = 0$$

The fine- scale mass matrix M in equation 42 defined as

$$\mathrm{M} \equiv \boldsymbol{Q}^T \boldsymbol{M}_A \boldsymbol{Q} = \boldsymbol{M}_A \boldsymbol{Q} = \boldsymbol{Q}^T \boldsymbol{M}_A \tag{43}$$

This will be proven by the complete expression for Q (Equation 36) and coarse- scale mass matrix M (Equation 30).

Thus, we can write Lagrangian 36 as

$$\boldsymbol{L}(\boldsymbol{d}, \dot{\boldsymbol{d}}; \boldsymbol{q}, \dot{\boldsymbol{q}}) = \frac{1}{2}\dot{\boldsymbol{d}}^T \boldsymbol{M}\dot{\boldsymbol{d}} + \frac{1}{2}\dot{\boldsymbol{q}}^T \mathrm{M}\dot{\boldsymbol{q}} - U(\boldsymbol{N}\boldsymbol{d} + \boldsymbol{Q}\boldsymbol{q}) + \boldsymbol{f}_{ext}^T \boldsymbol{N}\boldsymbol{d} + \boldsymbol{f}_{ext}^T \boldsymbol{Q}\boldsymbol{q} \tag{44}$$

This is the desired equation for us because we have eliminated cross terms such as d and q in the kinetic energy. The bridging scale is responsible for this elimination. Using the bridging scale makes us capable of decomposing total kinetic energy into the sum of the kinetic energy in the coarse scale plus that in the fine scale.

7.1.2. Multiscale equations of motion

We can derive equations of motion by the use of Lagrangian equation:

$$\frac{d}{dt}\left(\frac{\partial \boldsymbol{L}}{\partial \dot{\boldsymbol{d}}}\right) - \frac{\partial \boldsymbol{L}}{\partial \boldsymbol{d}} = 0 \tag{45}$$

$$\frac{d}{dt}\left(\frac{\partial \boldsymbol{L}}{\partial \dot{\boldsymbol{q}}}\right) - \frac{\partial \boldsymbol{L}}{\partial \boldsymbol{q}} = 0 \tag{46}$$

http://dx.doi.org/10.5772/61564

Substituting the Lagrangian Equation 44 into Equations 45 and 46 gives us

$$\boldsymbol{M}\ddot{\boldsymbol{d}} = -\frac{\partial U(\boldsymbol{d},\boldsymbol{q})}{\partial \boldsymbol{d}} \tag{47}$$

$$M\ddot{\boldsymbol{q}} = -\frac{\partial U(\boldsymbol{d},\boldsymbol{q})}{\partial \boldsymbol{q}} \tag{48}$$

For simplicity, we have ignored the external force. Using the chain rule, we can rewrite Equations 47 and 48 as

$$\boldsymbol{M}\ddot{\boldsymbol{d}} = \frac{\partial U}{\partial \boldsymbol{u}}\frac{\partial \boldsymbol{u}}{\partial \boldsymbol{d}} \tag{49}$$

$$M\ddot{\boldsymbol{q}} = -\frac{\partial U}{\partial \boldsymbol{u}}\frac{\partial \boldsymbol{u}}{\partial \boldsymbol{q}} \tag{50}$$

As we see, the derivation of potential energy U couples scales in both equations. We know that the variation of energy to displacement is the same type as force. These means we can conclude that the derivation of potential energy is the total forces on atoms:

$$\boldsymbol{f} = -\frac{\partial U(\boldsymbol{u})}{\partial \boldsymbol{u}} \tag{51}$$

Using Equations 51 and 38, we can rewrite Equations 49 and 50 as

$$\boldsymbol{M}\ddot{\boldsymbol{d}} = -\left(\frac{\partial \boldsymbol{u}}{\partial \boldsymbol{d}}\right)^{\boldsymbol{T}}\frac{\partial U}{\partial \boldsymbol{u}} = \boldsymbol{N}^{T}\boldsymbol{f} \tag{52}$$

$$M\ddot{\boldsymbol{q}} = -\left(\frac{\partial \boldsymbol{u}}{\partial \boldsymbol{q}}\right)^{\boldsymbol{T}}\frac{\partial U}{\partial \boldsymbol{u}} = \boldsymbol{Q}^{T}\boldsymbol{f} \tag{53}$$

where $M = Q^{T}M_{A}$. If the external forces in Equation 44 are kept in formulation, we will have

$$\boldsymbol{M}\ddot{\boldsymbol{d}} = \boldsymbol{N}^{\boldsymbol{T}}\left(\boldsymbol{f} + \boldsymbol{f}_{ext}\right) \tag{54}$$

$$M\ddot{\boldsymbol{q}} = \boldsymbol{Q}^T\left(\boldsymbol{f} + \boldsymbol{f}_{ext}\right) \tag{55}$$

Equation 55 can be rewritten as

$$\boldsymbol{Q}^T \boldsymbol{M}_A \ddot{q} = \boldsymbol{Q}^T\left(\boldsymbol{f} + \boldsymbol{f}_{ext}\right) \tag{56}$$

Q is a singular since multiplying a field by Q gives the fine- scale part of the field, and many different fields can have same fine- scale part. Therefore the solution of and equation like $Qu = u'$ for u is nonunique, and Q must be singular. It is concluded that Equation 56 does not have a unique solution, and we are free to choose any q that has the proper fine- scale part and thus satisfies Equation 56. One q that comply these conditions is that which exactly satisfies the following equation:

$$\boldsymbol{M}_A \ddot{q} = \boldsymbol{f} + \boldsymbol{f}_{ext} \tag{57}$$

This is nothing but a molecular dynamics simulation for the atomic displacements, with the atomic forces given by Equation 51. So we are able to solve this equation using MD simulation.

Anyway, now we have the equation of motions as

$$\boldsymbol{M}_A \ddot{q} = \boldsymbol{f} + \boldsymbol{f}_{ext} \tag{58}$$

$$\boldsymbol{M}\ddot{\boldsymbol{d}} = \boldsymbol{N}^T\left(\boldsymbol{f} + \boldsymbol{f}_{ext}\right) \tag{59}$$

There are some relevant comments:

- Equation 58 is a fine- scale equation of motion and a standard MD solver can be used to obtain the MD displacements q.
- Equation 59 is a coarse- scale equation of motion. This is an FE momentum equation, and to solve it, we should use finite elements method. It should be noted that for the coarse scale, the summation sign turns into integral sign.
- As we said before, q (MD solution) simulate u (real total displacement). Although we proved Equation 58 for q, it is obvious that u as real total displacement have to satisfy Equation 58 in each atomic position (X_α), this is found from Newton's second law. u is not necessarily equal to q; they just have the same behavior like solutions of a differential equations. Indeed, initial conditions determine their equality. Thus, if these two quantities have the same initial conditions, then (which they should) then they are equal forever.
- Equality of q and u and 38 gives

$$\boldsymbol{Nd} = \boldsymbol{Pq} \tag{60}$$

It means that the coarse scale is a projection of q, and instead of solving the FE equation, we can use q. However, as we will discuss, since we will eliminate the fine scale from the coarse scale, q will be limited to fine scale (not the entire domain), and thus it is not possible to calculate the coarse- scale solution everywhere via direct the projection of the MD displacements.

- Equations 58 and 59 converge to each other if FE nodal positions approach MD atomic positions.
- This convergence makes us to conclude FE equation (Equation 59) overlays both fine and coarse scales. This is the unwanted degree of freedom that prolongs spent time for FE solution, so we should eliminate the effects of the fine scales that lie in coarse- scale region.

7.2. Removing fine- scale degree of freedom in coarse- scale region

In the previous section, we saw the fine- scale equation of motion (Equation 58) and coarse-scale equation of motion (Equation 59) in areas among this two- scale overlay together, and this is in contrast with the orthogonality of these two scale, so we should eliminate the effect of fine scale from coarse scale. However why do we eliminate fine scale from coarse scale and not vice versa? The answer is because eliminating one scale from the other will enter another terms in the eliminated scale's equation of motion. Since we want to avoid wave reflection back into the MD domain and avoid heat retention within this region, we should add some terms to the fine- scale equation of motion, and it is the best opportunity for us to achieve this gold by eliminating the fine- scale effect from the coarse- scale domain. Thus, we limit our studies to the MD region and its equation of motion (Equation 58) with assumption f_{ext}=0:

$$\boldsymbol{M}_A\ddot{q} = \boldsymbol{f}(\boldsymbol{q}) \tag{61}$$

To find the area that shall be removed, we should consider the behavior of these two scales. In fine scale, we have a harmonic or nonlinear potential in atomic interactions. However, at coarse scale, we have a harmonic or linear potential energy in atomic interaction. Thus, we can conclude that there is a transition area among these two scales that belong in the MD region. This is the area that shall be eliminated, and instead of this elimination, we should add some boundary conditions for the MD region. Thus, we divide the MD region in two sections: linear and nonlinear. This division is called the "linearized MD equation of motion." Two find MD boundary condition, since we are transient from fine to coarse scale in the transition area, we use FE method and then Lattice mechanics. This means we are treating with transition area, like a coarse scale, and maybe we could say we are dividing the MD area into fine and coarse sections. We first study a chain of atoms and then consider the general state.

Consider a one-dimensional chain of atoms according to the following figure:

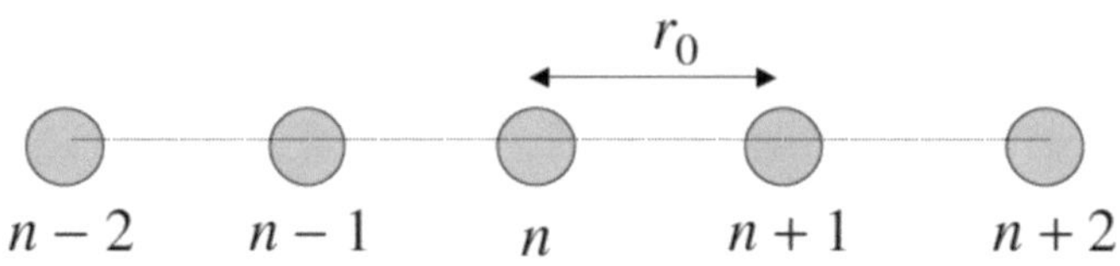

From the lattice mechanics, we have

$$f_n = -\frac{\partial U(u)}{\partial u_n}\Big|_{u=0} \tag{62}$$

$$K_{n-n'} = \frac{\partial f_n}{\partial u_{n'}}\Big|_{u=0} = -\frac{\partial^2 U(u)}{\partial u_n \partial u_{n'}}\Big|_{u=0} \tag{63}$$

where f_n denotes the total force on atom (unit cell) n and $K_{n-n'}$ is the springy factor. Using the Lagrange equation results in the equation of motion as follows:

$$m\ddot{u}_n(t) - \sum_{n'=n-\upsilon}^{n+\upsilon} K_{n-n'} u_{n'}(t) = f_n^{ext}(t) \tag{64}$$

This equation is validated only for harmonic lattice, and $f_n^{ext}(t)$ is the total external force acting on the unit cell n.

By dividing the MD region into fine and coarse scales, we have

$$q = \bar{u} + u' \tag{65}$$

where $\bar{u}$ denotes the coarse scale and u' denotes the fine scale of the MD region.

For the left side of Equation 61, we can write

$$M_A \ddot{q} = M_A \ddot{\bar{u}} + (M_A \ddot{u}') = f(\bar{u}) + \left(Ku' + f^{ext}\right) \tag{66}$$

where

$$\boldsymbol{K} = \frac{\partial \boldsymbol{f}}{\partial \boldsymbol{u}}\Big|_{\boldsymbol{u}'=0} \tag{67}$$

http://dx.doi.org/10.5772/61564

Moreover, f^{ext} is the total external force acting on the harmonic section of the MD by an unharmonic section of the MD (note that f^{ext} is different from f_{ext} we neglected in Equation 48; f^{ext} takes place within the MD section while f_{ext} lies outside of this section).

A comparison between Equations 66 and 61 gives

$$f(q) = f(\bar{u}) + \left(Ku' + f^{ext}\right) \tag{68}$$

In Equation 66, we have grouped terms based on fine scale and coarse scale. Because of the orthogonality of coarse and fine scales and that the timescale for the coarse scale is much larger than that of atomic vibrations present in the fine scale, it is reasonable to equate the corresponding parts from each side of equality sign, that is,

$$M_A \ddot{\bar{u}} = f(\bar{u}) \tag{69}$$

$$(M_A \ddot{u}') = \left(Ku' + f^{ext}\right) \tag{70}$$

Equation 70 is the one we were looking for; in other words, by removing the fine- scale degree Of freedom in the coarse- scale region, Equation 61 turns into Equation 70. Thus we use Equation 70 instead of Equation 61. However we still have not terminated; we should calculate f^{ext} as an issue problem. To do this, we use lattice mechanics.

In lattice mechanics, Equation 70 turns into Equation 71:

$$\ddot{u}_n'(t) = \sum_{n'=n\ 1}^{n} m_A^{-1} K_{n-n'} u_{n'}'(t) + m_A^{-1} f_n^{ext}(t) \tag{71}$$

Equation 70 is the one we were looking for; in other words, by removing the fine- scale degree Of freedom in the coarse- scale region, Equation 61 turns into Equation 70. Thus we use Equation 70 instead of Equation 61. However we still have not terminated; we should calculate f^{ext} as an issue problem. To do this, we use lattice mechanics.

In lattice mechanics, Equation 70 turns into Equation 71:

$$\ddot{u}_n'(t) = \sum_{n'=n-1}^{n} m_A^{-1} K_{n-n'} u_{n'}'(t) + m_A^{-1} f_n^{ext}(t) \tag{71}$$

where u_n is the displacement of unit cell n.

Again, consider the one- dimensional chain of atoms like the following figure:

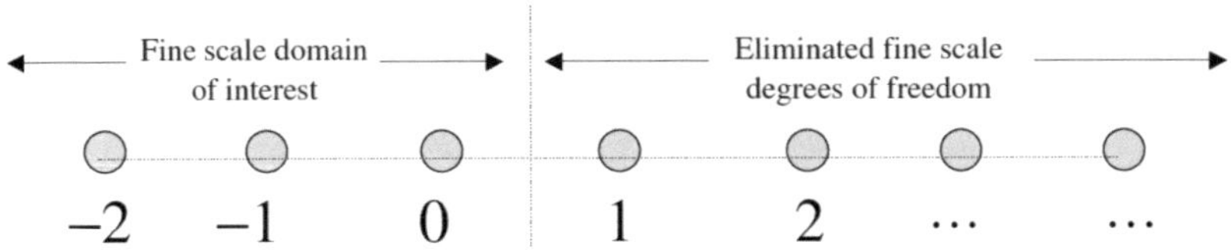

n=0 is the position that elimination has took place, and the result of that is f_n^{ext}. We can write the following:

$$f_n^{ext}(t) = \delta_{n,0} f_0^{ext}(t) \tag{72}$$

Applying discrete Fourier transform (DFT) to Equation 71, we have

$$\ddot{\hat{u}}'(p,t) = m_A^{-1}\hat{K}(p)\hat{u}'(p,t) + m_A^{-1} f_0^{ext}(t) \tag{73}$$

Using Laplace transform (LT) for Equation 73 and solving for resulting displacements $\hat{U}$ gives

$$\widehat{\boldsymbol{U}}'(p,s) = \hat{\boldsymbol{G}}(p,s)\left(m_A^{-1}\boldsymbol{F}_0^{ext}(s) + s\hat{u}'(0,p) + \dot{\hat{u}}'(0,p)\right) \tag{74}$$

where

$$\hat{\boldsymbol{G}}(p,s) = \left(s^2 - \hat{A}(p)\right)^{-1} \tag{75}$$

and

$$\hat{A}(p) = m_A^{-1}\hat{K}(p) \tag{76}$$

For f_0^{ext} we can write:

$$f_0^{ext}(t) = K_{-1}u'_1(t) \tag{77}$$

Taking LT from Equation 77 gives impedance force:

$$f_0^{imp}(s) = K_{-1}U'_{-1}(s) \tag{78}$$

So to find f_0^{ext}, we should calculate $\acute{U}_1(s)$, This requires to omit $F_0^{ext}(s)$ from Equation 74, so we first take the inverse discrete Fourier transform of Equation 61 in terms of p:

$$\boldsymbol{U}'_n(s) = \boldsymbol{G}_n(s) m_A^{-1} \boldsymbol{F}_0^{ext}(s) + \boldsymbol{R}_n^d(s) \tag{79}$$

where

$$\boldsymbol{R}_n^d(s) = s \sum_{n'=-N/2}^{\frac{N}{2}-1} G_{n-n'}(s) u'_{n'}(0) + \sum_{n'=-N/2}^{\frac{N}{2}-1} G_{n-n'}(s) \dot{u}'_{n'}(0) \tag{80}$$

and N is the total number of unit cells in the domain. Now we write Equation 66 for n=0,1:

$$\boldsymbol{U}'_0(s) = G_0(s) m_A^{-1} \boldsymbol{F}_0^{ext}(s) + \boldsymbol{R}_0^d(s) \tag{81}$$

$$\boldsymbol{U}'_1(s) = G_1(s) m_A^{-1} \boldsymbol{F}_0^{ext}(s) + \boldsymbol{R}_1^d(s) \tag{82}$$

Solving these two equations simultaneously can omit $F_0^{ext}(s)$ and gives

$$\boldsymbol{U}'_1(s) = G_1(s) G_0^{-1}(s) \left(\boldsymbol{U}'_0(s) - \boldsymbol{R}_0^d(s) \right) + \boldsymbol{R}_1^d(s) \tag{83}$$

We can rewrite Equation 83 as

$$\boldsymbol{U}'_1(s) = \boldsymbol{Q}(s) \left(\boldsymbol{U}'_0(s) - \boldsymbol{R}_0^d(s) \right) + \boldsymbol{R}_1^d(s) \tag{84}$$

where

$$\boldsymbol{Q}(s) = G_1(s) G_0^{-1}(s) \tag{85}$$

Now taking the inverse Laplace transform (ILT) of Equations 78 and 84 and convoluting the displacement, we obtain

$$f_0^{imp}(t)=\int_0^t \theta(t-\tau)\left(u_0^{'}(\tau)-\boldsymbol{R}_0^d(\tau)\right)d\tau+\boldsymbol{R}_0^f(t) \tag{86}$$

We know

$$u_0^{'}(t)=q_0(t)-\bar{u}_0(t) \tag{87}$$

And then

$$f_0^{imp}(t)=\int_0^t \theta(t-\tau)\left(q_0(\tau)-\bar{u}_0(\tau)-\boldsymbol{R}_0^d(\tau)\right)d\tau+\boldsymbol{R}_0^f(t) \tag{88}$$

Where the time history kernel $\theta(t)$ and random force $R_0^f(t)$ are defined to be

$$\theta(t)=L^{-1}\left(\boldsymbol{K}_{-1}\boldsymbol{Q}(s)\right) \tag{89}$$

$$\boldsymbol{R}_0^f(t)=\boldsymbol{K}_{-1}\boldsymbol{R}_1^d(t) \tag{90}$$

Other terms in Equation 71 is equal with $f(u^{'})$, and we can rewrite Equation 71 as

$$M_A\ddot{u}'=f(u')+f_0^{imp}(t) \tag{91}$$

We can turn $u^{'}$ to q as

$$\boldsymbol{M}_A\ddot{q}=\boldsymbol{f}(\boldsymbol{q})+f_0^{imp}(t) \tag{92}$$

Thus, the final coupled forms of the equations of motion are

$$\boldsymbol{M}\ddot{\boldsymbol{d}}=\boldsymbol{N}^T\boldsymbol{f}(\boldsymbol{u}) \tag{93}$$

$$\boldsymbol{M}_A\ddot{q}=\boldsymbol{f}(\boldsymbol{q})+f_0^{imp}(t) \tag{94}$$

$$f_0^{imp}(t) = \int_0^t \theta(t-\tau)\left(q_0(\tau) - \bar{u}_0(\tau) - \boldsymbol{R}_0^d(\tau)\right)d\tau + \boldsymbol{R}_0^f(t) \tag{95}$$

where the impedance force f_0^{imp} acts only on the boundary atom 0.

7.2.1. Elimination of fine- scale degrees of freedom: 3D generalization

In 3D state, each unit cell can be labeled with three indices, *l, m,* and *n* indicating the position along axes in the direction of the three primitive vectors of the crystal structure. The equation of motion can be rewritten as

$$\ddot{\boldsymbol{u}}_{l,m,n}(t) = \sum_{l'=l-1}^{l} \sum_{m'=m-\mu}^{m+\mu} \sum_{n'=n-\upsilon}^{n+\upsilon} \boldsymbol{M}_A^{-1}\boldsymbol{K}_{l-l',m-m',n-n'}\boldsymbol{u}_{l',m',n'}(t) + \boldsymbol{M}_A^{-1}\boldsymbol{f}_{l,m,n}^{ext}(t) \tag{96}$$

where μ and υ represent the range of the forces in the *m* and *n* coordinate directions. We consider the one dimensional boundary condition in our study. Our boundary is located in *l*=0, like the following figure:

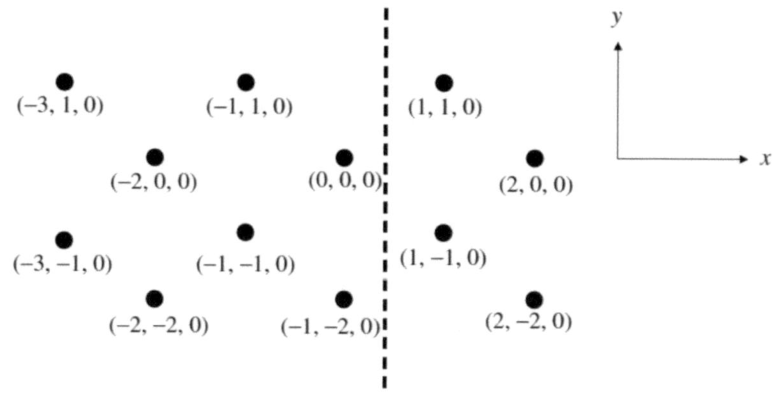

Like the one-dimensional chain, we can write

$$f_{l,m,n}^{ext}(t) = \delta_{l,0} f_{0,m,n}^{ext}(t) \tag{97}$$

By taking LT and DFT of Equation 96 and solving for $\hat{U}'$, we obtain

$$\hat{\boldsymbol{U}}'(p,q,r,s) = \hat{\boldsymbol{G}}(p,q,r,s)\left(\boldsymbol{M}_A^{-1}\hat{\boldsymbol{F}}_0^{ext}(p,q,r,s) + s\boldsymbol{u}'(p,q,r,0) + \dot{\boldsymbol{u}}'(p,q,r,0)\right) \tag{98}$$

where

$$\hat{A}(p,q,r) = M_A^{-1} K_{p,q,r} \tag{99}$$

and

$$\hat{\boldsymbol{G}}(p,q,r,s) = \left(s^2 \boldsymbol{I} - \hat{\boldsymbol{A}}(p,q,r)\right)^{-1} \tag{100}$$

For $f_{0,m,n}^{ext}$, we can write

$$f_{0,m,n}^{ext}(t) = \sum_{m'=m-\mu}^{m'=m+\mu} \sum_{n'=n-\upsilon}^{n'=n+\upsilon} K_{-1,m-m',n-n'} u'_{1,m',n'}(t) \tag{101}$$

Taking LT from Equation 101 gives impedance force:

$$f_{m,n}^{imp}(s) = \sum_{m'=m-\mu}^{m'=m+\mu} \sum_{n'=n-\upsilon}^{n'=n+\upsilon} K_{-1,m-m',n-n'} U'_{1,m',n'}(s) \tag{102}$$

Thus, to find $f_{0,m,n}^{ext}$ we should calculate $U'_{1,m',n'}(s)$. This requires to omit $\hat{F}_0^{ext}$ from Equation 74, so we first take inverse discrete Fourier transform of (98) in terms of p:

$$\tilde{\boldsymbol{U}}_l'(q,r,s) = \boldsymbol{M}_A^{-1} \tilde{\boldsymbol{G}}_l(q,r,s) \hat{\boldsymbol{F}}_0^{ext}(q,r,s) + \tilde{\boldsymbol{R}}_l^d(q,r,s) \tag{103}$$

where

$$\tilde{\boldsymbol{R}}_l^d(q,r,s) = \sum_{l'=-L/2}^{\frac{L}{2}-1} \tilde{\boldsymbol{G}}_{l-l'}(q,r,s)\left(s\boldsymbol{U}_{l'}'(q,r,0) + \dot{\boldsymbol{U}}_{l'}'(q,r,0)\right) \tag{104}$$

Here L is the total number of unit cells in the x-direction. Letting l=0,1 and eliminating $\hat{F}_0^{ext}$ we have:

$$\tilde{\boldsymbol{U}}'_1(q,r,s) = \tilde{\boldsymbol{Q}}(q,r,s)\left(\tilde{\boldsymbol{U}}_0'(q,r,s) - \tilde{\boldsymbol{R}}_0^d(q,r,s)\right) + \tilde{\boldsymbol{R}}_1^d(q,r,s) \tag{105}$$

where

$$\tilde{Q}(q,r,s)=\tilde{G}_1(q,r,s)\tilde{G}_0^{-1}(q,r,s) \tag{106}$$

Taking IFT of 105 and using convolution property of the DFT, we gain

$$U'_{1,m,n}(s)=\sum_{m'=-M/2}^{\frac{M}{2}-1}\sum_{n'=-N/2}^{\frac{N}{2}-1}Q_{m-m',n-n'}(s)\left(U'_{0,m',n'}(s)-R^d_{0,m',n'}(s)\right)+R^d_{1,m,d}(s) \tag{107}$$

where upon plane l we can write:

$$R^d_{l,m,n}(s)=\sum_{l'=-L/2}^{\frac{L}{2}-1}\sum_{m'=-M/2}^{\frac{M}{2}-1}\sum_{n'=-N/2}^{\frac{N}{2}-1}G_{l-l',m-m',n-n'}(s)\left(sU'_{l',m',n'}(0)+\dot{U}'_{l',m',n'}(0)\right) \tag{108}$$

Taking ILT of Equation 108 gives

$$R^d_{l,m,n}(t)=\sum_{l'=-L/2}^{\frac{L}{2}-1}\sum_{m'=-M/2}^{\frac{M}{2}-1}\sum_{n'=-N/2}^{\frac{N}{2}-1}\left(\dot{g}_{l-l',m-m',n-n'}(t)u'_{l',m',n'}(0)+g_{l-l',m-m',n-n'}(t)\dot{u}'_{l',m',n'}(0)\right) \tag{109}$$

where

$$g_{l,m,n}(t)=L^{-1}\left(G_{l,m,n}(s)\right) \tag{110}$$

$$\dot{g}_{l,m,n}(t)=L^{-1}\left(sG_{l,m,n}(s)\right) \tag{111}$$

Substituting Equation 107 into Equation 102 and applying ILT, the impedance force boundary condition obtains as

$$f^{imp}_{m,n}(t)=\sum_{m'=-M/2}^{\frac{M}{2}-1}\sum_{n'=-N/2}^{\frac{N}{2}-1}\int_0^t\theta_{m-m',n-n'}(t-\tau)\left(u'_{0,m',n'}(\tau)-R^d_{0,m',n'}(\tau)\right)d\tau+R^f_{0,m,n}(t) \tag{112}$$

where the time history kernel $\theta(t)$ is defined to be

$$\theta_{m,n}(t)=L^{-1}\left(\Theta_{m,n}(s)\right) \tag{113}$$

$$\boldsymbol{\Theta}_{m,n}\left(s\right)=\sum_{m'=m-\mu}^{m+\mu}\sum_{n'=n-\upsilon}^{n+\upsilon}\boldsymbol{K}_{-1,m-m',n-n'}\boldsymbol{Q}_{m',n'}(s) \tag{114}$$

and

$$\boldsymbol{R}^{f}_{0,m,n}\left(t\right)=\sum_{l'=-L/2}^{\frac{L}{2}-1}\sum_{m'=-M/2}^{\frac{M}{2}-1}\sum_{n'=-N/2}^{\frac{N}{2}-1}\boldsymbol{K}_{-1,m-m',n-n'}\boldsymbol{R}^{d}_{l,m,n}(t) \tag{115}$$

where R^{d} is given by Equation 109.

For simplicity, we rewrite Equation 112 as

$$\boldsymbol{f}^{imp}_{m,n}\left(t\right)=\sum_{m'=-n_c}^{n_c}\sum_{n'=-n_c}^{n_c}\int_0^t\theta_{m-m',n-n'}(t-\tau)\left(\boldsymbol{u}'_{0,m',n'}\left(\tau\right)-\boldsymbol{R}^{d}_{0,m',n'}(\tau)\right)d\tau+\boldsymbol{R}^{f}_{0,m,n}(t) \tag{116}$$

where n_c refers to a maximum number of atomic neighbors, which will be used to compute the impedance force. Certainly, all of this n_c atoms are located in the MD region with linear or harmonic behavior.

As we evolved in Equation 92, we can write the final coupled form of the equations of motion as:

$$\boldsymbol{M}_{A}\ddot{\boldsymbol{q}}=\boldsymbol{f}\left(\boldsymbol{q}\right)+f^{imp}_{0,m,n} \tag{117}$$

$$\boldsymbol{M}\ddot{d}=\mathrm{N}^{\boldsymbol{T}}\boldsymbol{f}(\boldsymbol{u}) \tag{118}$$

where

$$\boldsymbol{f}^{imp}_{m,n}\left(t\right)=\sum_{m'=-n_c}^{n_c}\sum_{n'=-n_c}^{n_c}\int_0^t\theta_{m-m',n-n'}(t-\tau)\left(\boldsymbol{q}_{0,m',n'}\left(\tau\right)-\bar{\boldsymbol{u}}_{0,m',n'}(\tau)-\boldsymbol{R}^{d}_{0,m',n'}(\tau)\right)d\tau+\boldsymbol{R}^{f}_{0,m,n}(t) \tag{119}$$

Knowing that

$$\boldsymbol{u}'_{0,m',n'}\left(\tau\right)=\boldsymbol{q}_{0,m',n'}\left(\tau\right)-\bar{\boldsymbol{u}}_{0,m',n'}(\tau) \tag{120}$$

Note:

1. The impedance force f^{imp} became an appliance for us to vanish all fine- scale information that cannot be represented in the continuum. $f^{imp}=0$ means we do not have any unharmonic area in the MD region.

2. R^f and R^d are called random or stochastic forces. These forces arise from the initial conditions in the eliminated fine- scale degrees of freedom. These terms represent thermally motivated displacements and forces exerted on the reduced MD system by the removed fine- scale degrees of freedom.

Author details

Farzad Ebrahimi* and Ebrahim Heidari

*Address all correspondence to: febrahimy@eng.ikiu.ac.ir

Department of Mechanical Engineering, Faculty of Engineering, Imam Khomeini International University, Qazvin, Iran

References

[1] Allen, Matthew J., Vincent C. Tung, and Richard B. Kaner. "Honeycomb carbon: a review of graphene." Chemical reviews 110.1 (2009): 132 145.

[2] Geim, Andre Konstantin. "Graphene: status and prospects." science 324.5934 (2009): 1530 1534.

[3] Rao, C. N. R, et al. "Graphene: The New Two-Dimensional Nanomaterial." Angewandte Chemie International Edition 48.42 (2009): 7752-7777.

[4] Stankovich, Sasha, et al. "Synthesis and exfoliation of isocyanate-treated graphene oxide nanoplatelets." Carbon 44.15 (2006): 3342 3347.

[5] Stankovich, Sasha, et al. "Synthesis of graphene-based nanosheets via chemical reduction of exfoliated graphite oxide." Carbon 45.7 (2007): 1558 1565.

[6] Ferrari, Andrea C. "Raman spectroscopy of graphene and graphite: disorder, electron–phonon coupling, doping and nonadiabatic effects." Solid state communications 143.1 (2007): 47 57.

[7] Katsnelson, M. I., and K. S. Novoselov. "Graphene: new Bridge between condensed matter physics and quantum electrodynamics." Solid State Communications 143.1 (2007): 3 13.

[8] Bunch, J. Scott, et al. "Electromechanical resonators from graphene sheets." Science 315.5811 (2007): 490 493.

[9] Atalaya, Juan, Andreas Isacsson, and Jari M. Kinaret. "Continuum elastic modeling of graphene resonators." Nano Letters 8.12 (2008): 4196 4200.

[10] Sakhaee-Pour, A., M. T. Ahmadian, and A. Vafai. "Applications of single-layered graphene sheets as mass sensors and atomistic dust detectors." Solid State Communications 145.4 (2008): 168 172.

[11] Ramanathan, T., et al. "Functionalized graphene sheets for polymer nanocomposites." Nature nanotechnology 3.6 (2008): 327 331.

[12] Saremi, F., et al. "Adsorption of carbon monoxide on a (6, 6) armchair carbon nanotube: ab initio study." Journal of Physical and Theoretical Chemistry Islamic Azad University of Iran 4.4 (2008): 235 238.

[13] Ekinci, K. L. "Electromechanical transducers at the nanoscale: actuation and sensing of motion in nanoelectromechanical systems (NEMS)." small 1.8 9 (2005): 786 797.

[14] Eom, Kilho, et al. "Nanomechanical resonators and their applications in biological/ chemical detection: nanomechanics principles." Physics Reports 503.4 (2011): 115 163.

[15] Arlett, J. L., E. B. Myers, and M. L. Roukes. "Comparative advantages of mechanical biosensors." Nature nanotechnology 6.4 (2011): 203 215.

[16] Poot, Menno, and Herre SJ van der Zant. "Mechanical systems in the quantum regime." Physics Reports 511.5 (2012): 273 335.

[17] Peng, H. B., et al. "Ultrahigh frequency nanotube resonators." Physical Review Letters 97.8 (2006): 087203.

[18] Sakhaee-Pour, A., M. T. Ahmadian, and A. Vafai. "Potential application of single-layered graphene sheet as strain sensor." Solid State Communications 147.7 (2008): 336 340.

[19] Sheehan, Madoc E., and Paul N. Sharratt. "Molecular dynamics methodology for the study of the solvent effects on a concentrated Diels-Alder reaction and the separation of the post-reaction mixture." Computers and chemical engineering 22 (1998): S27-S33.

[20] Kresse, Georg, and Jürgen Hafner. "Ab initio molecular dynamics for liquid metals." Physical Review B 47.1 (1993): 558.

[21] Gumbsch, Peter, and Rowland M. Cannon. "Atomistic aspects of brittle fracture." MRS Bulletin 25.05 (2000): 15 20.

[22] Müser, Martin H. "Towards an atomistic understanding of solid friction by computer simulations." Computer physics communications 146.1 (2002): 54 62.

[23] Horstemeyer, M. F., M. I. Baskes, and S. J. Plimpton. "Computational nanoscale plasticity simulations using embedded atom potentials." Theoretical and applied fracture mechanics 37.1 (2001): 49 98.

[24] Abraham, Farid F., et al. "Spanning the continuum to quantum length scales in a dynamic simulation of brittle fracture." Europhysics Letters 44.6 (1998): 783.

[25] Broughton, Jeremy Q., et al. "Concurrent coupling of length scales: methodology and application." Physical review B 60.4 (1999): 2391.

[26] Tadmor, Ellad B., Michael Ortiz, and Rob Phillips. "Quasicontinuum analysis of defects in solids." Philosophical magazine A 73.6 (1996): 1529 1563.

[27] Tadmor, E. B., Rob Phillips, and M. Ortiz. "Mixed atomistic and continuum models of deformation in solids." Langmuir 12.19 (1996): 4529 4534.

[28] Shenoy, V. B., et al. "Quasicontinuum models of interfacial structure and deformation." Physical Review Letters 80.4 (1998): 742.

[29] Miller, R., et al. "Quasicontinuum models of fracture and plasticity." Engineering Fracture Mechanics 61.3 (1998): 427 444.

[30] Miller, R., et al. "Quasicontinuum simulation of fracture at the atomic scale." Modelling and Simulation in Materials Science and Engineering 6.5 (1998): 607.

[31] Ortiz, M., et al. "Mixed Atomistic–Continuum models of material behavior: The art of transcending atomistics and informing continua." Mrs Bulletin 26.03 (2001): 216 221.

[32] Wagner, Gregory J., and Wing Kam Liu. "Coupling of atomistic and continuum simulations using a bridging scale decomposition." Journal of Computational Physics 190.1 (2003): 249 274.

[33] Geim, Andre Konstantin. "Graphene: status and prospects." science 324.5934 (2009): 1530 1534.

[34] Hirth, John P., and Jens Lothe. "Theory of dislocations." (1982).

[35] Sholl, David S. "Structural Nanocrystalline Materials. Fundamentals and Applications. Von Carl C. Koch, Ilya A. Ovid'ko, Sudipta Seal und Stan Veprek." Angewandte Chemie 120.6 (2008): 1022 1022.

[36] Zhu, T., and Ju Li. "J. Ultra-strength materials Prog." Mater. Sci 55 (2010): 710 757.

[37] Lee, Changgu, et al. "Measurement of the elastic properties and intrinsic strength of monolayer graphene." science 321.5887 (2008): 385 388.

[38] Liu, Fang, Pingbing Ming, and Ju Li. "Ab initio calculation of ideal strength and phonon instability of graphene under tension." Physical Review B 76.6 (2007): 064120.

[39] Kim, Kwanpyo, et al. "Ripping graphene: preferred directions." Nano letters 12.1 (2011): 293 297.

[40] Sen, Dipanjan, et al. "Tearing graphene sheets from adhesive substrates produces tapered nanoribbons." Small 6.10 (2010): 1108 1116.

[41] Meyer, Jannik C., et al. "Direct imaging of lattice atoms and topological defects in graphene membranes." Nano letters 8.11 (2008): 3582 3586.

[42] Warner, Jamie H., et al. "Dislocation-driven deformations in graphene." Science 337.6091 (2012): 209 212.

[43] Kim, Kwanpyo, et al. "Grain boundary mapping in polycrystalline graphene." ACS nano 5.3 (2011): 2142 2146.

[44] Huang, P. Y., et al. "Grains and grain boundaries in single-layer graphene atomic patchwork quilts." Nature 469 (2011): 389 392.

[45] Ruiz-Vargas, Carlos S., et al. "Softened elastic response and unzipping in chemical vapor deposition graphene membranes." Nano Letters 11.6 (2011): 2259 2263.

[46] Warner, Jamie H., et al. "Dislocation-driven deformations in graphene." Science 337.6091 (2012): 209 212.

[47] Ong, Zhun-Yong, Eric Pop, and Junichiro Shiomi. "Reduction of phonon lifetimes and thermal conductivity of a carbon nanotube on amorphous silica." Physical Review B 84.16 (2011): 165418.

[48] Kang, Kwangu, et al. "Lifetimes of optical phonons in graphene and graphite by time-resolved incoherent anti-Stokes Raman scattering." Physical Review B 81.16 (2010): 165405.

[49] Qiu, Bo, and Xiulin Ruan. "Reduction of spectral phonon relaxation times from suspended to supported graphene." Applied Physics Letters 100.19 (2012): 193101.

[50] Tohei, Tetsuya, et al. "Debye temperature and stiffness of carbon and boron nitride polymorphs from first principles calculations." Physical Review B 73.6 (2006): 064304.

[51] Nicklow, R., N. Wakabayashi, and H. G. Smith. "Lattice dynamics of pyrolytic graphite." Physical Review B 5.12 (1972): 4951.

[52] Nihira, Takeshi, and Tadao Iwata. "Temperature dependence of lattice vibrations and analysis of the specific heat of graphite." Physical Review B 68.13 (2003): 134305.

[53] Chen, Shanshan, et al. "Raman measurements of thermal transport in suspended monolayer graphene of variable sizes in vacuum and gaseous environments." ACS nano 5.1 (2010): 321 328.

[54] Balandin, Alexander A. "Thermal properties of graphene and nanostructured carbon materials." Nature materials 10.8 (2011): 569 581.

[55] Chen, Shanshan, et al. "Thermal conductivity of isotopically modified graphene." Nature materials 11.3 (2012): 203 207.

http://dx.doi.org/10.5772/61564

[56] Pettes, Michael Thompson, et al. "Influence of polymeric residue on the thermal conductivity of suspended bilayer graphene." Nano Letters 11.3 (2011): 1195 1200.

[57] Anthony, T. R., et al. "Thermal diffusivity of isotopically enriched C 12 diamond." Physical Review B 42.2 (1990): 1104.

[58] Berman, R. "Thermal conductivity of isotopically enriched diamonds." Physical Review B 45.10 (1992): 5726.

[59] Pop, Eric. "Energy dissipation and transport in nanoscale devices." Nano Research 3.3 (2010): 147 169.

[60] Chen, Z., et al. "Thermal contact resistance between graphene and silicon dioxide." Applied Physics Letters 95.16 (2009): 161910.

[61] Koh, Yee Kan, et al. "Heat conduction across monolayer and few-layer graphenes." Nano Letters 10.11 (2010): 4363 4368.

[62] Mak, Kin Fai, Chun Hung Lui, and Tony F. Heinz. "Measurement of the thermal conductance of the graphene/SiO_2 interface." Applied Physics Letters 97.22 (2010): 221904 221904.

[63] Liao, Albert D., et al. "Thermally limited current carrying ability of graphene nanoribbons." Physical Review Letters 106.25 (2011): 256801.

Chapter 2

Modeling and Control of a Smart Single-Layer Graphene Sheet

A. Ghorbanpour Arani and F. Ebrahimi

Additional information is available at the end of the chapter

http://dx.doi.org/10.5772/61277

Abstract

In this study, a smart single-layer graphene sheet (SLGS) is analytically modeled and its buckling is controlled using coupled polyvinylidene fluoride (PVDF) nanoplates. A voltage is applied to the PVDF nanoplate in thickness direction in order to control the critical load of the SLGS. Electric potential distribution is assumed as a combination of a half-cosine and linear variation in order to satisfy the Maxwell equation. The exact analysis is performed for the case when all four ends are simply supported and in free electrical boundary condition. The nonlocal governing equations are derived through Hamilton's principle and energy method based on a nonlocal Mindlin plate theory. The detailed mathematical derivations are presented and numerical investigations are performed, while the emphasis is placed on investigating the effect of several parameters such as small-scale coefficient, stiffness of the internal elastic medium, graphene length, mode number, and external electric voltage on the buckling smart control of the SLGS in detail. It is explicitly shown that the imposed external voltage is an effective controlling parameter for buckling of the SLGS. Numerical results are presented to serve as benchmarks for design and smart control of nanodevices.

Keywords: Smart control, Mindlin plate theory, buckling analysis, coupled system, Pasternak model

1. Introduction

In recent years, nanostructural carbon materials have received considerable interest from the scientific community due to their superior properties. Among carbon-based nanomaterials, single-layered graphene sheet (SLGS) is defined as a flat monolayer of carbon atoms tightly packed into a two-dimensional honeycomb lattice, in which carbon atoms bond covalently with their neighbors [1,2]. Graphene sheets (GSs) possess extraordinary properties, such as strong mechanical strength (Young's modulus = 1.0 TPa), large thermal conductivity (thermal conductivity = 3000 W / Km), excellent electric conductivity (electric conductivity up to 6000 S / cm), high surface area, and unusual optical properties [3-5]. These unique properties have made GSs one of the most promising materials in many applications such as nanosensors, nano-electronics, nanocomposites, batteries, nano-oscillators, nano-actuators, nanoresonators, nano-optomechanical systems, supercapacitors, fuel cells, solar cells, and hydrogen storage [6, 7]. On the other hand, polyvinylidene fluoride (PVDF) is a piezoelectric polymer. It has superior properties, such as excellent dimensional stability, abrasion and corrosion resistance, high strength, and capability of maintaining its mechanical properties at elevated temperature [8]. In contrast to GS, PVDF is a smart material, since it has piezoelectric property. The specific characteristic of piezoelectric materials is their ability to produce an electric field when deformed, and undergo deformation when subjected to an electric field. The coupling nature of piezoelectric materials has wide applications in electromechanical and electric devices, such as electromechanical actuators, sensors, and transducers [9].

Understanding mechanical behaviors of GSs is a key step for designing many nano-electro-mechanical system (NEMS) devices. Especially, stability response of GSs as NEMS component has great importance. Many studies have been carried out on the basis of nonlocal elasticity theory, which was initiated in the papers of Eringen [10-12]. He regarded the stress state at a given point as a function of the strain states of all points in the body, while the local continuum mechanics assumes that the stress state at a given point depends uniquely on the strain state at the same point. Pradhan and Murmu [13] studied small-scale effect on the buckling analysis of SLGS embedded in an elastic medium based on nonlocal plate theory. They found that the buckling loads of SLGS are strongly dependent on the small-scale coefficients and the stiffness of the surrounding elastic medium. Explicit analytical expressions for the critical buckling stresses in a monolayer GS based on nonlocal elasticity were investigated by Ansari and Rouhi [14]. They concluded that with the appropriate selection of the nonlocal parameter, the nonlocal relations are capable of yielding excellent results from the static deflection of monolayer GS under a uniformly distributed load. Also, their results showed that the importance of the small length scale is dependent on the boundary conditions of monolayer GS. Akhavan et al. [15] introduced exact solutions for the buckling analysis of rectangular Mindlin plates subjected to uniformly and linearly distributed in-plane loading on two opposite edges simply supported resting on elastic foundation. Their results indicated that the buckling load parameter increases as the thickness-to-length ratio decreases. Hashemi and Samaei [16] proposed an analytical solution for the buckling analysis of rectangular nanoplates based on the nonlocal Mindlin plate theory. They graphically presented the effects of small length scale on buckling loads for different geometrical parameters.

Samaei et al. [17] employed nonlocal Mindlin plate theory to analyze buckling of SLGS embedded in an elastic medium. They found that the effects of small length scale and surrounding elastic medium are significant to the mechanical behavior of nanoplates or SLGS and cannot be ignored. Furthermore, they showed that the nonlocal assumptions present larger buckling loads and stiffness of elastic medium in comparison to classical plate theory.

With respect to developmental works on mechanical behavior analysis of SLGS, it should be noted that none of the research mentioned above [13-17] have considered coupled double-nanoplate system. Herein, Murmu and Adhikari [18] analyzed nonlocal vibration of bonded double-nanoplate systems. Their study highlighted that the small-scale effects considerably influence the transverse vibration of NDNPS. Besides, they elucidated that the increase of the stiffness of the coupling springs in nonlocal double-nanoplate system (NDNPS) reduces the small-scale effects during the asynchronous modes of vibration. Also, nonlocal buckling behavior of bonded double-nanoplate system was studied by Murmu et al. [19] who showed that the nonlocal effects in coupled system are higher with increasing values of the nonlocal parameter for the case of synchronous buckling modes than in the asynchronous buckling modes. Moreover, their analytical results indicated that the increase of the stiffness of the coupling springs in the double-GS system reduces the nonlocal effects during the asynchronous modes of buckling. Both papers [18, 19] have considered Winkler model for simulation of elastic medium between two nanoplates. In this simplified model, a proportional interaction between pressure and deflection of SLGSs is assumed, which is carried out in the form of discrete and independent vertical springs. Whereas, Pasternak suggested considering not only the normal stresses but also the transverse shear deformation and continuity among the spring elements, and its subsequent applications for developing the model for buckling analysis proved to be more accurate than the Winkler model. To the best of our knowledge, none of works in the literature have taken into account the Pasternak model for coupled system. This study aims to couple SLGS with PVDF nanoplate by an elastic medium which is simulated by the Pasternak model.

The use of smart materials has received considerable attention due to their higher potential applicability for the mechanical behavior control in various research areas. These materials can produce control forces to structural elements according to the applied voltage. A number of researchers utilized the piezoelectric materials to control mechanical behavior. For example, analysis of composite plates with piezoelectric actuators for vibration controls using layerwise displacement theory was performed by Han and Lee [20]. Piezoelectric control of composite plate vibration was carried out by Pietrzakowski [21], who considered effect of electric potential distribution. Recently, Liao et al. [22] examined nonlinear vibration of the piezoelectric nanobeams based on the nonlocal theory by considering external voltage. They assumed an electric potential as a combination of cosine and linear variation. They observed that the positive/negative voltage decreases/increases the linear and nonlinear frequencies of the nanobeam. None of the aforementioned studies [20-22] have considered a nanostructure (e.g., SLGS) whose vibrational behavior is controlled by an elastically bonded smart material (e.g., PVDF nanoplate).

However, to date, no report has been found in the literature on buckling analysis and smart control of SLGS using elastically coupled PVDF nanoplate based on nonlocal Mindlin plate theory. Motivated by these considerations, in order to improve optimum design of nanostructures, we aim to investigate the buckling smart control of SLGS based on nonlocal Mindlin plate theory. Herein, SLGS is elastically coupled with PVDF nanoplate and controlled by applying external electric voltage in thickness direction of PVDF nanoplate. The influences of external electric voltage, small-scale parameter, elastic medium, length of SLGS, and axial half wave number on buckling behavior of SLGS have been taken into account.

2. Review of nonlocal piezoelasticity and Mindlin plate theory

2.1. Nonlocal piezoelasticity

Based on the theory of nonlocal piezoelasticity, the stress tensor and the electric displacement at a reference point depend not only on the strain components and electric-field components at same position but also on all other points of the body. The nonlocal constitutive behavior for the piezoelectric material can be given as follows [22]:

$$\sigma_{ij}^{nl}(x)=\int_{v}\alpha(|x-x'|,\tau)\sigma_{ij}^{l}dV(x'),\qquad \forall x\in V \tag{1}$$

$$D_{k}^{nl}=\int_{v}\alpha(|x-x'|,\tau)D_{k}^{l}dV(x'),\qquad \forall x\in V \tag{2}$$

where σ_{ij}^{nl} and σ_{ij}^{l} are, respectively, the nonlocal stress tensor and local stress tensor; D_k^{nl} and D_k^{l} are the components of the nonlocal and local electric displacement; $\alpha(|x-x'|,\tau)$ is the nonlocal modulus; $|x-x'|$ is the Euclidean distance; and $\tau=e_0a/l$ is defined such that l is the external characteristic length, e_0 denotes constant appropriate to each material, and a is the internal characteristic length of the material. Consequently, e_0a is a constant parameter which is obtained with molecular dynamics, experimental results, experimental studies, and molecular structure mechanics. The constitutive equation of the nonlocal elasticity can be written as [20]:

$$(1-\mu\nabla^2)\sigma_{ij}^{nl}=\sigma_{ij}^{l}, \tag{3}$$

where the parameter $\mu=(e_0a)^2$ denotes the small-scale effect on the response of structures in nanosize, and ∇^2 is the Laplacian operator in the above equation. Similarly, Eq. (2) can be written as [22]:

$$(1-\mu\nabla^2)D_{k}^{nl}=D_{k}^{l}. \tag{4}$$

2.2. Mindlin plate theory

Based on the Mindlin plate theory, the displacement field can be expressed as [15-17]:

$$\begin{aligned}u_x(x,y,z,t)&=z\psi_x(x,y,t),\\ u_y(x,y,z,t)&=z\psi_y(x,y,t),\\ u_z(x,y,z,t)&=w(x,y,t),\end{aligned} \tag{5}$$

where $\psi_x(x, y)$ and $\psi_y(x, y)$ are the rotations of the normal to the mid-plane about x- and y-directions, respectively.

The von Kármán strains associated with the above displacement field can be expressed in the following form:

$$\varepsilon_{xx} = z\frac{\partial \psi_x}{\partial x}, \varepsilon_{yy} = z\frac{\partial \psi_y}{\partial y}, \gamma_{yz} = \frac{\partial w}{\partial y} + \psi_y, \gamma_{xz} = \frac{\partial w}{\partial x} + \psi_x, \gamma_{xy} = z(\frac{\partial \psi_x}{\partial y} + \frac{\partial \psi_y}{\partial x}), \tag{6}$$

where $(\varepsilon_{xx}, \varepsilon_{yy})$ are the normal strain components and $(\gamma_{yz}, \gamma_{xz}, \gamma_{xy})$ are the shear strain components.

3. Modeling of the problem

Consider a coupled SLGS–PVDF nanoplate system as depicted in Fig. 1, in which geometrical parameters of length L , width b, and thickness h are indicated. The SLGS and PVDF nanoplate are coupled by an elastic medium which is simulated by Pasternak foundation. As is well-known, this foundation model is capable of both transverse shear loads (k_g) and normal loads (k_w). The PVDF nanoplate is subjected to external electric voltage ϕ in thickness direction which is used for buckling smart control of SLGS.

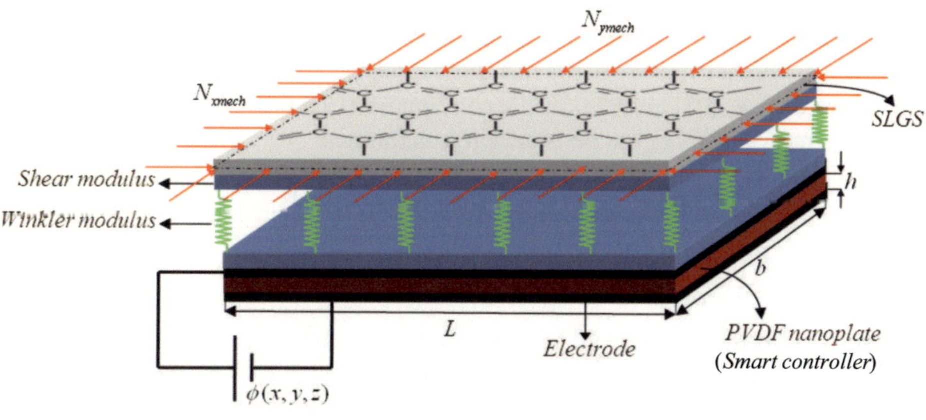

Figure 1. SLGS coupled by a Pasternak foundation with PVDF nanoplate subjected to applied external electric potential in thickness direction.

3.1. Constitutive relations for PVDF nanoplate

In a piezoelectric material, application of an electric field will cause a strain proportional to the mechanical field strength, and vice versa. The constitutive equation for stresses σ and strains ε matrix on the mechanical side, as well as flux density D and field strength E matrix on the electrostatic side, may be arbitrarily combined as follows [23]:

$$\begin{Bmatrix} \sigma_{xx} \\ \sigma_{yy} \\ \sigma_{yz} \\ \sigma_{zx} \\ \sigma_{xy} \end{Bmatrix} = \begin{bmatrix} C_{11}^{PVDF} & C_{12}^{PVDF} & 0 & 0 & 0 \\ C_{21}^{PVDF} & C_{22}^{PVDF} & 0 & 0 & 0 \\ 0 & 0 & C_{44}^{PVDF} & 0 & 0 \\ 0 & 0 & 0 & C_{55}^{PVDF} & 0 \\ 0 & 0 & 0 & 0 & C_{66}^{PVDF} \end{bmatrix} \begin{Bmatrix} \varepsilon_{xx} \\ \varepsilon_{yy} \\ \gamma_{yz} \\ \gamma_{xz} \\ \gamma_{xy} \end{Bmatrix} - \begin{bmatrix} 0 & 0 & e_{31} \\ 0 & 0 & e_{32} \\ 0 & e_{24} & 0 \\ e_{15} & 0 & 0 \\ 0 & 0 & 0 \end{bmatrix} \begin{Bmatrix} E_{xx} \\ E_{yy} \\ E_{zz} \end{Bmatrix}, \tag{7}$$

$$\begin{Bmatrix} D_{xx} \\ D_{yy} \\ D_{zz} \end{Bmatrix} = \begin{bmatrix} 0 & 0 & 0 & e_{15} & 0 \\ 0 & 0 & e_{24} & 0 & 0 \\ e_{31} & e_{32} & 0 & 0 & 0 \end{bmatrix} \begin{Bmatrix} \varepsilon_{xx} \\ \varepsilon_{yy} \\ \gamma_{yz} \\ \gamma_{xz} \\ \gamma_{xy} \end{Bmatrix} + \begin{bmatrix} \in_{11} & 0 & 0 \\ 0 & \in_{22} & 0 \\ 0 & 0 & \in_{33} \end{bmatrix} \begin{Bmatrix} E_{xx} \\ E_{yy} \\ E_{zz} \end{Bmatrix}, \tag{8}$$

where C_{ij}^{PVDF}, e_{ij}, and $\in_{ij}$ denote elastic, piezoelectric, and dielectric coefficients, respectively.

Also, electric field $E_i\,(i=x,\ y,\ z)$ in terms of electric potential (ϕ) is given as follows [23]:

$$E = -\nabla\Phi. \tag{9}$$

The electric potential distribution in the thickness direction of the PVDF nanoplate in the form proposed by references [21] and [24] as the combination of a half-cosine and linear variation which satisfies the Maxwell equation is adopted as follows:

$$\Phi(x,y,z,t) = -\cos(\frac{\pi z}{h})\phi(x,y,t) + \frac{2zV_0}{h}e^{i\Omega t} \tag{10}$$

where $\phi(x,\ y,\ t)$ is the time and spatial distribution of the electric potential caused by bending which must satisfy the electric boundary conditions, V_0 is external electric voltage, and Ω is the natural frequency of the SLBNS which is zero for buckling analysis.

The strain energy of the PVDF nanoplate can be expressed as:

$$U = \frac{1}{2}\int_{\Omega_0}\int_{-h/2}^{h/2}\left(\sigma_{xx}\varepsilon_{xx} + \sigma_{yy}\varepsilon_{yy} + \sigma_{xy}\gamma_{xy} + \sigma_{xz}\gamma_{xz} + \sigma_{yz}\gamma_{yz} - D_{xx}E_{xx} - D_{yy}E_{yy} - D_{zz}E_{zz}\right) \tag{11}$$

Combining Eqs. (6)–(11) yields:

$$U=\frac{1}{2}\int_{\Omega_0}\left(M_{xx}\frac{\partial\psi_x}{\partial x}+M_{yy}\frac{\partial\psi_y}{\partial x}+M_{xy}\left(\frac{\partial\psi_x}{\partial y}+\frac{\partial\psi_y}{\partial x}\right)+Q_{xz}\left(\frac{\partial w_0}{\partial x}+\psi_x\right)+Q_{yz}\left(\frac{\partial w_0}{\partial y}+\psi_y\right)\right)dxdy$$
$$-\frac{1}{2}\int_{\Omega_0}\int_{-h/2}^{h/2}D_{xx}\left(\cos(\frac{\pi z}{h})\times\frac{\partial\phi}{\partial x}\right)+D_{yy}\left(\cos(\frac{\pi z}{h})\times\frac{\partial\varphi}{\partial x}\right)-D_{zz}\left(\frac{\pi}{h}\sin(\frac{\pi z}{h})\phi+\frac{2V_0}{h}\right)dzdxdy, \quad (12)$$

where the stress resultant-displacement relations can be written as:

$$\left\{(N_{xx},N_{yy},N_{xy}),(M_{xx},M_{yy},M_{xy})\right\}=\int_{-h/2}^{h/2}\left\{\sigma_{xx},\sigma_{yy},\tau_{xy}\right\}(1,z)dz, \quad (13)$$

$$\left\{Q_{xx},Q_{yy}\right\}=K\int_{-h/2}^{h/2}\left\{\tau_{xz},\tau_{yz}\right\}dz, \quad (14)$$

in which K is shear correction coefficient.

The external work due to surrounding elastic medium can be written as:

$$W=\frac{1}{2}\int_0^L q\,w_2 dx, \quad (15)$$

where q is related to Pasternak foundation. Finally, using Hamilton's principles lead to the following governing equations:

$$\frac{\partial M_{xx}^{PVDF}}{\partial x}+\frac{\partial M_{xy}^{PVDF}}{\partial y}-Q_{xx}^{PVDF}=0, \quad (16)$$

$$\frac{\partial M_{xy}^{PVDF}}{\partial x}+\frac{\partial M_{yy}^{PVDF}}{\partial y}-Q_{yy}^{PVDF}=0, \quad (17)$$

$$\frac{\partial Q_{xx}^{PVDF}}{\partial x}+\frac{\partial Q_{yy}^{PVDF}}{\partial y}-\left(1-\mu\nabla^2\right)\left[k_w\left(w_1-w_2\right)-k_g\nabla^2\left(w_1-w_2\right)+(N_{xm}+N_{xe})\frac{\partial^2 w}{\partial x^2}+(N_{ym})\frac{\partial^2 w}{\partial y^2}=0\right], \quad (18)$$

$$\int_{-h/2}^{h/2}\left[\cos(\frac{\pi z}{h})\frac{\partial D_{xx}}{\partial x}+\cos(\frac{\pi z}{h})\frac{\partial D_{yy}}{\partial x}+\frac{\pi}{h}\sin(\frac{\pi z}{h})D_{zz}\right]dz=0, \quad (19)$$

in which k_w and k_g are Winkler and shear coefficients of Pasternak medium, respectively. Also, mechanical force are zero (i.e. $N_{xm}=N_{ym}=0$) and electrical force is $N_{xe}=2e_{11}V_0$. Substituting Eqs.

(7) and (8) into Eqs. (13) and (14), the stress resultant-displacement relations and electric displacement for PVDF nanoplate can be obtained as follows:

$$(1-\mu\nabla^2)M_{xx}^{PVDF} = \frac{24he_{31}\phi + \pi h^3 C_{11}^{PVDF}\frac{\partial\psi_{xx}}{\partial x} + \pi h^3 C_{12}^{PVDF}\frac{\partial\psi_{yy}}{\partial y}}{12\pi}, \tag{20}$$

$$(1-\mu\nabla^2)M_{yy}^{PVDF} = \frac{24he_{32}\phi + \pi h^3 C_{11}^{PVDF}\frac{\partial\psi_{yy}}{\partial y} + \pi h^3 C_{12}^{PVDF}\frac{\partial\psi_{xx}}{\partial x}}{12\pi}, \tag{21}$$

$$(1-\mu\nabla^2)M_{xy}^{PVDF} = \frac{h^3 C_{66}^{PVDF}}{12}\left(\frac{\partial\psi_{xx}}{\partial y} + \frac{\partial\psi_{yy}}{\partial x}\right), \tag{22}$$

$$(1-\mu\nabla^2)Q_{xx}^{PVDF} = \frac{Kh\left(-2e_{15}\frac{\partial\phi}{\partial x} + C_{55}^{PVDF}\pi(\psi_{xx} + \frac{\partial w}{\partial y})\right)}{\pi}, \tag{23}$$

$$(1-\mu\nabla^2)Q_{yy}^{PVDF} = \frac{Kh\left(-2e_{24}\frac{\partial\phi}{\partial x} + C_{44}^{PVDF}\pi(\psi_{yy} + \frac{\partial w}{\partial y})\right)}{\pi}, \tag{24}$$

$$\int_{-h/2}^{h/2}\left[(1-\mu\nabla^2)D_{xx}\right]\cos(\frac{\pi z}{h})\,dz = \left(\int_{-h/2}^{h/2} e_{15}\cos(\frac{\pi z}{h})dz\right)\left(\frac{\partial w}{\partial x} + \psi_{xx}\right) + \left(\int_{-h/2}^{h/2} \epsilon_{11}\cos(\frac{\pi z}{h})^2\,dz\right)\frac{\partial\phi}{\partial x}, \tag{25}$$

$$\int_{-h/2}^{h/2}\left[(1-\mu\nabla^2)D_{yy}\right]\cos(\frac{\pi z}{h})\,dz = \left(\int_{-h/2}^{h/2} e_{24}\cos(\frac{\pi z}{h})dz\right)\left(\frac{\partial w}{\partial y} + \psi_{yy}\right) + \left(\int_{-h/2}^{h/2} \epsilon_{22}\cos(\frac{\pi z}{h})^2\,dz\right)\frac{\partial\phi}{\partial x}, \tag{26}$$

$$\begin{aligned}\int_{-h/2}^{h/2}\left[(1-\mu\nabla^2)D_{zz}\right]\frac{\pi}{h}\sin(\frac{\pi z}{h})\,dz = &\left(\int_{-h/2}^{h/2} e_{31}\frac{\pi}{h}\sin(\frac{\pi z}{h})zdz\right)\frac{\partial\psi_{xx}}{\partial x} + \left(\int_{-h/2}^{h/2} e_{32}\frac{\pi}{h}\sin(\frac{\pi z}{h})zdz\right)\frac{\partial\psi_{yy}}{\partial y}\\ &-\left(\int_{-h/2}^{h/2}\epsilon_{33}\left(\frac{\pi}{h}\sin(\frac{\pi z}{h})\right)^2 dz\right)\phi.\end{aligned} \tag{27}$$

3.2. Constitutive relations for SLGS

The SLGS is subjected to uniform compressive edge loading along x and y axis. In order to obtain the governing equations of SLGS, the procedure outlined above for PVDF nanoplate can be expressed repeatedly by ignoring piezoelectric coefficient in Eq. (7) and electric displacement (Eq. (8)).

$$\frac{\partial M_{xx}^{GS}}{\partial x}+\frac{\partial M_{xy}^{GS}}{\partial y}-Q_{xx}^{GS}=0, \tag{28}$$

$$\frac{\partial M_{xy}^{GS}}{\partial x}+\frac{\partial M_{yy}^{GS}}{\partial y}-Q_{yy}^{GS}=0, \tag{29}$$

$$\frac{\partial Q_{xx}^{GS}}{\partial x}+\frac{\partial Q_{yy}^{GS}}{\partial y}-\left(1-\mu\nabla^2\right)\left[k_w\left(w_2-w_1\right)-k_g\nabla^2\left(w_2-w_1\right)+(N_{xm})\frac{\partial^2 w}{\partial x^2}+(N_{ym})\frac{\partial^2 w}{\partial y^2}=0\right]. \tag{30}$$

It is noted that the superscript PVDF nanoplate in section 3.1 can be changed to *GS* in this section.

4. Solution procedure

Steady-state solutions to the governing equations of the plate motion and the electric potential distribution which relate to the simply supported boundary conditions and zero electric potential along the edges of the surface electrodes can be assumed as [17,21]:

$$\begin{Bmatrix}\psi_{x1}(x,y,t)\\ \psi_{y1}(x,y,t)\\ w_1(x,y,t)\\ \psi_{x2}(x,y,t)\\ \psi_{y2}(x,y,t)\\ w_2(x,y,t)\\ \phi_2(x,y,t)\end{Bmatrix}=\begin{Bmatrix}\bar{\psi}_{x1}\cos(\frac{m\pi x}{L})\sin(\frac{n\pi y}{b})\\ \bar{\psi}_{y1}\sin(\frac{m\pi x}{L})\cos(\frac{n\pi y}{b})\\ \bar{w}_1\sin(\frac{m\pi x}{L})\sin(\frac{n\pi y}{b})\\ \bar{\psi}_{x2}\cos(\frac{m\pi x}{L})\sin(\frac{n\pi y}{b})\\ \bar{\psi}_{y2}\sin(\frac{m\pi x}{L})\cos(\frac{n\pi y}{b})\\ \bar{w}_2\sin(\frac{m\pi x}{L})\sin(\frac{n\pi y}{b})\\ \bar{\phi}_2\sin(\frac{m\pi x}{L})\sin(\frac{n\pi y}{b})\end{Bmatrix}. \tag{31}$$

As mentioned above, it is assumed that the SLGS plate is free from any transverse loadings. Uniform compressive edge loading along x and y axis are $N_{xm} = -P$ and $N_{ym} = -kP$, respectively. Substituting Eq. (31) into Eqs. (16)–(19) and (28)–(30) yields:

$$\left\{\begin{bmatrix} L_{11} & L_{12} & L_{13} & L_{14} & 0 & 0 & 0 \\ L_{21} & L_{22} & L_{23} & 0 & 0 & 0 & 0 \\ L_{31} & L_{32} & L_{33} & 0 & 0 & 0 & 0 \\ L_{41} & 0 & 0 & L_{44} & L_{45} & L_{46} & L_{47} \\ 0 & 0 & 0 & L_{54} & L_{55} & L_{56} & L_{57} \\ 0 & 0 & 0 & L_{64} & L_{65} & L_{66} & L_{67} \\ 0 & 0 & 0 & L_{74} & L_{75} & L_{76} & L_{77} \end{bmatrix} + P\xi_{mn}\eta_{mn} \begin{bmatrix} 1 & 0 & 0 & 0 & 0 & 0 & 0 \\ 0 & 0 & 0 & 0 & 0 & 0 & 0 \\ 0 & 0 & 0 & 0 & 0 & 0 & 0 \\ 0 & 0 & 0 & 0 & 0 & 0 & 0 \\ 0 & 0 & 0 & 0 & 0 & 0 & 0 \\ 0 & 0 & 0 & 0 & 0 & 0 & 0 \\ 0 & 0 & 0 & 0 & 0 & 0 & 0 \end{bmatrix}\right\} \begin{bmatrix} \bar{w}_1 \\ \bar{\psi}_{x1} \\ \bar{\psi}_{y1} \\ \bar{w}_2 \\ \bar{\psi}_{x2} \\ \bar{\psi}_{y2} \\ \bar{\phi}_2 \end{bmatrix} = \begin{bmatrix} 0 \\ 0 \\ 0 \\ 0 \\ 0 \\ 0 \\ 0 \end{bmatrix}, \tag{32}$$

where $\xi_{mn} = 1 + \mu\left(\left(\frac{m\pi}{L}\right)^2 + \left(\frac{n\pi}{b}\right)^2\right)$, $\eta_{mn} = \left(\frac{m\pi}{L}\right)^2 + k\left(\frac{n\pi}{b}\right)^2$ and L_{ij} are defined in Appendix A. Finally, buckling load of the system (P) can be calculated by solving the above equation.

5. Numerical results

In this section, buckling smart control of SLGS using elastically bonded PVDF nanoplate is discussed so that the effects of nonlocal parameter, mode number, Pasternak foundation, and SLGS length on the buckling of the SLGS are also considered. For this purpose, buckling load ratio is defined as follows:

$$\text{Buckling Load Ratio} = \frac{\text{Buckling load from nonlocal theory } (P_{nl})}{\text{Buckling load from local theory } (P_l)}$$

The orthotropic mechanical properties of SLGS with thickness $h = 0.34\ nm$ are listed in Table 1 [25]. Also, mechanical and electrical characteristics of PVDF nanoplate with thickness $h = 2\ nm$ are presented in Table 1 [26], which are taken from the macroscopic piezoelectric materials. In fact, the polymeric piezoelectric nanomaterials' characteristics are size-dependent and determined by experiments or molecular dynamic (MD) simulation [22]. To the best of the author's knowledge, the experimental values of piezoelectric nanomaterials' properties are not available in the literature. Therefore, in order to quantitatively analyze the present work, the characteristics of PVDF nanoplate are assumed in macroscale.

The developed nonlocal theory to date is incapable of determining the small scaling parameter e_0a. However, Eringen [11] proposed $e_0a = 0.39\ nm$ by matching the dispersion curves using nonlocal theory for plane waves and Born–Karman model of lattice dynamics. For CNT, the

SLGS	PVDF
E_1=1765(GPa)	C_{11}=238.24 (GPa)
E_2=1588(GPa)	C_{22}=23.60 (GPa)
υ_{12}=0.3	C_{12}=3.98 (GPa)
υ_{21}=0.3	C_{44}=2.15 (GPa)
	C_{55}=4.40 (GPa)
	C_{66}=6.43 (GPa)
	$e_{31}=-0.13\,(C/m^2)$
	$e_{32}=-0.145\,(C/m^2)$
	$e_{24}=-0.276\,(C/m^2)$
	$e_{15}=-0.009\,(C/m^2)$
	$e_{32}=-0.145\,(C/m^2)$
	$\in_{11}=1.1068\times10^{-8}\,(F/m)$

Table 1. Material properties of PVDF and SLGS.

e_0a is found to be less than 2 nm [27]. Small-scale effects on the buckling behavior of SLGS and PVDF nanoplate were carried out analytically by assuming a range of values $0<e_0a<2\ nm$, since its exact value is not known [27].

In the absence of similar publications in the literature covering the same scope of the problem, one cannot directly validate the results found here. However, the present work could be partially validated based on a simplified analysis suggested by Samaei et al. [17], Pradhan [28], Murmu and Pradhan [29], and Hashemi and Samaei [16] on buckling of the SLGS for which the coupled PVDF nanoplate in this paper was ignored. For this purpose, an SLGS with $E=1\ TPa$, $\nu=0.3$, $h=0.34\ nm$, and $5<L<50\ nm$ [17] as well as a nanoplate with $E=1765\ GPa$, $\nu=0.25$, $h=0.34\ nm$, and $5<L<50\ nm$ [16] are considered.

On the same basis and assuming Mindlin plate theory for buckling of SLGS embedded in a Pasternak foundation, the results obtained here are compared with those of Samaei et al. [17]. The results are shown in Fig. 2, in which buckling load ratio versus nonlocal parameter is plotted for $K_g=0$ and $K_g=20$. It is noted that in this figure the shear modulus is normalized the same as [17], $K_g=k_gL^2/D$ in which $D=Eh^3/12(1-\nu^2)$. As can be seen, the two analyses agree well and show similar results. A comparison between the buckling analysis of SLGS using the theories of classical plate [29], higher-order shear deformation [28] and Mindlin plate [16] is presented in Table 2. In this table, critical buckling load for different values of nonlocal parameter and aspect ratio of length to thickness is shown. As can be seen, the present results based on Mindlin plate theory closely match with those reported by Hashemi and Samaei [16], Pradhan [28], and Murmu and Pradhan [29].

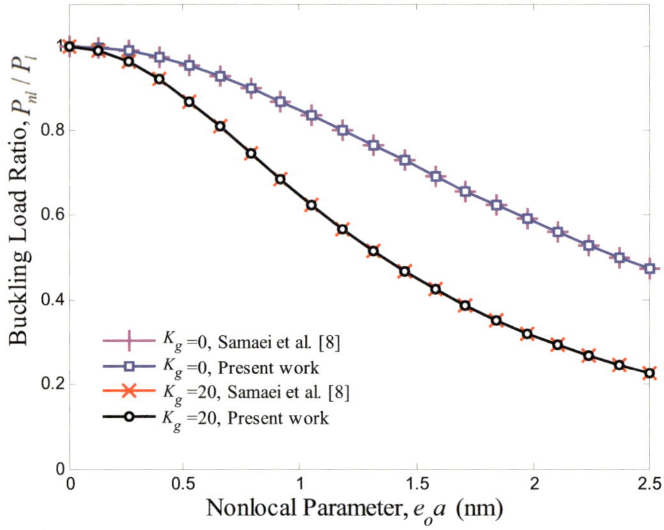

Figure 2. Comparison of buckling load ratio versus nonlocal parameter for $K_g = 0$ and $K_g = 20$.

L/h	μ	Dimensionless critical buckling load from higher order plate theory [28]	Dimensionless critical buckling load from classical plate theory [29]	Dimensionless critical buckling load from Mindlin plate theory (present work)	Dimensionless critical buckling load from Mindlin plate theory [16]
100	0.0	9.8791	9.8671	9.8671	9.8671
	0.5	9.4156	9.4031	9.4029	9.4028
	1.0	8.9947	8.9807	8.9803	8.9801
	1.5	8.6073	8.5947	8.5939	8.5939
	2.0	8.2537	8.2405	8.2393	8.2393
20	0.0	9.8177	9.8067	9.8067	9.8067
	0.5	9.3570	9.3455	9.3455	9.3451
	1.0	8.9652	8.9528	8.9527	8.9522
	1.5	8.5546	8.5421	8.5420	8.5419
	2.0	8.2114	8.1900	8.1898	8.1898

Table 2. Comparison between the buckling analysis of SLGS using the theories of classical plate, higher-order shear deformation and Mindlin plate.

The effect of the external electric voltage (V_0) on the buckling load ratio with respect to nonlocal parameter (μ) is demonstrated in Fig. 3. It is shown that applying positive electric potential can increase the buckling load ratio of the SLGS and vice versa. This is because the imposed positive and negative voltages generate the axial compressive and tensile forces in the PVDF nanoplate, respectively. Meanwhile, the effect of external voltage becomes more prominent at higher μ. Hence, the imposed external voltage is an effective controlling parameter for buckling of the SLGS which is coupled by a smart PVDF nanoplate. It is also concluded that increasing the μ decreases the buckling load ratio. This is due to the fact that the increase of nonlocal parameter decreases the interaction force between graphene sheet atoms, and that leads to a softer structure. Figure 4 illustrates the effect of mode number of SLGS on the variation of buckling load ratio versus μ. As can be seen, buckling load ratio decreases with increasing mode numbers. Also, the small-scale effects on the buckling load ratio become more distinguished at higher modes. Obviously, the difference between the buckling load ratios of the SLGS is larger at higher nonlocal parameters. Furthermore, the buckling load ratio for all mode numbers decreases by increasing the μ.

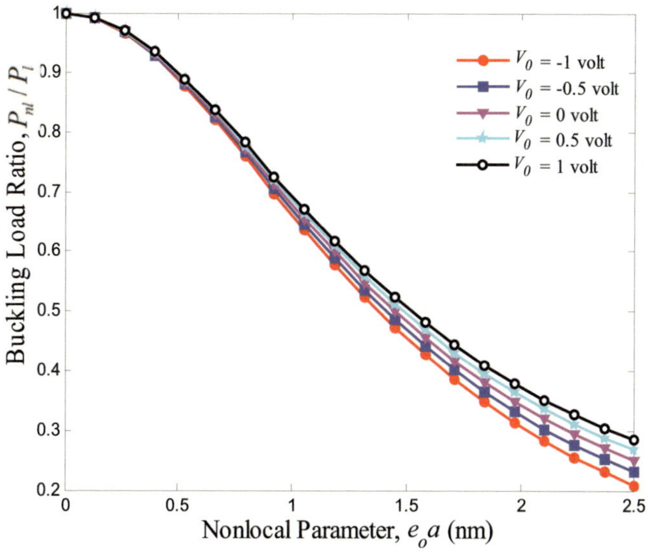

Figure 3. The effect of the external voltage on the buckling load ratio versus nonlocal parameter.

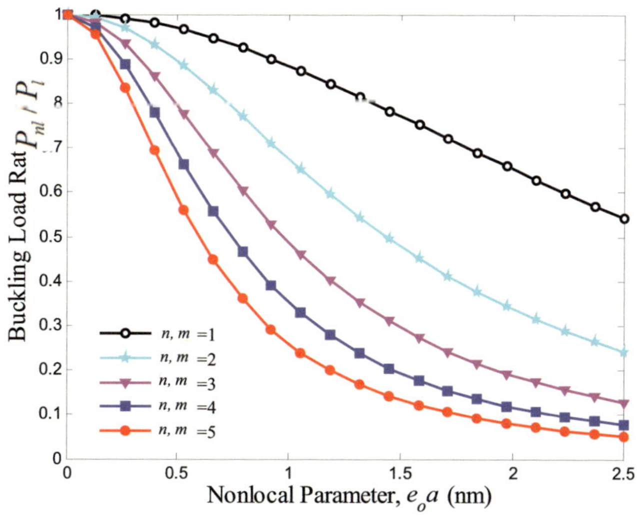

Figure 4. The effect of mode number on the buckling load ratio versus nonlocal parameter.

The effects of the SLGS length and the imposed external voltage to the PVDF nanoplate on the buckling load ratio are shown in Fig. 5. It is noted that the length of the SLGS is considered between $10<L<40$ *nm*, since the maximum length of the graphene sheet taken is 45.2896 *nm* in the literature by Sakhaee Pour [30], Pradhan and Murmu [13], and Samaei et al. [17]. As length of the graphene sheet increases, the buckling load ratio increases. It is also observed that for a given length, the SLGS, applying negative external voltage to PVDF nanoplate, will buckle first as compared to the SLGS with positive one.

Figure 6 depicts the effects of axial half wave number (m) and external voltage on the buckling load ratio of the SLGS. It is obvious that the buckling load ratio decreases sharply with increasing m. As can be seen, with the increase of external voltage, buckling load ratio is increased. Moreover, the effect of V_0 is not considerable for $m<2$. It means that the external voltage effect decreases with decrease of the axial half wave number.

Buckling smart control of SLGS using PVDF nanoplate versus shear modulus parameter (K_g) is plotted in Fig. 7. The obtained results show that at a given K_g, when the imposed external voltage changes from -1 V to 1 V, the buckling load ratio increases. It is also worth mentioning that the influences of V_0 at higher K_g values are more apparent than at lower K_g's. As the shear modulus parameter of the coupled system increases, generally, the buckling load ratio reduces and approaches a constant value. This is because increasing shear modulus parameter increases the structure stiffness.

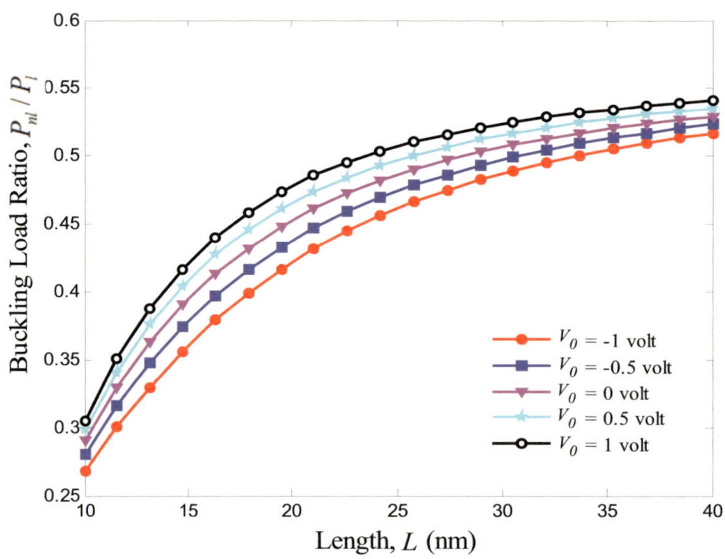

Figure 5. The effect of the external voltage on the buckling load ratio versus graphene length.

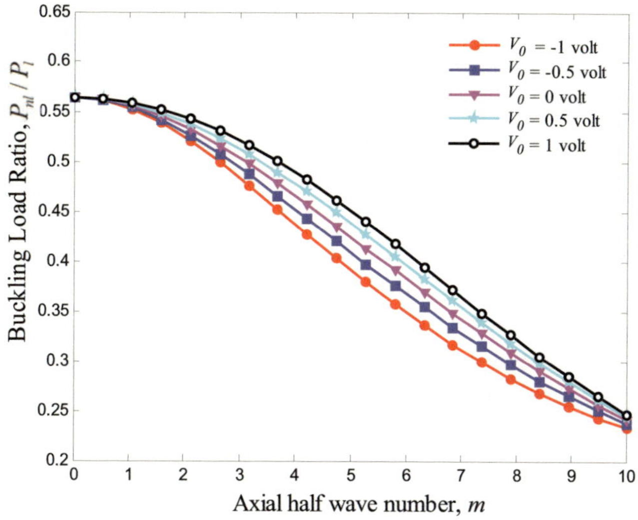

Figure 6. The effect of the external voltage on the buckling load ratio versus axial half wave number.

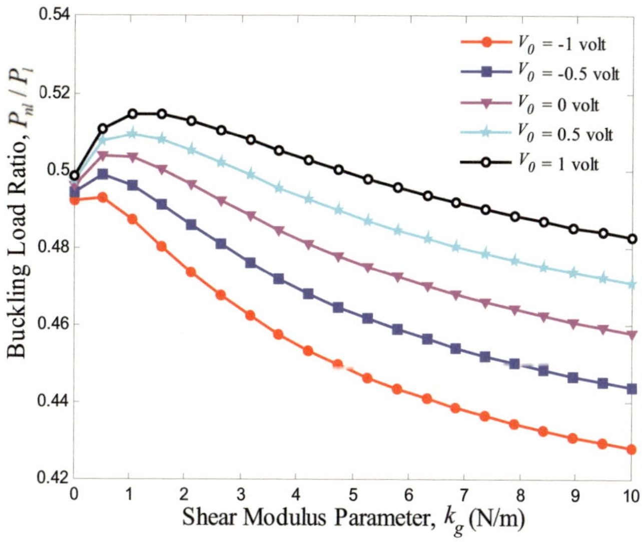

Figure 7. The effect of the external voltage on the buckling load ratio versus shear modulus parameter.

6. Conclusion

Buckling response of graphene sheets has applications in designing many NEMS/MEMS devices such as strain sensor, mass and pressure sensors, and atomic dust detectors. Buckling smart control of the SLGS using elastically bonded PVDF nanoplate which is subjected to external voltage is the main contribution of the present paper. The elastic medium between SLGS and PVDF nanoplate is simulated by a Pasternak foundation. The governing equations are obtained based on nonlocal Mindlin plate theory so that the effects of small-scale, elastic medium coefficient, mode number, and graphene length are discussed. The results indicate that the imposed external voltage is an effective controlling parameter for buckling of the SLGS. It is found that the effect of external voltage becomes more prominent at higher nonlocal parameter and shear modulus. It is also observed that for a given length, the SLGS with negative external voltage will buckle first as compared to the SLGS with positive one. The results of this study are validated as far as possible by the buckling of SLGS in the absence of PVDF nanoplate, as presented by [16, 17, 28, and 29]. Finally, it is hoped that the results presented in this paper would be helpful for study and design of bonded systems based on smart control and electromechanical systems.

Appendix A

$$L_{11} = -\frac{KC_{55}^{GS} m^2\pi^2 h}{L^2} - \frac{KC_{44}^{GS} n^2\pi^2 h}{b^2} + k_w + k_g\left(\frac{m^2\pi^2}{L^2} + \frac{n^2\pi^2}{b^2}\right) -$$

$$\mu\left(-\frac{k_w m^2\pi^2}{L^2} - k_g\left(\frac{m^4\pi^4}{L^4} + \frac{m^2\pi^4 n^2}{L^2 b^2}\right) - \frac{k_w n^2\pi^2}{b^2} - k_g\left(\frac{m^2\pi^4 n^2}{L^2 b^2} + \frac{n^4\pi^4}{b^4}\right)\right)$$

$$L_{12} = -\frac{KC_{55}^{GS} m\pi h}{L}$$

$$L_{13} = -\frac{KC_{44}^{GS} n\pi h}{b}$$

$$L_{14} = -k_g\left(\frac{m^2\pi^2}{L^2} + \frac{n^2\pi^2}{b^2}\right) - k_w -$$

$$\mu\left(\frac{k_w m^2\pi^2}{L^2} + K_g\left(\frac{m^4\pi^4}{L^4} + \frac{m^4\pi^4 n^2}{L^2 b^2}\right) + \frac{k_w n^2\pi^2}{b^2} + k_g\left(\left(\frac{m^2\pi^4 n^2}{L^2 b^2} + \frac{n^4\pi^4}{b^4}\right)\right)\right)$$

$$L_{21} = -\frac{KC_{55}^{GS} m\pi h}{L}$$

$$L_{22} = -\frac{C_{11}^{GS} m^2\pi^2 h^3}{12L^2} + \frac{C_{66}^{GS} n^2\pi^2 h^3}{12b^2} + KC_{55}^{GS} h$$

$$L_{23} = -\frac{m\pi^2 n h^3}{12Lb}(C_{12}^{GS} + C_{66}^{GS})$$

$$L_{31} = -\frac{KC_{44}^{GS} n\pi h}{b}$$

$$L_{32} = -\frac{m\pi^2 n h^3}{12Lb}(C_{12}^{GS} + C_{66}^{GS})$$

$$L_{33}=-\frac{C_{66}^{GS}m^2\pi^2h^3}{12L^2}-\frac{C_{11}^{GS}n^2\pi^2h^3}{12b^2}-KC_{44}^{GS}h$$

$$L_{41}=-k_w-k_g\left(\frac{m^2\pi^2}{L^2}+\frac{n^2\pi^2}{b^2}\right)-$$

$$\mu(\left(\frac{k_wn^2\pi^2}{b^2}+k_g\left(\frac{m^2\pi^4n^2}{L^2b^2}+\frac{n^4\pi^4}{b^4}\right)+\frac{k_wm^2\pi^2}{L^2}+k_g\left(\frac{m^4\pi^4}{L^4}+\frac{m^2\pi^4n^2}{L^2b^2}\right)\right)$$

$$L_{44}=-\frac{KC_{55}^{PVDF}m^2\pi^2h}{L^2}-\frac{KC_{44}^{PVDF}n^2\pi^2h}{b^2}-\frac{2e_{31}V_0m^2\pi^2}{L^2}+k_w+k_g\left(\frac{m^2\pi^2}{L^2}+\frac{n^2\pi^2}{b^2}\right)-$$

$$\mu\left(\frac{2e_{31}V_0m^4\pi^4}{L^4}-\frac{k_wm^2\pi^2}{L^2}-k_g\left(\frac{m^4\pi^4}{L^4}+\frac{m^2\pi^4n^2}{L^2b^2}\right)+\frac{2e_{31}V_0m^2\pi^4n^2}{L^2b^2}-\frac{k_wn^2\pi^2}{b^2}-k_g\left(\frac{m^2\pi^4n^2}{L^2b^2}+\frac{n^4\pi^4}{b^4}\right)\right)$$

$$L_{45}=-\frac{KC_{55}^{PVDF}m\pi h}{L}$$

$$L_{46}=-\frac{KC_{44}^{PVDF}n\pi h}{b}$$

$$L_{47}=\frac{2Khe_{15}m^2\pi}{L^2}+\frac{2Khe_{24}n^2\pi}{b^2}$$

$$L_{54}=-\frac{KC_{55}^{PVDF}m\pi h}{L}$$

$$L_{55}=-\frac{C_{11}^{PVDF}m^2\pi^2h^3}{12L^2}-\frac{C_{66}^{PVDF}n^2\pi^2h^3}{12b^2}-KC_{55}^{PVDF}h$$

$$L_{56}=-\frac{m\pi^2nh^3}{12Lb}(C_{12}^{PVDF}+C_{66}^{PVDF})$$

$$L_{57}=\frac{2he_{31}m+2Khe_{15}m}{L}$$

$$L_{64}=-\frac{KC_{44}^{PVDF}n\pi h}{b}$$

$$L_{65}=-\frac{m\pi^2nh^3}{12Lb}(C_{12}^{PVDF}+C_{66}^{PVDF})$$

$$L_{66}=-\frac{C_{66}^{PVDF}m^2\pi^2h^3}{12L^2}-\frac{C_{11}^{PVDF}n^2\pi^2h^3}{12b^2}-KC_{44}^{PVDF}h \tag{33}$$

$$L_{67}=\frac{2he_{32}n}{b}+\frac{2Khe_{24}n}{b}$$

$$L_{74}=-\frac{4m^2h^2b^2e_{15}\pi+4n^2h^2L^2e_{24}\pi}{2L^2b^2h}$$

$$L_{75}=-\frac{4mh^2b^2e_{15}L+4Lb^2e_{31}mh^2}{2L^2b^2h}$$

$$L_{76}=-\frac{4nh^2L^2e_{24}b+4L^2be_{32}nh^2}{2L^2b^2h}$$

$$L_{77}=-\frac{h^2m^2\pi^2\in_{11}b^2+\in_{33}\pi^2L^2b^2+h^2n^2\pi^2\in_{22}L^2}{2L^2b^2h}$$

Author details

A. Ghorbanpour Arani[1] and F. Ebrahimi[2*]

*Address all correspondence to: febrahimy@eng.ikiu.ac.ir

1 Faculty of Mechanical Engineering, Institute of Nanoscience & Nanotechnology University of Kashan, Kashan, Islamic Republic of Iran

2 Department of Mechanical Engineering, Faculty of Engineering, Imam Khomeini International University, Qazvin, Islamic Republic of Iran

References

[1] M. Yang, A. Javadi, H. Li, Sh. Gong, *Biosens Bioelec.* 26 (2010) 560–565.

[2] B. Arash, Q. Wang, K.M. Liew, *Comput Meth Appl Mech Engin.* 223 (2012) 1–9.

[3] J.S. Oh, Ta. Hwang, G.Y. Nam, J.P. Hong, A.H. Bae, S.I. Son, G.H. Lee, H.k. Sung, H.R. Choi, J.C. Koo, J.D. Nam, *Thin Solid Film* (2011) In Press.

[4] C. Chen, W. Fu, C. Yu, *Mat Lett.* 82 (2012) 132–136.

[5] C. Baykasoglu, *Comput Mat Sci.* 55 (2012) 228–236.

[6] Y. Qian, C. Wang, Z.G. Le, *Appl Surf Sci.* 257 (2011) 10758–10762.

[7] E. Jomehzadeh, A.R. Saidi, *Comput Mat Sci.* 50 (2011) 1043–1051.

[8] A.A. Mosallaie Barzoki, A. Ghorbanpour Arani, R. Kolahchi, M.R. Mozdianfard, *Appl Math Model.* 36 (2012) 2983-2995.

[9] A. Ghorbanpour Arani, R. Kolahchi, A.A. Mosallaie Barzoki, A. Loghman, *Appl Math Model.* 36 (2012) 139–157.

[10] A.C. Eringen, *Int J Engin Sci.* 10 (1972) 1–16.

[11] A.C. Eringen, *J Appl Phys.* 54 (1983) 4703–4710.

[12] A.C. Eringen, *Nonlocal Continuum Field Theories*, Springer-Verlag, New York, 2002.

[13] S.C. Pradhan, T. Murmu, *Physica E.* 42 (2010) 1293–1301.

[14] R. Ansari, H. Rouhi, *Solid State Comm.* 152 (2012) 56–59.

[15] H. Akhavan, Sh. Hosseini Hashemi, H. Rokni Damavandi Taher, A. Alibeigloo, Sh. Vahabi, *Comput Mat Sci.* 44 (2009) 968–978.

[16] Sh. Hosseini Hashemi, A. Tourki Samaei, *Physica E.* 43 (2011) 1400–1404.

[17] A.T. Samaei, S. Abbasion, M.M. Mirsayar, *Mech Res Comm*. 38 (2011) 481– 485.

[18] T. Murmu, S. Adhikari, *Compos Part B: Engin*. 42 (2011) 1901–1911.

[19] T. Murmu, J. Sienz, J. Adhikari, C. Arnold, *J Appl Phys*. 110 (2011) 084316-9.

[20] J.H. Han, I. Lee, *Compos Part B: Engin 29B* (1998) 621–632.

[21] M. Pietrzakowski, *Comput Struct*. 86 (2008) 948–954

[22] L.L. Ke, Y.Sh. Wang, Zh.D. Wang, *Compos Struct*. 96 (2012) 2038–2047.

[23] A. Ghorbanpour Arani, R. Kolahchi, A.A. Mosallaie Barzoki, *Appl Math Model*. 35 (2011) 2771–2789.

[24] S.T. Quek, Q. Wang, *Smart Mater Struct*. 9 (2000) 859–867.

[25] S.C. Pradhan, A. Kumar, *Comput Mat Sci*. 50 (2010) 239–245.

[26] A. Salehi-Khojin, N. Jalili, *Compos Sci Technol*. 68 (2008) 1489–1501.

[27] Q. Wang, *J Appl Phys*. 98 (2005) 124301.

[28] S.C. Pradhan, *Phys Lett*. A 373 (2009) 4182–4188.

[29] S.C. Pradhan, T. Murmu, *Comput Mat Sci*. 47 (2009) 268–274.

[30] A. Sakhaee-Pour, Comput Mat Sci. 45 (2009) 266–270.

Chapter 3

Review of Graphene Technology and Its Applications for Electronic Devices

Ashok K. Sood, Isaac Lund, Yash R. Puri, Harry Efstathiadis, Pradeep Haldar, Nibir K. Dhar, Jay Lewis, Madan Dubey, Eugene Zakar, Priyalal Wijewarnasuriya, Dennis L. Polla and Michael Fritze

Additional information is available at the end of the chapter

http://dx.doi.org/10.5772/61316

Abstract

Graphene has amazing abilities due to its unique band structure characteristics defining its enhanced electrical capabilities for a material with the highest characteristic mobility known to exist at room temperature. The high mobility of graphene occurs due to electron delocalization and weak electron–phonon interaction, making graphene an ideal material for electrical applications requiring high mobility and fast response times. In this review, we cover graphene's integration into infrared (IR) devices, electro-optic (EO) devices, and field effect transistors (FETs) for radio frequency (RF) applications. The benefits of utilizing graphene for each case are discussed, along with examples showing the current state-of-the-art solutions for these applications.

Graphene has many outstanding properties due to its unique bonding and subsequently band gap characteristics, having electronic carriers act as "massless" Dirac-Fermions. The material characteristics of graphene are anisotropic, having phenomenal characteristic within a single sheet and diminished material characteristics between sheet with increasing sheet number and grain boundaries. We will discuss the integration of graphene into many electronic device applications.

Graphene has the highest mobility values measured in a material at room temperature, allowing integration into fast response time devices such as a high electron mobility transistor (HEMT) for RF applications. Graphene has shown promise in IR detectors by utilizing graphene in thermal-based detection applications.

Keywords: Graphene Electronics (GE), Graphene Field Effect Transistor (GFET), Graphene Radio-Frequency (RF) Devices, Electro-Optical (EO) Devices, Infra-Red (IR) Detectors

1. Introduction

Graphene is a two-dimensional (2D) analogue of graphite (carbon, or C) material that has exceptional characteristics derived from the bonding characteristics of C bonding sheets. C has four valence electrons, with three of these electrons participating in σ-bonding with its closest neighbors, creating a honeycomb structure. [1] The fourth of these valence electrons occupies an orbital perpendicular to the one-dimensional (1D) sheet creating delocalized π-bonding, as shown in Figure 1, which allows for the creation of a two-dimensional electron gas (2DEG) with high mobility within the sheets. [1, 2]

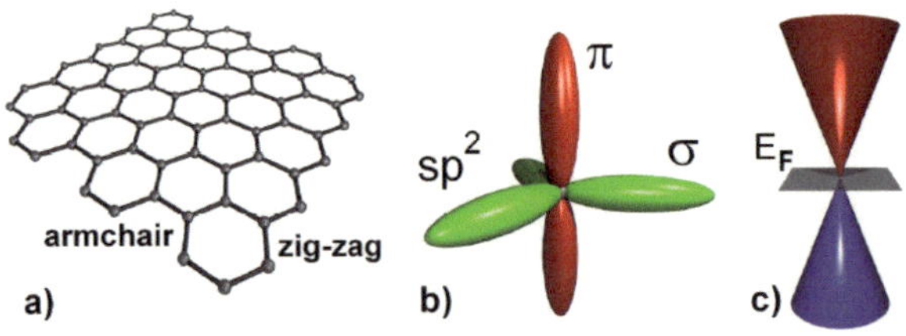

Figure 1. Graphene geometry, bonding, and a related band diagram [1].

The delocalization of the π-bonding electrons allows for the graphene sheets to have high mobility, up to 15,000–200,000 cm^2/Vs, limited by interactions with the substrate, any contaminant particles, or from itself during bilayer growth. [1, 3–7] This makes cleanliness, grain size, and substrate interference very important issues for growing and using graphene for high mobility and ultrafast applications.

In this review, we are going to focus on the important electrical properties of graphene; however, we should mention some of its other properties for completeness. Due to the 2D nature of a graphene crystal, a single flake will exhibit a large breaking strength of ≈40 N/m due to the absence of slip planes associating the fracture strength of graphene with the strong bonding of c–c in a hex ring. [8] The isolation of electrons from phonons also contributes to the high room temperature thermal conductivity of ≈5,000 W/mK. [9] Along with its high breaking strength graphene is also very pliable with a Young's modulus ≈1.0 TPa and an elastic strain of up to 20%. [8] These values were partially expected on the basis of previous studies of carbon nanotubes and graphite; although the higher values observed in graphene can be attributed to the crystal defects in samples obtained by micromechanical cleavage.

There are even more intriguing material characteristics of graphene such as shrinkage with increasing T at all T due to membrane phonons dominating in 2D. [10] Also, graphene exhibits simultaneously high pliability (folds and pleats are commonly observed) and brittleness (it fractures like glass at high strains). [11] Equally unprecedented is the observation that the one-atom-thick film is impermeable to gases, including helium. [12]

For electronic applications the structure of graphene creates a semi-metal with a direct Fermi-Dirac band structure, as shown in Figure 1, having charge carriers interacting as Dirac Fermions (with zero-effective mass) that allows for ballistic transport of up to a micron at room temperature. [13– 15] The Fermi-Dirac cone as shown in Figure 1c, however, is modified either by the addition of multiple layers as shown in Figure 2iii, the addition of two layers and doping from contaminant particles (metal or polymer particles lying on the surface) shown in 2iv, or contaminants doping a single layer as shown in Figure 2ii. [16] The contaminant-induced doping would move the Fermi level either up or down, the Dirac cone causing a rounding of the k states resulting in a decrease in the mobility of the current carriers (electron or holes). [16] This, along with the thickness restriction for graphene, creates large resistance and chemical inertness, unless chemically doped and functionalized, making its use for pure conductive applications less attractive. [16, 17]

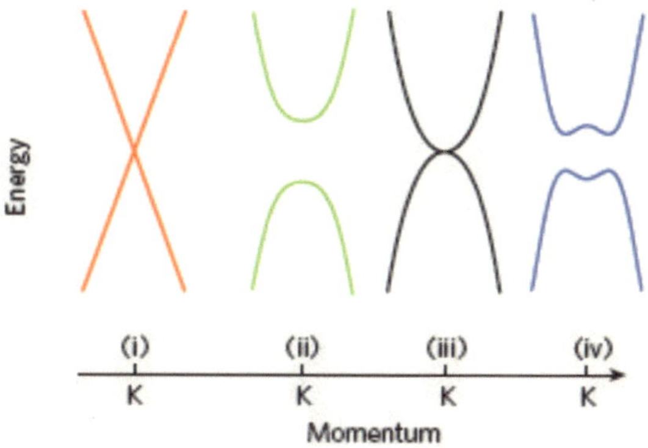

Figure 2. (i) Diagram showing the Dirac Fermi cone; (ii) the modification of the k states by chemical or geometry restrictive doping; (iii) the modification of the k states by bilayer graphene; (iv) and finally, the modification of the k states in doped bilayer graphene. [16]

For applications such as the channel in a field effect transistor, graphene provides an interesting solution since it can be doped electrostatically and has extremely high mobility allowing for quick response. [18] The replacement of Si by graphene for logic gates might be considered due to the high potential switching speed; however, the absence of a band gap means that a relatively large band gap would have to be induced through a variety of doping or other symmetry breaking mechanisms. [18] The absence of a band gap in graphene limits voltage and power gains that is achieved through operation of a device in the saturation regime, along with having a low Ion/Ioff ratio. [16] To overcome this, several doping strategies as shown in Figure 3 have been proposed and tested including: electrostatic doping, chemical doping, and stress or geometry restricted doping by breaking the graphene periodicity (and band properties). [18]

The induction of a band gap has been attempted by multiple groups creating transistors with low on/off ratio and high mobility with a tradeoff between on/off ratio and mobility possible through graphene functionalization techniques.[1] This makes graphene more desirable for

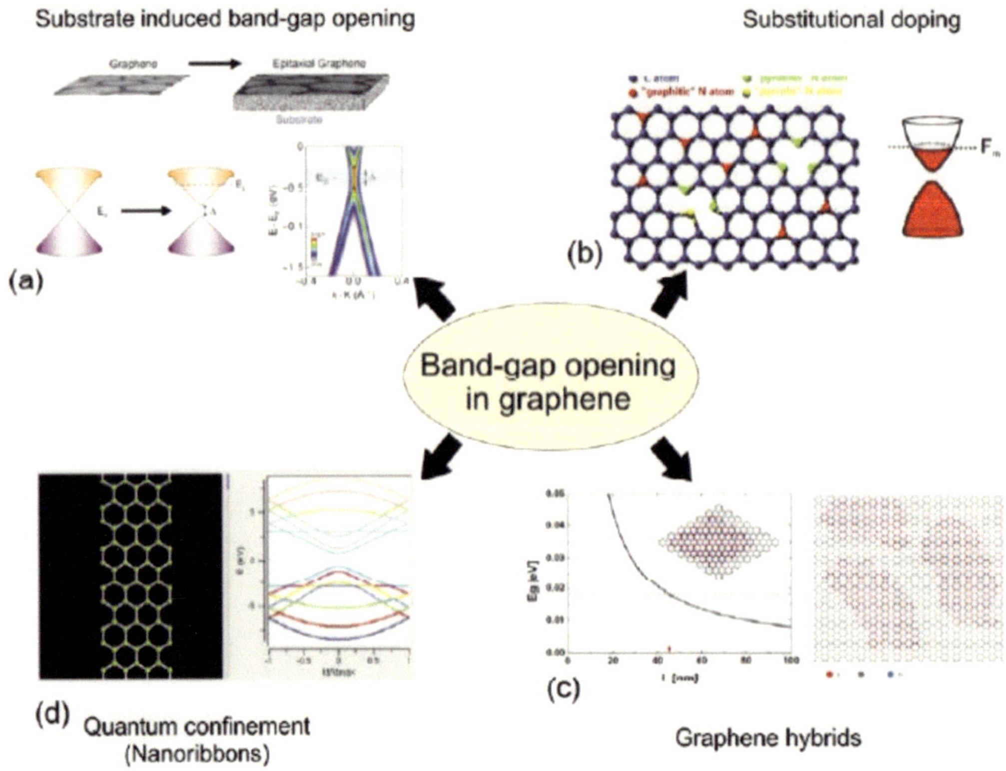

Figure 3. Diagram showing multiple mechanisms for inducing a band gap in graphene [19].

applications that require fast response times, but not necessarily big on/off ratios such as RF electronics and IR detectors.

2. Graphene fabrication

Since graphene's performance is very susceptible to contamination and structural defects (such as folds, grain boundaries, and pin holes) from processing or the transfer process, a review of graphene growth techniques should be done to determine the benefits and drawbacks of each. [1] Due to the sensitivity of graphene, the growth method must be chosen with the required quality, processing, scale, and device architecture in mind, making exfoliation good for small test structures but inadequate for a repeatable semiconductor targeted process. As shown in Figure 4 there are five major pathways for creating graphene sheets: exfoliation from bulk graphite, unzipping through etching a carbon nanotube, growth from sublimation and reconstruction of carbon from a carbide surface, epitaxial growth from a carbide forming catalyst layer by utilizing condensation during cooling, and the epitaxial growth utilizing a non-carbide forming catalyst layer.

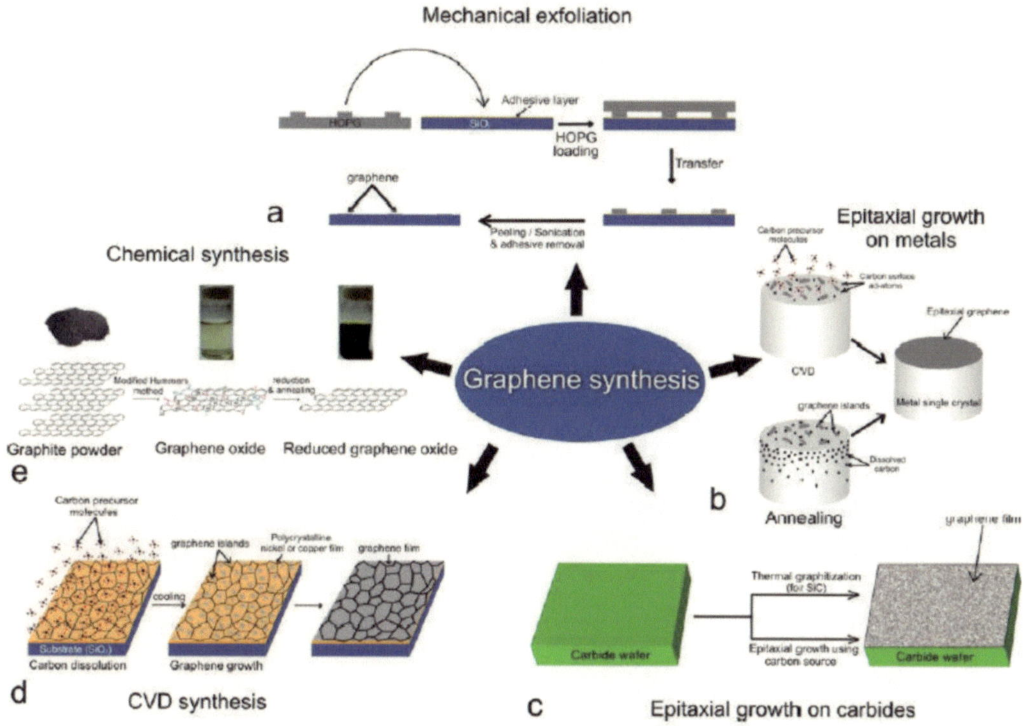

Figure 4. Fabrication schemes for the large scale synthesis of graphene sheets [19].

2.1. Exfoliation

As of 2014, exfoliation methods produced graphene with the lowest number of defects and the highest electron mobility by the pioneers of graphene, Novoselov and Geim, using the adhesive tape method to isolate graphene from graphite.[1, 13, 20] The most common exfoliation method utilizes an adhesive tape to pull graphene films off a graphite crystal, which are subsequently thinned down by further strips of tape and finally rubbed against the desired substrate. This rather crude method creates a random array of single and double layer graphene flakes on the desired substrate that has been a key driver for investigating the many properties of graphene. Since graphene is susceptible to creating folds during this process, it cannot be produced with high accuracy and repeatability, so other mechanical and chemical exfoliation processes have been investigated. To address the difficulties of the scotch tape method, one group tried to exfoliate graphene from highly ordered pyrolytic graphite (HOPG) utilizing a sharp single-crystal diamond wedge penetrating into the graphite source to exfoliate layers. [21] This method has problems with defect initiation through shear stress and the reliable placement of the graphene flakes after exfoliation.

The other main exfoliation method is to utilize liquid-based techniques to create a dispersion of graphene or graphene oxide flakes that are drop-casted or ink-jet-printed, and in the case of graphene oxide subsequently reduced. Liquid exfoliation can be accomplished through the use of solvents or ionic liquids with similar surface tension to graphene, which when sonicated exfoliate the bulk graphite into graphene sheets that can be subsequently centrifuged to create a supernatant and dispersed. [22–25] Probably the oldest known method for producing graphene is through the production of graphite oxide using Hummers' method, sonicating to create a dispersion and then reduction of the graphene oxide either through the introduction of hydrazine at elevated temperature or through the introduction of a quick burst of energy introduced either through a light burst as shown in Figure 5 (flash or laser) or a temperature spike. [21, 26, 27]

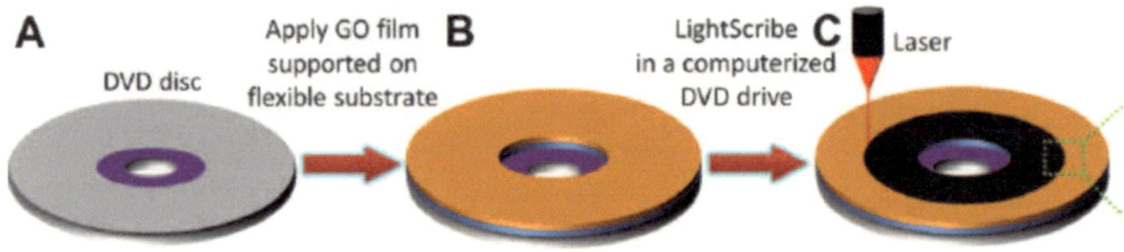

Figure 5. Reduction of graphene oxide using a LightScribe laser writing system ion, a standard DVD writer [28],

One of the more interesting liquid exfoliation methods utilizes sonicating graphite at the interface of two immiscible liquids, most notably heptane and water, producing macro-scale graphene films. [29] The graphene sheets are adsorbed to the high energy interface between the heptane and the water, where they are kept from restacking. [29] The graphene remains at the interface and the solvents may then be evaporated isolating the graphene flakes. [29]

Straightforward mechanical exfoliation methods have been able to produce high-quality graphene flakes that have been very beneficial for the investigation of the amazing characteristics of graphene, while liquid exfoliation (and reduction) methods have been utilized for the production of transparent conducting oxides, conductive inks, and electrodes for Li-ion batteries and super capacitors. Mechanical exfoliation, however, cannot be reliably scaled up to provide the reliable placement and large area high-quality graphene sheets desired for transistor and device applications.

2.2. Carbon nanotube unzipping

As shown in Figure 6, graphene can be created by cutting open carbon nanotubes. [7] In one such method, multi-walled carbon nanotubes are cut open in a solution by action of potassium permanganate and sulfuric acid. [30] In another method, graphene nanoribbons were produced by plasma etching of nanotubes partly embedded in a polymer film. [30] This method is useful for producing nanoribbons of graphene that induces a band gap in graphene through geometry breaking, which will be discussed in Section 3. However, the placement of the nanotubes on an integratable chip has been problematic, and thus this method once again is only good for the production of test structures to probe graphene characteristics.

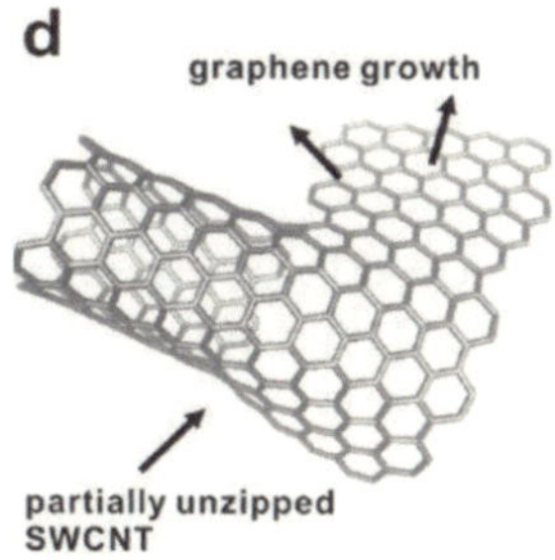

Figure 6. Image showing the unzipping of a carbon nanotube to produce graphene sheets [21].

2.3. Sublimation and reconstruction from carbide

Heating silicon carbide (SiC) or other carbide materials (TaC, NbCm ZrC, HfC, TiC) to high temperatures (>1,100°C) under low pressures (~10^{-6} torr) boils off the Si (from either the Si face or underlying Si from the C face) and reconstructs the C into a single layer graphene film, although multi-layer graphene has been produced through this approach as well. [19, 32] This process produces epitaxial graphene with dimensions dependent upon the size of the wafer.

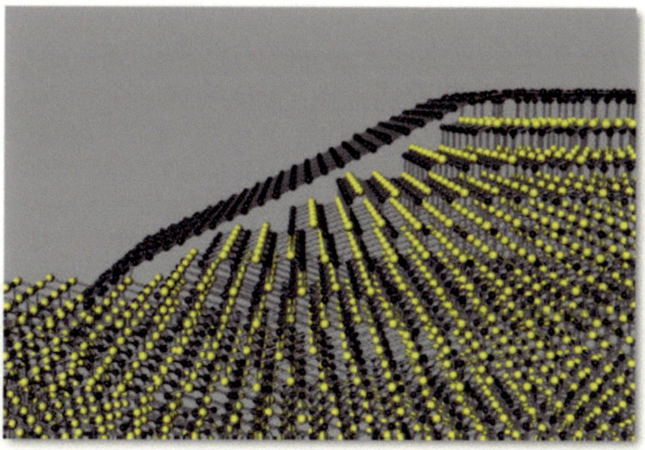

Figure 7. Bonding of graphene at a SiC step edge [33].

The particular face of the SiC used for graphene formation, silicon- or carbon-terminated, highly influences the thickness, mobility, and carrier density of the resulting graphene, with the best results coming from a step edge in SiC that produces "floating" graphene attached to the SiC on the top and the bottom of the step edge as shown in Figure 7. [33] There has also been some work utilizing Ni and Cu bilayers to catalyze the production of graphene from SiC achieving growth at higher pressures and lower temperature. [34] The benefit of using graphene produced from SiC is that SiC is easily integratable with microelectronics processing technologies. The SiC is not desired for most electronics applications, making it desirable to transfer the graphene from its SiC substrate to a more standard substrate such as Si. The sublimation of graphene from SiC also creates a Si_2O_3 insulating under layer that could assist with the transfer process. Under high temperatures, a large variety of intercalant species can also be placed between the graphene and SiC layer that can potentially help with the exfoliation or the electrical modification/isolation. [19] Under normal conditions, the graphene SiC interface forms a Schottky contact; however, it has been shown that the oxide can be transformed to a nitrogen underlayer through a thermal annealing process in a nitride atmosphere modifying the electronic characteristics between the two. [35]

2.4. Growth through condensation after carbide formation

Graphene growth utilizing a carbonaceous source material (such as methane introduced through a CVD process) differs from material to material with the carbon solubility in the metal and the growth conditions determining the deposition mechanism as shown by the phase diagrams in Figure 8. [7] For carbide producing metallic substrates (such as Ni), graphene growth occurs through a precipitation process during cooling from the carbide. [7] The solubility of C in the metal (Ni for example) is higher at higher temperatures, and thus during the furnace cooling phase carbon diffuses out of its Ni host. [7] The process of forming graphene on Ni has the fundamental limitation that single and few layered graphene is obtained over few to tens of micron regions and not homogeneously over the entire substrate. [7] The lack of control over the number of layers is attributed to the difference in out-diffusion of C from the grains and the grain boundaries of Ni creating non-homogeneous growth conditions.

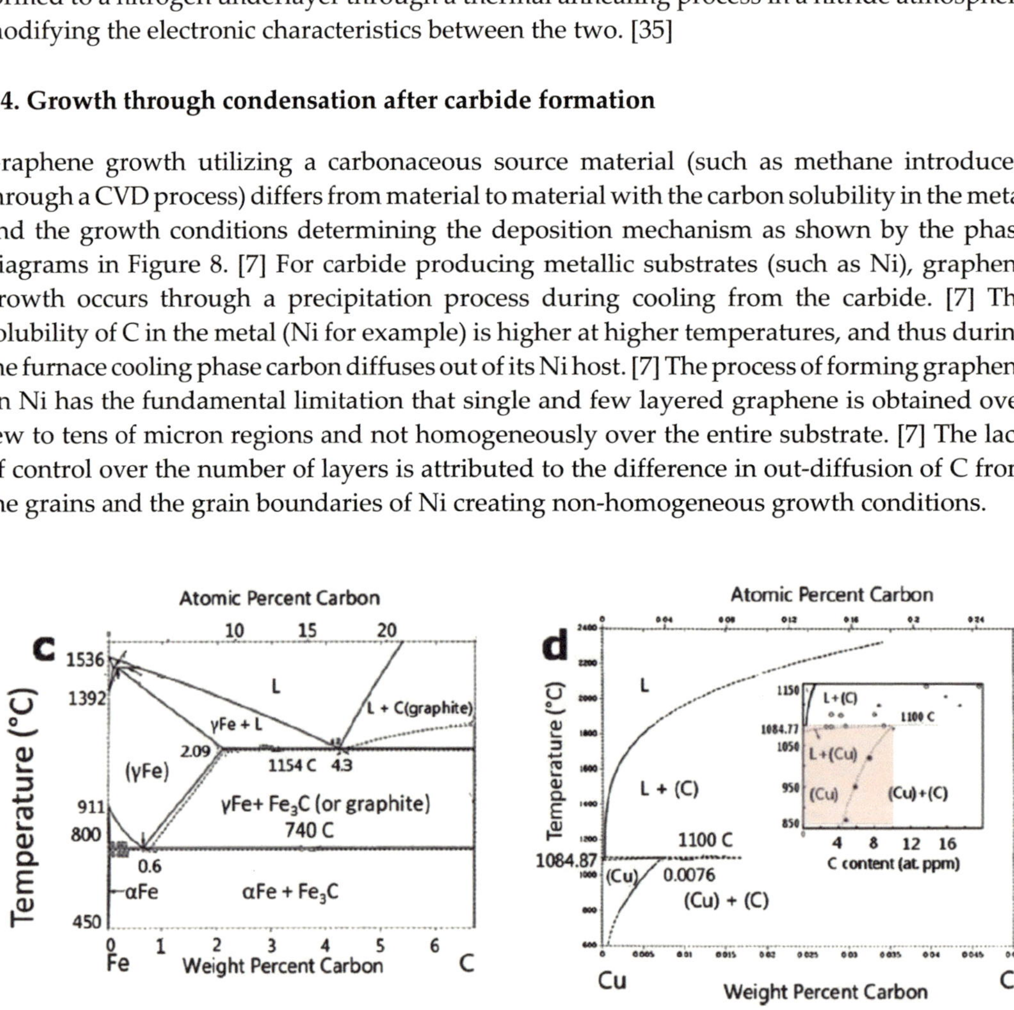

Figure 8. The phase diagram for a carbide creating catalyst (Fe) and a non-carbide creating catalyst (Cu) [7].

2.5. Epitaxial growth of graphene on a non-carbide forming catalyst

Epitaxy refers to the deposition of a crystalline overlayer on a crystalline substrate, where there is registry between the two. In some cases, epitaxial graphene layers are coupled to surfaces weakly enough (by Van der Waals forces) to retain the 2D electronic band structure of isolated graphene. [31, 32] It is commonly accepted that the production of graphene through the surface absorption of carbon on a non-carbide producing metal (such as Cu or Ir) is an epitaxial process due to the registry between the underlying Cu (or Ir) crystal structure and the graphene layer.

Exceptional high-quality single layer graphene growth over large areas have been recently achieved on polycrystalline copper foils.[7] The growth on Cu or Ir is simple and straightforward due to the metallic substrates not having a stable carbide material, thus the decomposition of C is only reliant upon the grain orientation. [36] For example, Cu is an FCC lattice with three dominant grain orientations Cu(100), Cu(110), and Cu(111) along with high index facets which are made up of combinations of low index facets. [36] Cu(100), Cu(110), and Cu (111) have cubic, rectangular, and hexagonal geometries making the Cu(111) grain orientation able to support epitaxial growth. [36] Thus, grain growth on Cu(111) grains tend to be monolayered graphene sheets while Cu(100) and Cu(110) geometries prevent C diffusion causing compact multilayered C islands to form with higher index facets replicating the performance of the lower index grains. [36]

Despite the ability for Cu and other metal substrates to grow high-quality graphene flakes for device applications, graphene has to first be transferred onto a semiconducting or insulating substrate when using this growth method. [37] The transference process usually involves spinning on a polymer, etching off the catalyst metal layer, then transferring the graphene onto the desired substrate by placing it on the substrate, and finally etching off the substrate. [37] Both of these processes can produce contaminants on the graphene layer, reducing the mobility by adding scattering centers in the sheet. [1] Groups have been working on ways to reliably reduce these contamination effects; one group has utilized Ti sputtering along with a Ti etch to remove any remaining Cu, while another group has shown that by first spinning on a lift off resist before a normal polymer backing layer produces a much cleaner graphene layer. [38, 39] High-quality graphene has also been shown to be grown between a GaN and Ni interface where the Ni can be peeled off and the graphene layer is left on the GaN substrate. [40]

3. Graphene doping

Due to the chemical nature of the graphene with its zero band gap, mobility related to the delocalization of the π-bonding orbitals, and lattice periodicity, the doping of graphene can be achieved either through the breaking of lattice periodicity or the electrostatic confinement of the delocalized p_z orbitals. [19] There have been several mechanisms proposed and tested that have been effective in shifting the Fermi energy to either p- or n-type regions of the band structure and the creation of p-n junctions at the interface. [19] It should be noted that by breaking the lattice symmetry, the electronic states and band edges are modified as shown in Figure 2, decreasing the mobility. With chemical functionalization, scattering is introduced into the graphene flakes, also decreasing the mobility.

Since graphene is a self-contained sheet with no real interface layer, it should be noted that process integration with oxide dielectrics as shown in Figure 9 can be difficult due to trapped

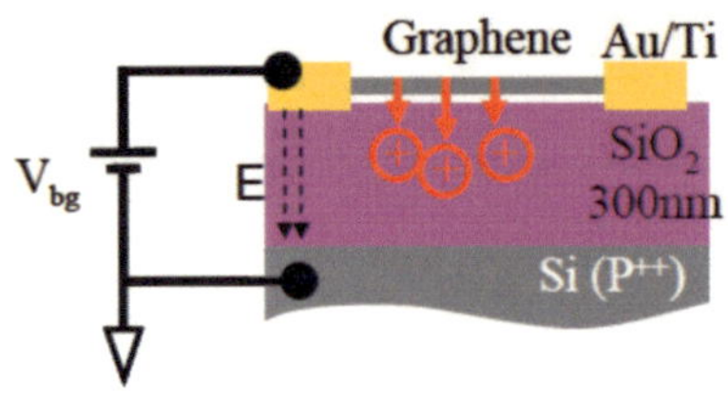

Figure 9. Image showing trapped charges at graphene oxide interfaces [41].

impurities at the interface in terms of creating floating gates for voltage-controlled variable gate transistors. [41] This effect is relevant for device applications of graphene films. It should also be noted that the introduction of trap centers and doping sites through contamination will degrade the continuity of the 2DEG at the trap or dopant site, causing electron and hole pooling to occur. [16, 17, 42]

3.1. Electrostatic doping

Electrostatic doping as shown in Figure 10 can be controlled through a variety of methods; some use floating gates with oxide buffer layers and others use direct gate contacts to locally modify the Fermi level allowing for the voltage-controlled operation of the graphene device. [43] Most electrostatic gating is accomplished through a horizontal device architecture to preserve the mobility of graphene for ultrafast devices. With both direct and indirect contact, electrostatic gating can be accomplished by utilizing metals with two different work functions; polymers with different end groups as shown in Figure 11; and finally, layered materials with different opposing band gaps with the higher band gap being the acceptor and the lower being the donor as shown in Figure 12. [19] Metals with dissimilar work functions are normally integrated into a horizontal device with many different combinations to choose from. [44, 45] The amount of gap opening is defined by the difference between the two metal work functions and the induced electric field decreasing down the length of the sheet making the contact placement critical. [44, 45]

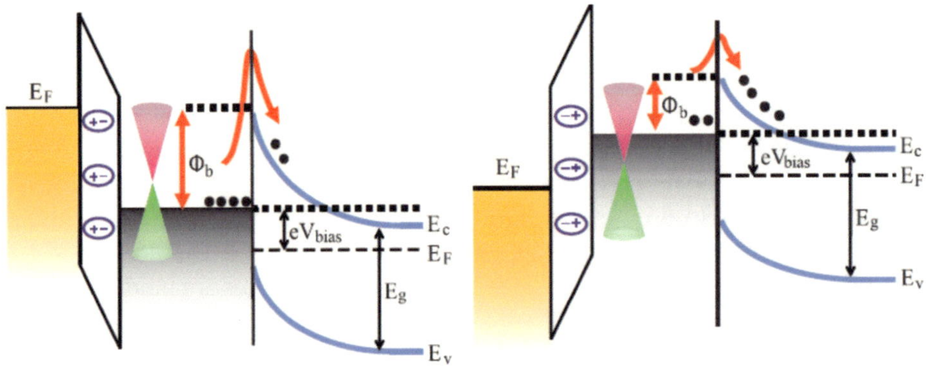

Figure 10. Diagram showing charge injection and Fermi modification of a graphene Schottky contact [43].

http://dx.doi.org/10.5772/61316

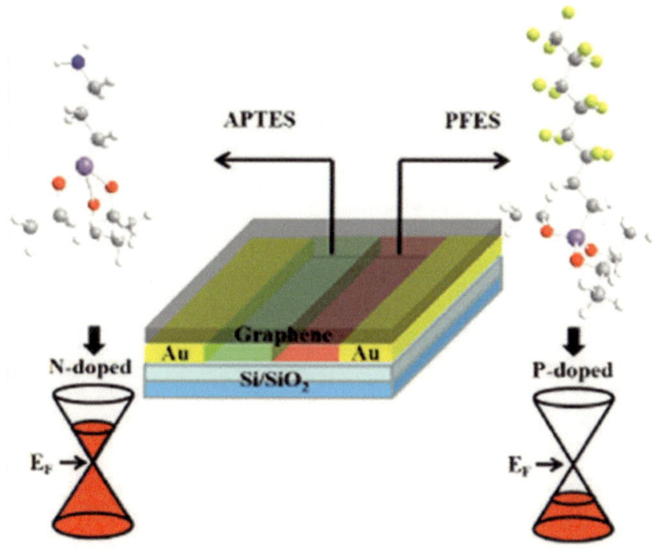

Figure 11. Diagram showing the doping of graphene utilizing different polymer end groups [18].

For polymers, the use of different functional groups can electrostatically dope a horizontal graphene sheet with an isolated amine group (isolated nitrogen atom as in nitric acid) n-doping the graphene sheet while fluorine is well known as a good electron acceptor so a polymer containing an isolated fluorine end group p-dopes the polymer as shown in Figure 11. [18] For the polymer electrostatic doping technique, similar atomic dopants are utilized for the chemical doping regime with atoms lower than group V providing n-type doping and elements higher than group V creating p-type dopants (this will create environmental sensitivity within an exposed graphene sheet due to the oxygen and hydroxide adatoms p-doping the graphene). [18, 46]

Finally, for electrostatic doping, the utilization of other 2D materials as shown in Figure 12 can be used by vertical device integration with either a homojunction-based device or a heterojunction-based device. For a homojunction-based device, graphene is utilized in a double layer with electrostatic doping coming from a layer above one graphene sheet with a lower band gap (such as tungsten diselenide WSe_2) and one below the other graphene sheet with a higher band gap (such as molybdenum disulfide MoS_2) creating an electric field between the two 2D materials with different band gaps and electrostatically doping the graphene as shown in Figure 12. [48] Since the electrostatic potential outside a sheet with a band gap will only induce a shift in the Fermi energy for graphene, a heterojunction can be formed between the junction of the 2D materials with a tuning of the upper and lower contacts required. [48] The use of 2D stacked devices is interesting but it should be remembered that many of these layers have not been shown to be readily deposited on top of the other, requiring transfer techniques that can induce defects, transfer contaminants, and have alignment issues between the lattices creating different properties across the lattice due to misalignment as shown in Figure 13. [47, 49]

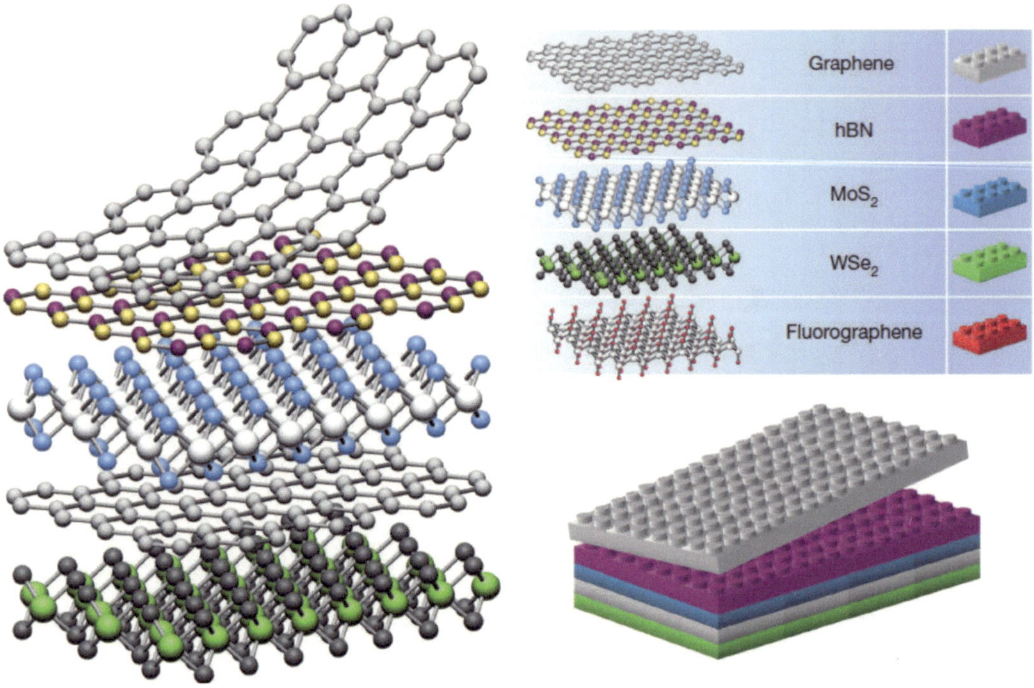

Figure 12. Diagram showing the stacking of multiple Van der Waals materials in order to create unique and tunable electrical properties [47]

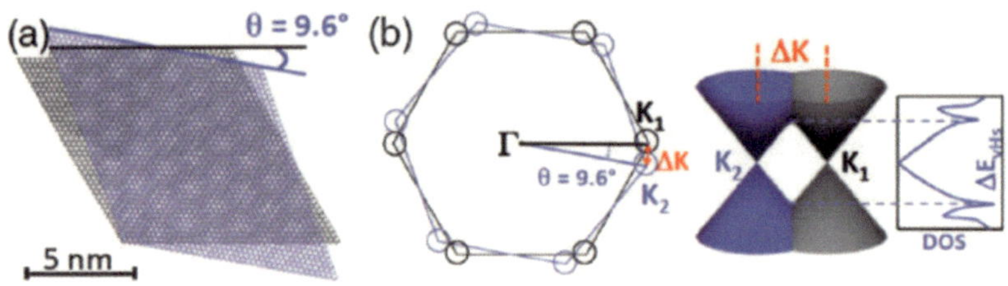

Figure 13. Image showing that the misalignment of 2D materials can electrically isolate the two sheets by separating the Dirac cones [50].

It has been shown that a twist angle between two graphene sheets above 2° electrically isolates the two graphene sheets from one another except at certain twist angles as shown in Figure 13. [49] Most 2D materials have also been shown to have intrinsic doping due to vacancies and edge defects that create more problems for device integration. [49] It should be noted that the mobilities in graphene on boron nitride (BN) substrates have been measured up to 140,000 cm^2/Vs, which is very close to completely suspended graphene grown from a SiC step edge, showing the validity of using 2D heterostructures for device integration and isolation. [32, 42]

http://dx.doi.org/10.5772/61316

3.2. Chemical doping

As mentioned briefly in the electrostatic doping section, chemical dopants can be utilized to modify the electrical characteristics of graphene, modifying the Fermi energy to create p- or n-type doping as shown in Figure 14. [19, 46] The chemical doping mechanism of graphene works by having the dopant bond either ionically or covalently to the delocalized p_z orbital. [46] The covalent bonding of a dopant with graphene occurs through modification of the delocalized p_z orbitals to electrostatically hold an adatom onto the surface, which modifies the band structure by binding the electrons in the p_z orbitals, thus creating a required energy (a band gap) for conduction. [46] The chemical dopant can be ionically bonded to a single carbon atom by breaking a c–c bond and attaching to that bonding spot, which breaks the graphene symmetry introducing a scattering defect and a band gap opening. [46, 51] The amount of surface adatoms is reliant upon the dopant and the type of bonding with ionic bonding and larger electronegativity being able to obtain a stronger bond, higher dopant concentration, and higher band gap shifting. [46] However, it should be noted that the higher the doping, the more scattering and the lower the mobility, leading chemical doping to be typically done on vertical devices with a small cross section and thus small diffusion length. [19, 46]

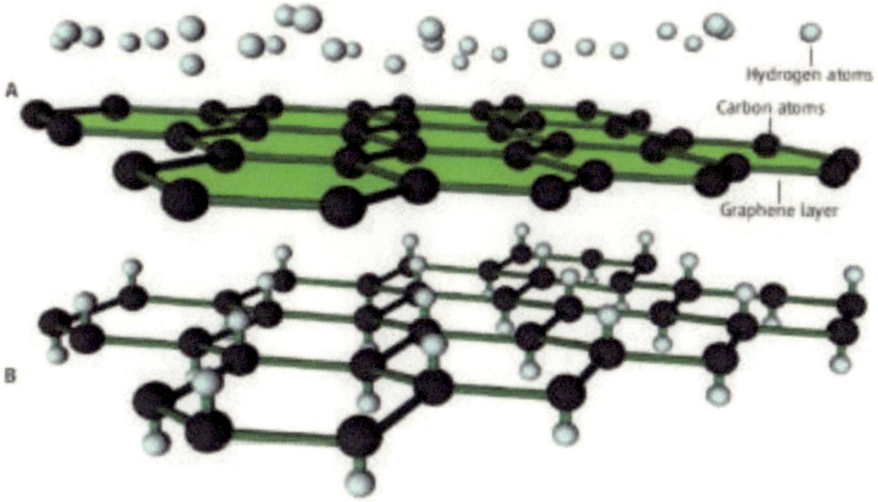

Figure 14. The functionalization scheme of graphene utilizing a H_2 plasma [46].

3.3. Geometry restriction

The final way to dope graphene is by breaking the lattice periodicity of graphene as shown in Figure 15. [19, 53] This can be done by reducing the size of a graphene sheet in one direction so that the Fermi levels from the periodic boundary conditions are refined, providing doping through a quantum confinement effect. [53] Quantum confinement occurs when the material dimensions are below the Bohr radius, which for graphene is at 10 nm. [13, 53] This has been shown to be accomplished through the patterning of graphene into ribbons with one dimension restricted to under 10 nm, thus opening a gap of 2.5–3.0 eV in theory and 0.5 eV experimentally. [19, 53] Graphene with a size in either x or y under 10 nm is known as a graphene nanoribbon and it suffers, like many other graphene synthesis techniques, from a lack of a reliable production technique. [53] Traditional semiconductor line definition techniques cannot reliably get a line definition below 20 nm, with large problems creating lines with acceptable line edge roughness. For graphene nanoribbons, the resistance induced through scattering from the line edge roughness is coupled with a lack of graphene conformality, not knowing whether the line definition will create "zig-zag" or "arm-chair" end terminations that provide different conductivity values. [19, 53] The difference between "zig-zag" and "arm-chair" end terminations is shown in Figure 16 and the difference in conductivities between the two create a discrepancy when designing a device utilizing multiple nanoribbons or multiple devices utilizing a graphene nanoribbon. [54]

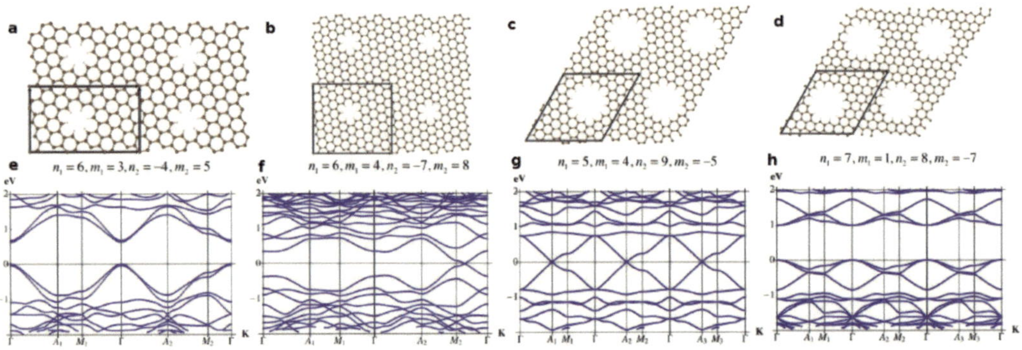

Figure 15. Different defined graphene sheet edge states and the associated band diagrams showing opening according to edge definitions [52].

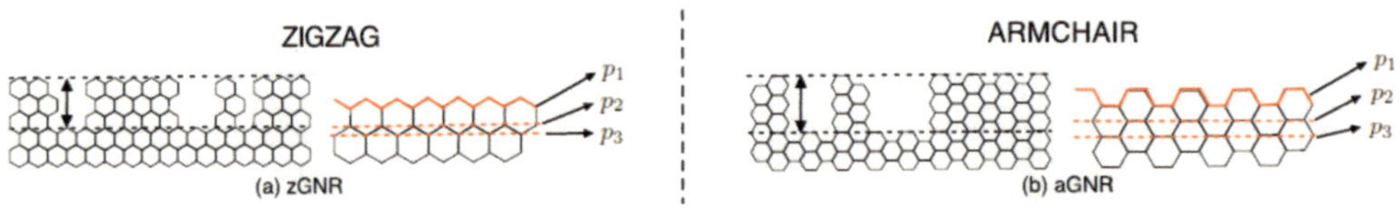

Figure 16. A picture showing the difference between zig-zag and armchair graphene end terminations [54].

The induced line edge roughness produces many scattering defect reducing the lattice periodicity, obliterating the induced band gap, and decreasing the mobility ultimately limiting the usefulness of graphene nanoribbon formation. [16, 19] Thus, to achieve useful devices from geometry restricted graphene, a reliable method of patterning graphene with low line edge roughness and uniform width must be developed.

4. Graphene Field Effect Transistors (GFETs) and Radio Frequency (RF) electronics

There has been an interest in looking at graphene for nanoelectronics applications due to its high intrinsic mobility allowing for greater switching speed. [18] The main problem with the

http://dx.doi.org/10.5772/61316

integration of graphene into three-terminal devices is the lack of a high Ion/Ioff current, which for regular metal oxide semiconductor field effect transistors (MOSFETs) is on the order of 10^4–10^7, while for most graphene devices is on the order of 10. [16, 55] This makes graphene-based devices more attractive for the replacement of RF-based devices that are currently dominated by high electron mobility transistors (HEMT) that require a Ion/Ioff ratio of around 30. [16, 55] In addition to this, RF electronics require current saturation to obtain voltage and power gains of around 30, which for graphene means the creation of a band gap through one of the doping mechanisms described above. [16] Saturation current is normally attained through the saturation of charge carrier velocity; however, due to the high mobility of the graphene layer the velocity saturation is unattainable without going to extremely high source drain voltage. [16] Therefore, saturation must be obtained through current pinchoff effects and voltage gain as shown in Figure 17, which can be created in graphene through band gap formation. [16] Even with the formation of a band gap, graphene does not exhibit a saturation current at zero doping due to the Fermi-Dirac cone shape, but the band gap does allow the creation of current pinchoff due to electrical band gap modulation via the source and drain contacts. With this background, we are going to address in the subsequent section several issues that hinder the integration of graphene into FETs that can be utilized for RF applications and review the state-of-the-art technology in terms of GFETs for RF electronics.

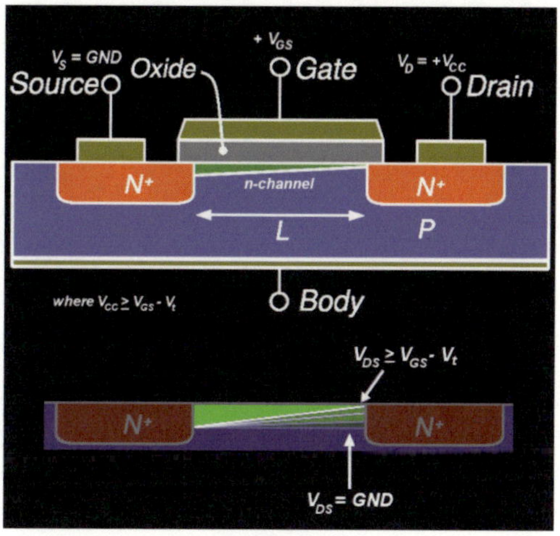

Figure 17. Diagram showing current pinchoff in a Si MOSFET.

4.1. Short channel effects

Graphene normally has a grain size of several to tens of microns with the desire to use a single grain as the channel material to avoid scattering at grain edges (also a factor in current MOSFET structures that is why single crystalline Si is used as a substrate). [16] This creates many of the short channel effects commonly seen in MOSFETS such as drain induced barrier lowering, surface scattering, velocity saturation, impact ionization, and hot-electron effects. [16] Specifically for graphene, drain-induced barrier lowering, surface scattering, and hot electron effects are all in play. Surface scattering is due to the intrinsic susceptibility of graphene to surface contamination and scattering sites, while the hot electron and barrier lowering effects affect graphene due to the pinchoff formation needed to have the large Ion/Ioff ratio required for typical electronics applications and to create large enough voltage and power gains for RF applications.

4.2. Metallic doping by source drain contacts in graphene

Graphene is a self-contained electronic sheet showing no classical band bending interactions when coupled to a metallic contact as shown in Figure 18. This creates an abrupt transition in the vacuum level, creating a barrier that any carrier would have to tunnel through, creating charge buildup at the band edges and large contact resistances. [56]

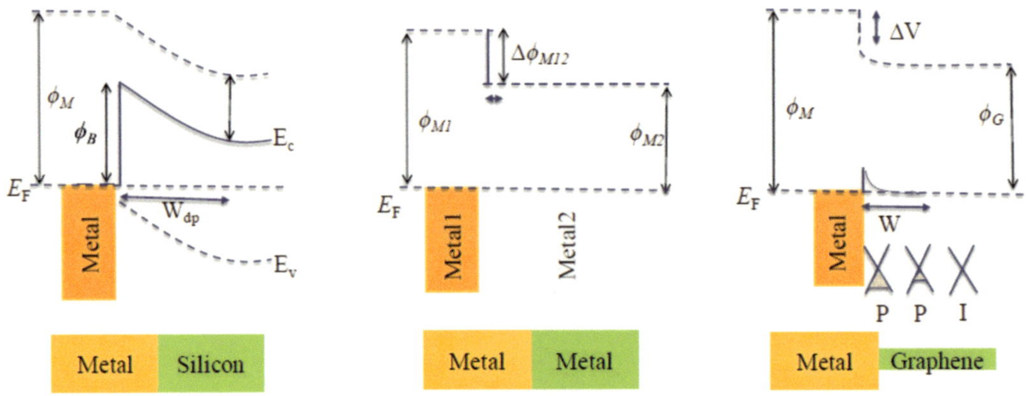

Figure 18. Classical band diagrams for a metal-silicon interface, a metal-metal interface, and a metal graphene interface.

In conjunction to this challenge is the relative inertness of a graphene sheet, making good electrical contacts difficult to realize and mainly occurring at grain edges. [56] This creates a situation where the bulk of the contact sits over the graphene electrostatically doping it, while also trying to realize good adhesion creating a search for a metals with good adhesion to graphene along with the correct Fermi level. [56] To achieve this goal, a double or triple metal stack is commonly used with an oxygen scavenger interfacing the graphene (normally Ti), followed by one or a couple of Fermi level contacts (Au, Pd, Ni). [56] The metallic doping effect, however, can be utilized for some interesting devices such as one using asymmetric contacts to create an internal electric field making an IR detector through the photothermoelectric effect, or using large gap superconducting contacts to confine electrons and holes in a graphene sheet to enhance bolometric response. [45, 57]

4.3. Dielectric deposition and trap states

As stated in Section 1 and Section 3, graphene is a self-contained layer without any dangling bonds, thus adhesion and interfaces with graphene are a challenge. Multiple groups have been experimenting with different types of oxides with either an aluminum deposition and oxidation or a nitrogen dioxide surface pretreatment prior to a hafnium oxide, silicon dioxide, or aluminum oxide deposition. [58] The dielectric which seems to work the best (but has not yet been implemented in a complementary metal oxide semiconductor (CMOS) fabrication process) is another 2D self-contained dielectric BN with which graphene has shown mobilities of 140,000 cm^2/Vs, which is very close to completely suspended graphene grown from a SiC step edge, demonstrating low interaction and good isolation between the two substrates. [42]

4.4. FET design and gate coupling

To overcome some of the short channel issues and problems with graphene integration into common process flows, a wafer bonding type of integration has been suggested as shown in Figure 19. [59] This allows for the separation of the drain and gate contacts, which reduces coupling and alleviates some of the issues with drain induced barrier lowering. [60]

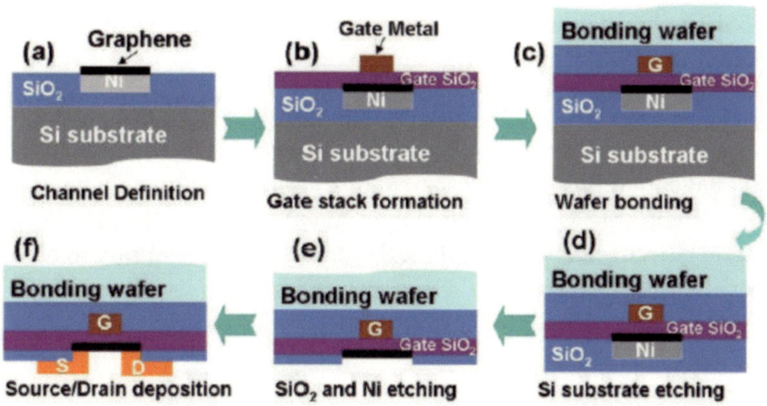

Figure 19. Wafer bonding with subsequent source drain contact deposition [59].

Gate coupling is a significant issue with graphene FETs due to the large gate voltages needed to create sufficient barriers for high Ion/Ioff ratios, the metallic characteristic of the graphene layer, the thin gate oxide needed to ensure good gate control and reduced gate potential for smaller electrical field propagation, and finally the dielectric breakdown strength. [60] All of these needs show that a thin high-k gate with opposing gate and source drain contact geometry is desired as shown in Figure 19.

Graphene devices have a very thin cross section where the active electric field can affect one another. It has been shown that by using tapered contacts as shown in Figure 20, the amount of source drain coupling is reduced due to electric field reduction. [60] This is especially effective if utilizing a back gate design as shown later in Figure 21, or a large gate that could overlap the source and drain contacts on the opposing side of the devices channel. [55, 60]

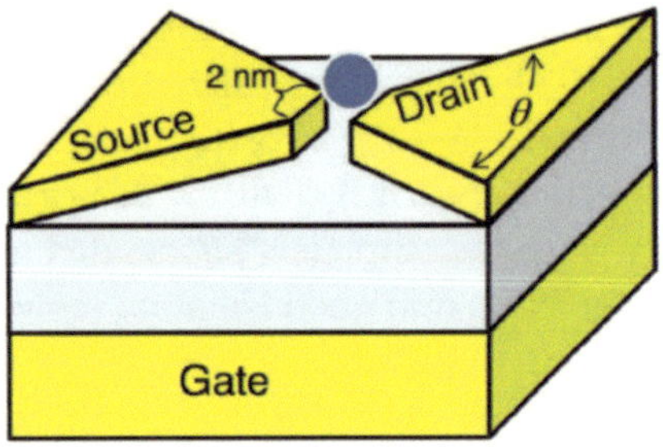

Figure 20. Tapered contacts used on opposing sides of the gate to reduce [60].

4.5. Graphene FET and RF electronic performance

In order to create a device that allows for the opening of a band gap required for current saturation and appropriate voltage and current gains, several device geometries have been proposed. [16, 55] The main mechanisms for increasing graphene performance in FETs is to increase gate coupling with graphene layer and to optimize the graphene dielectric interface to reduce scattering and make the conduction and valence states continuous. [16]

One possible device geometry shown in Figure 21 utilizes a bilayer graphene channel with a large backgate voltage to induce an electric field of 1.7 V/nm that opens a band gap in bilayer graphene of 80 meV with the Mexican hat shape shown in Figure 2iv. [16, 55] The band gap creates a saturation current due to pinchoff at the drain contact resulting in a voltage gain of 35, which is relevant for RF electronics.

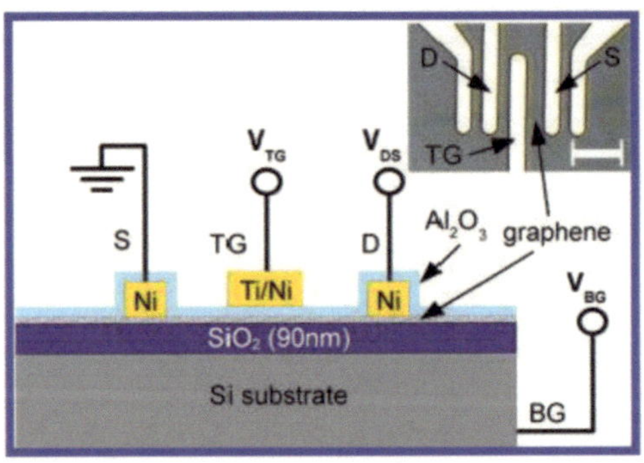

Figure 21. Diagram of a B-bilayer graphene FET with back contact to create a pinchoff region and voltage gain [55].

This mechanism works much better for bilayer graphene than monolayer graphene as bilayer graphene more easily forms a pinchoff region. To demonstrate this, the amount of voltage gain in such a graphene FET is graphed as contour plots with voltage gain axis on the right hand side of the graph as shown in Figure 22. [55]

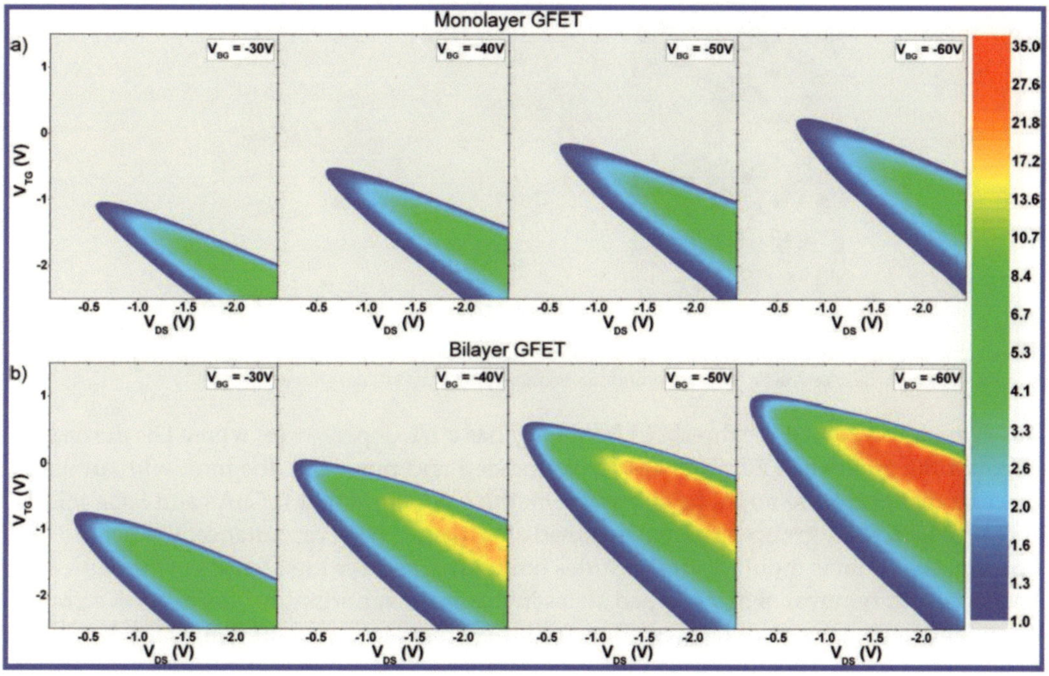

Figure 22. Contour maps of voltage gain in a single layer and bilayer graphene channel with modification of back gate voltage [55].

Current designs for graphene FETs are shown in Figure 23, with the back gated design commonly used to overcome any doping in the graphene channel due to substrate, atmospheric, or dielectric effects. [16] The back gate and top gate design are the most common since these allow for the shifting of the Dirac point to zero through an induced electric field and proper gate modulation. [16]

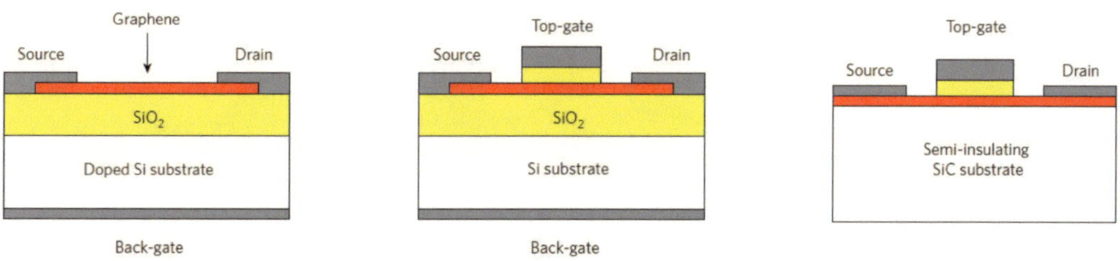

Figure 23. Image showing the most common designs for GFETs [16]

Utilizing a three-terminal top gate design of CVD graphene grown on a SiC substrate, one group was able to achieve a 350 GHz cutoff frequency, utilizing a channel length of 40 nm as shown in Figure 24. [61]

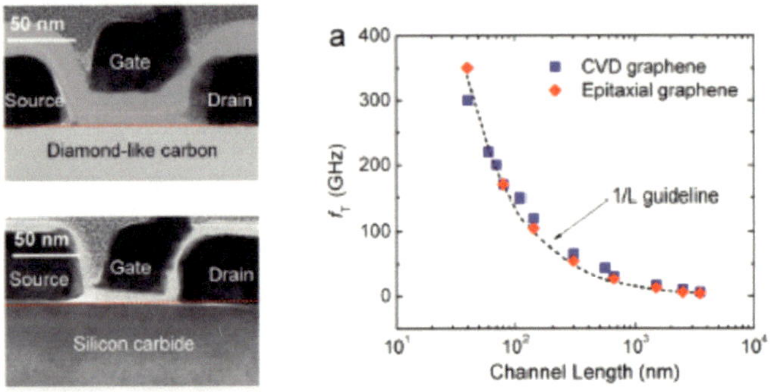

Figure 24. Image showing the threshold frequency versus gate length for the device architectures shown on the left, the epitaxial graphene is on the SiC substrate, and the frequency shows a 1/L dependence [61].

This group showed that the threshold frequency has a 1/L dependence, where L is the channel length of the graphene FET. This has been modeled and pushed to the limit with an understanding that graphene might be able to break the 1THz limit that InGaAs and SiGe HEMTs can't break. [62] One group theoretically tuned all of the parasitic capacitances that would limit the graphene channel mobility; this includes removing Schottky interactions at the source and drain contacts, removal of any trapped states in the oxide, ignoring any electron/hole pooling effects, and having the gate voltage perfectly coupled to channel potential, allowing for a GFET that operates at 1.5 THz. [62] This GFET is optimized to have zero gain due to the current saturation in the 50 nm channel. [62] By allowing for current saturation in the GFET, a voltage gain can be engineered in the graphene channel; however, this would deteriorate the operating frequency of the GFET as shown in Figure 25. [62]

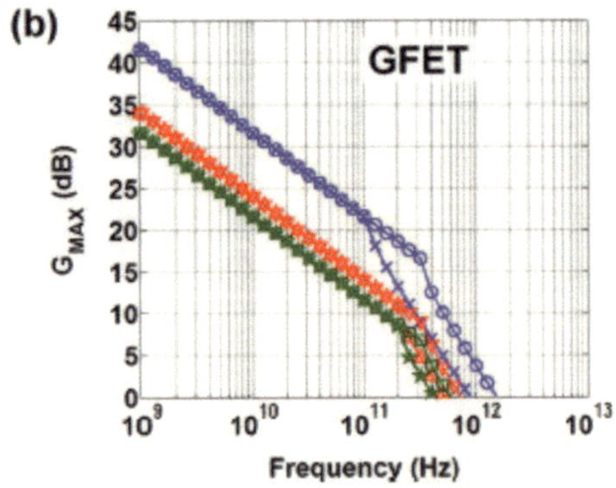

Figure 25. An image showing the threshold frequency for each possible gain in a GFET for systems with different amounts of tuning of parasitic resistance; the blue line has no parasitic resistance [62].

5. Graphene use in Electro-Optical (EO) devices

One of the interesting applications for graphene is its use in EO devices and lasers. Graphene can absorb wavelengths from the visible to the mid-IR with wavelength modulation enabled through electrostatic gating. [63–67] The electrostatic gating interacts with light either by modulating the band gap width up to a certain wavelength working as an absorption modulating element, or it modifies the graphene surface plasmon modes that interact with light. [63–67] The last example is how graphene was utilized for mode locking a laser. [63–67] The problem with utilizing graphene for pure optical devices is due to its inherent thinness only absorbing 2.3% of the incident light per monolayer. [63–67] This makes it more desirable to integrate graphene with other electro-optical components such as photonic cavities or plasmonic waveguides with an example shown in Figure 26. [63–67]

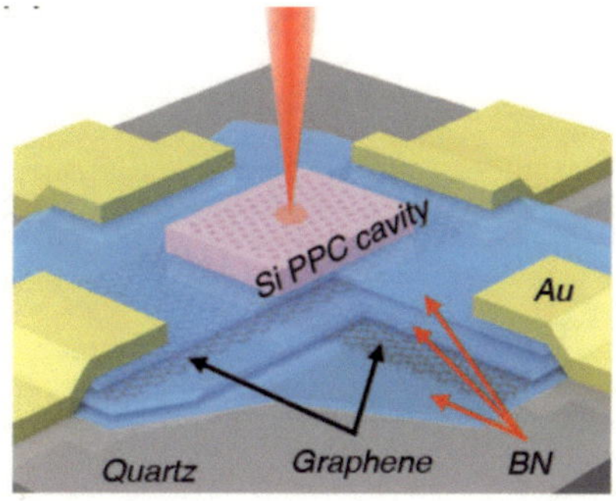

Figure 26. Integration of a tunable graphene capacitor with an EO modulator [63].

The EO modulator pictured in Figure 26 was created through the coupling of Si plasmonic nanocavities to a tunable graphene capacitor made from stacked layers of graphene and BN dielectric film. [63] The top and bottom graphene layers are electrostatically doped differently from one another with varying voltages for optical modulation of absorbed light. [63] The modulator worked up to 1.2 GHz frequency, which was limited by the RC time constant of the capacitor. [63]

Although on its own graphene is not practical for use as a waveguide or modulator, it can be combined with already active materials to increase the performance of such devices.

6. Graphene Infrared (IR) detectors

IR detectors can be separated into two separate categories: thermal-based IR detection and photon-based detection. [68] In thermal-based detectors, the incident IR radiation is absorbed, raising the temperature of the material. [68] The raised temperature affects some temperature-dependent property of the material; for pyrometers this is a change in electrical polarization, while for bolometers, this is a change in materials resistance. [68] Another more recent study utilized the photothermoelectric effect in graphene to create a net electric field due to electron diffusion into dissimilar metal contacts. [45] Photon-based detectors utilize band gap-based detection with the arriving photon being absorbed and utilized to promote electron hole pairs to create a photocurrent. [68] The photon-based detectors can be tuned to certain wavelengths by creating a quantum well structure. [68] Photon-based IR absorbers are characterized by having fast absorption response, but usually require cooling due to thermal effects, while thermal-based IR detectors have high responsivity over a large wavelength and can be utilized at room temperature but normally have slow absorption response. [68] This is where utilizing a graphene-based sensing element is attractive due to the high mobility with little temperature sensitivity making it ideal for IR detectors. [2]

Several groups have attempted to integrate graphene into IR detectors. The groups have tried both photon- and the thermal-based absorption methods. [45, 69–74] For photon-based absorption methods, the main focus has been the opening of a band gap through geometric modification. [45, 69] One group utilized bilayer graphene to open a small band gap that is sensitive to thermalization requiring cooling to 5 K for operation. [69]

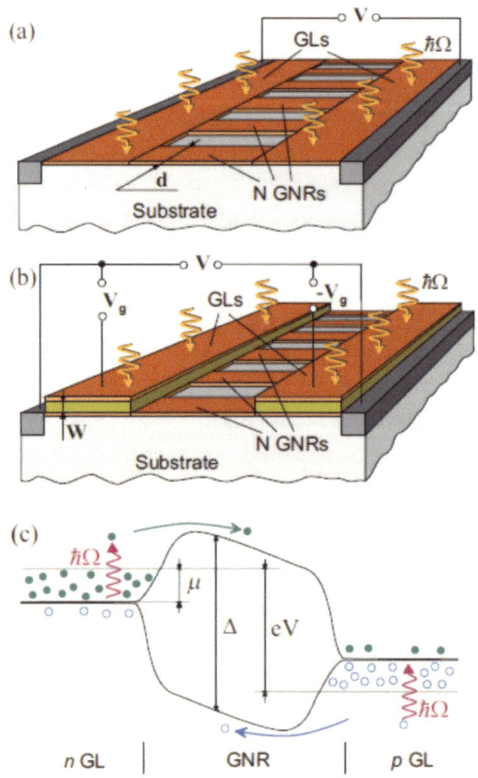

Figure 27. The utilization of graphene nanoribbons to open a small band gap that is enhanced through the use of p- and n-type graphene contacts [45]

Another group utilized an array of aligned graphene nanoribbons as shown in Figure 27 to open up a small band gap that has significant difficulties in fabrication and noise properties from the nanoribbon edges. [45] Groups that have tried thermal-based IR detectors seem to have created more novelty, with one group utilizing multiple vertically aligned graphene flakes, while another group utilized a resonant structure of two graphene sheets separated by a dielectric to tune the photon wavelength of absorption as shown in Figure 28. Finally, another group utilized the photothermoelectric effect as shown in Figure 29 to induce an electric current in graphene due to electric gating or dissimilar metal contacts. [45, 70, 71] The bolometer utilizing vertically aligned graphene sheets used distance-based tunneling between sheets for the bolometric effect, which is sensitive to contamination between sheets and alignment of the graphene flakes making reproduction difficult. [71]

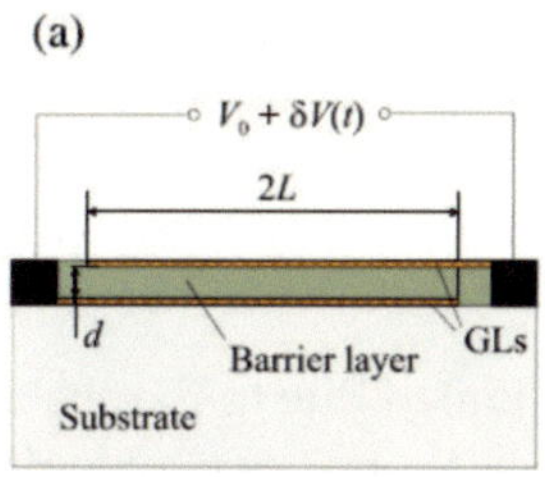

Figure 28. Phonon resonance-based IR detector [70].

The resonance-based IR detector shown in Figure 28 utilizes the phonon resonance of two separate graphene sheets separated by a dielectric allowing for the tuning of wavelength detection based upon separation distance, but the fabrication is difficult requiring pristine graphene and no trapped states in the oxide that would both modify the resonant frequency and could possibly contaminate the detector out of detection range. [70]

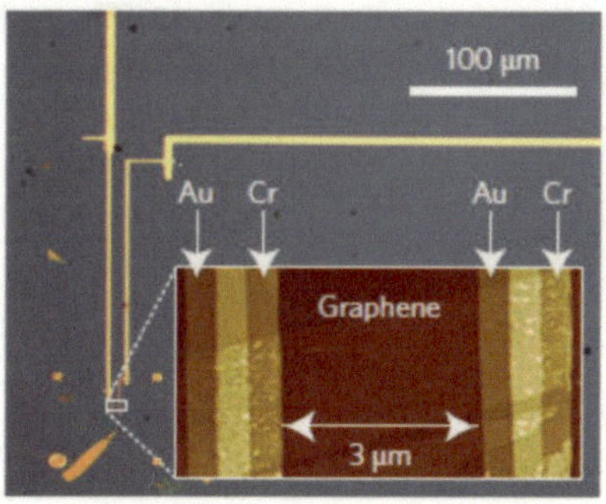

Figure 29. Image of a detector based upon the photothermoelectric effect [45].

The photothermoelectric effect detector shown in Figure 29 is relatively straight forward with contamination only affecting the speed of the detector and the noise only susceptible to trap states of the insulating oxide that the graphene is transferred onto. [45]

7. Conclusion

We have shown graphene to have many amazing properties due to its unique bonding and subsequently band gap characteristics, having electronic carriers act as "massless" Dirac-Fermions. The material characteristics of graphene are anisotropic, having phenomenal characteristic within a single sheet and diminished material characteristics between sheet with increasing sheet number and grain boundaries. This restricts the applications of graphene to technology that is consistent with miniaturization such as microelectronics. Therefore the integration of graphene into several electronic device applications was reviewed.

Graphene has the highest mobilities values measured in a material at room temperature making integration into fast response time devices such as a HEMT for RF applications. It has been shown that although the integration of graphene is challenging due to mobility degradation due to surface contamination in the graphene and trapped states in the oxide dielectric, a graphene RF detector with an overall response frequency of 300 Ghz was achieved utilizing a three-terminal design on a SiC substrate with a channel length of 40 nm.

Graphene use in optical devices is limited due to the absorption of 2.3% of incident light per layer making graphene's use for optical devices a tradeoff between getting enough layer for good optical absorption and modulation versus restricting number of layers for fast carrier propagation. On its own, graphene is not practical for use as a waveguide or modulator but when it is combined with already active materials, it increases the performance of such devices thus an EO modulator utilizing a stacked graphene-BN capacitor along with a Si microcavity array displays the ability to modulate light at a rate of 1.2 GHz.

Graphene for IR detectors has shown some promising results utilizing graphene in thermal-based detection regimes since photon-based absorption regimes all require inducing a band gap, adding complexity and reliability issues. The unique thermal-based properties of graphene either in a traditional bolometric type of device or one based upon current produced from the photoelectric effect allowed for the creation of a graphene IR detector with sensitivity to a 2.5 THz (119μm) laser.

Author details

Ashok K. Sood[1*], Isaac Lund[1], Yash R. Puri[1], Harry Efstathiadis[2], Pradeep Haldar[2], Nibir K. Dhar[3], Jay Lewis[4], Madan Dubey[5], Eugene Zakar[5], Priyalal Wijewarnasuriya[5], Dennis L. Polla[6] and Michael Fritze[7]

*Address all correspondence to: aksood@magnoliaoptical.com

1 Magnolia Optical Technologies, Inc., Woburn, MA, USA

2 College of Nanoscale Science and Engineering (CNSE), Albany, NY, USA

3 Night Vision & Electronic Sensors Directorate, VA, USA

4 DARPA/MTO, Arlington, VA, USA

5 Army Research Laboratory, Adelphi, MD, USA

6 University of Minnesota, Minneapolis, MN, USA

7 Senior Fellow, Potomac Institute for Policy Studies, Arlington, VA, USA

References

[1] Max C. Lemme, "Current Status of Graphene Transistors", Solid State Phenomena, Volumes 156-158, Pages 499-509, 2010.

[2] Harold Y. Hwang, Yoh Iwasa, Masashi Kawasaki, Bernhard Keimer, Naoto Nagaosa, and Yoshinori Tokura, "Emergent Phenomena at Oxide Interfaces", Nature Materials, Volume 11, Pages 103-113, 2012.

[3] Soon-Yong Kwon, Cristian V. Ciobanu, Vania Petrova, Vivek B. Shenoy, Javier Bareno, Vincent Gambin, Ivan Petrov, and Suneel Kodambaka, "Growth of semiconducting graphene on palladium", Nano Letters, Volume 9, Number 12, Pages 3985-90, December 2009.

[4] Peter W. Sutter, Jan-Ingo Flege, and Eli A. Sutter, "Epitaxial graphene on ruthenium", Nature Materials, Volume 7, Pages 406-411, May 2008.

[5] Johann Coraux, Alpha T. N'Diaye, Carsten Busse, and Thomas Michely, "Structural Coherency of Graphene on Ir(111)", Nano Letters, Volume 8, Number 2, Pages 565-570, 2008.

[6] Xuesong Li, Weiwei Cai, Jinho An, Seyoung Kim, Junghyo Nah, Dongxing Yang, Richard Piner, Aruna Velamakanni, Inhwa Jung, Emanuel Tutuc, Sanjay K. Banerjee, Luigi Colombo, and Rodney S. Ruoff, "Large-Area Synthesis of High-Quality and Uniform Graphene Films on Copper Foils", Science, Volume 324, Pages 1312-14, June 2009.

[7] Cecilia Mattevi, Hokwon Kima, and Manish Chhowalla, "A review of chemical vapour deposition of graphene on copper", Journal of Material Chemistry, Issue 21, Pages 3324-3334, 2011.

[8] Changgu Lee, Xiaoding Wei, Jeffrey W. Kysar, and James Hone, "Measurement of the Elastic Properties and Intrinsic Strength of Monolayer Graphene" Science, Volume 321, Pages 385-388, 2008.

[9] Alexander A. Balandin, Suchismita Ghosh, Wenzhong Bao, Irene Calizo, Desalegne Teweldebrhan, Feng Miao, and Chun Ning Lau, "Superior Thermal Conductivity of Single-Layer Graphene", Nano Letters, Volume 8, Number 3, Pages 902-907, 2008.

[10] Wenzhong Bao, Feng Miao, Zhen Chen, Hang Zhang, Wanyoung Jang, Chris Dames, and Chun Ning Lau, "Ripple Texturing of Suspended Graphene Atomic Membranes", Nature Nanotechnology, Volume 4, Pages 562-566, 2009.

[11] Tim J. Booth, Peter Blake, Rahul R. Nair, Da Jiang, Ernie W. Hill, Ursel Bangert, Andrew Bleloch, Mhairi Gass, Kostya S. Novoselov, M. I. Katsnelson, and A. K. Geim, "Macroscopic Graphene Membranes and Their Extraordinary Stiffness", Nano Letters, Volume 8, Issue 8, Pages 2442-2446, 2008.

[12] J. Scott Bunch, Scott S. Verbridge, Jonathan S. Alden, Arend M. van der Zande, Jeevak M. Parpia, Harold G. Craighead, and Paul L. McEuen, "Impermeable Atomic Membranes from Graphene Sheets", Nano Letters, Volume 8, Issue 8, Pages 2458-2462, 2008.

[13] Antonio H. Castro Neto, Francisco Guinea, Nuno M. R. Peres, Konstantin S. Novoselov and Andre K. Geim, "The electronic properties of graphene", Reviews of Modern Physics, Volume 81, Pages 109-162, January–March 2009.

[14] Jian-Hao Chen, Chaun Jang, Shudong Xiao, Masa Ishigami, and Michael S. Fuhrer, "Intrinsic and extrinsic performance limits of graphene devices on SiO_2", Nature Nanotechnology, Volume 3, Pages 206-209, 2008.

[15] Xu Du, Ivan Skachko, Anthony Barker, and Eva Y. Andrei, "Approaching ballistic transport in suspended graphene", Nature Nanotechnology, Volume 3, Pages 491-495, 2008.

[16] Frank Schwierz, "Graphene Transistors", Nature Nanotechnology, Volume 5, Pages 487-496, July 2010.

[17] A. K. Geim, "Graphene: Status and Prospects", Science, Volume 324, Number 5934, Pages 1530-1534, June 2009.

[18] Hossein Sojoudi, Jose Baltazar, Laren M. Tolbert, Clifford L. Henderson, and Samuel Graham, "Creating graphene p-n junctions using self-assembled monolayers", ACS Applied Material Interfaces, Volume 4, Pages 4781-6, 2012.

[19] Deep Jariwala, Anchal Srivastava, and Puklickel Ajayan, "Graphene synthesis and band gap opening", Journal of Nanoscience and Nanotechnology, Volume 11, Issue 8, Pages 6621-41, 2011.

[20] http://www.graphene.manchester.ac.uk/explore/the-story-of-graphene/a-discovery-at-manchester/

[21] Zheng Yan, Zhiwei Peng, Gilberto Casillas, Jian Lin, Changsheng Xiang, Haiqing Zhou, Yang Yang, Gedeng Ruan, Abdul-Rahman O. Raji, Errol L. G. Samuel, Robert H. Hauge, Miguel Jose Yacaman, and James M. Tour, "Rebar Graphene", ACS Nano, Volume 8, Issue 5, Pages 5061-5068, 2014.

[22] Jonathan N. Coleman, Mustafa Lotya, Arlene O'Neill, Shane D. Bergin, Paul J. King, Umar Khan, Karen Young, Alexandre Gaucher, Sukanta De, Ronan J. Smith, Igor V. Shvets, Sunil K. Arora, George Stanton, Hye-Young Kim, Kangho Lee, Gyu Tae Kim, Georg S. Duesberg, Toby Hallam, John J. Boland, Jing Jing Wang, John F. Donegan, Jaime C. Grunlan, Gregory Moriarty, Aleksey Shmeliov, Rebecca J. Nicholls, James M. Perkins, Eleanor M. Grieveson, Koenraad Theuwissen, David W. McComb, Peter D. Nellist, and Valeria Nicolosi, "Two-Dimensional Nanosheets Produced by Liquid Exfoliation of Layered Materials", Science, Volume 331, Number 6017, Pages 568-571, February 4 2011.

[23] Peter Sutter, "Epitaxial graphene: How silicon leaves the scene", Nature Materials Volume 8, Issue 3, Pages 171-172, 2009

[24] Taisuke Ohta, Aaron Bostwick, Jessica L. McChesney, Thomas Seyller, Karsten Horn, and Eli Rotenberg, "Interlayer Interaction and Electronic Screening in Multilayer Graphene Investigated with Angle-Resolved Photoemission Spectroscopy", Physical Review Letters, Volume 98, Page 206802, 2007.

[25] Aaron Bostwick, Taisuke Ohta, Jessica L. McChesney, Konstantin V. Emtsev, Thomas Seyller, Karsten Horn, Eli Rotenberg, "Symmetry breaking in few layer graphene films" New Journal of Physics, Volume 9, Issue 10, 385, 2007

[26] Rahul Mukherjee, Abhay Varghese Thomas, Ajay Krishnamurthy, and Nikhil Koratkar, "Photothermally Reduced Graphene as High-Power Anodes for Lithium-Ion Batteries", ACS Nano, Volume 6, Issue 9, Pages 7867-8, 2012.

[27] Hongbin Feng, Rui Cheng, Xin Zhao, Xiangfeng Duan, and Jinghong Li, "A low-temperature method to produce highly reduced graphene oxide", Nature Communications 4, Article number 1539, 2013.

[28] Maher F. El-Kady, Veronica Strong, Sergey Dubin, and Richard B. Kaner, "Laser scribing of high-performance and flexible graphene-based electrochemical capacitors", Science, Volume 335, Number 6074, Pages 1326-30, March 16 2012.

[29] Keith R. Paton, Eswaraiah Varrla, Claudia Backes, Ronan J. Smith, Umar Khan, Arlene O'Neill, Conor, Boland, Mustafa, Lotya, Oana M. Istrate, Paul King, Tom Higgins, Sebastian Barwich, Peter May, Pawel Puczkarski, Iftikhar Ahmed, Matthias Moebius, Henrik Pettersson, Edmund Long, João Coelho, Sean E. O'Brien, Eva K. McGuire, Beatriz Mendoza Sanchez, Georg S. Duesberg, Niall McEvoy, Timothy J. Pennycook, Clive Downing, Alison Crossley, Valeria Nicolosi, and Jonathan N. Coleman, "Scalable production of large quantities of defect-free few-layer graphene by shear exfoliation in liquids", Nature Materials, Volume 13, Pages 624-630, 2014.

[30] Daniel R. Lenski, and Michael S. Fuhrer, "Raman and optical characterization of multilayer turbostratic graphene grown via chemical vapor deposition", Journal of Applied Physics, Volume 110, 013720, 2011.

[31] Victor Calado, Shou-En Zhu, Srijit Goswami, Qiang Xu, Kenji Watanabe, Takashi Taniguchi, Guido C. A. M. Janssen, and Lieven Vandersypen, "Ballistic transport in graphene grown by chemical vapor deposition", Applied Physics Letter, Volume 104, 023103, 2014.

[32] Valeria Alzari, Daniele Nuvoli, Roberta Sanna, Sergio Scognamillo, Massimo Piccinini, Jose Maria Kenny, Giulio Malucelli, and Alberto Marian "In situ production of high filler content graphene-based polymer nanocomposites by reactive processing", Journal of Materials Chemistry, Volume 21, Issue 24, Pages 8727-8733, 2011.

[33] Jens Baringhaus, Ming Ruan, Frederik Edler, Antonio Tejeda, Muriel Sicot, Amina Taleb-Ibrahimi, An-Ping Li, Zhigang Jiang, Edward H. Conrad, Claire Berger, Chris-

toph Tegenkamp, and Walt A. de Heer, "Exceptional ballistic transport in epitaxial graphene nanoribbons", Nature, Volume 506, Pages 349–354, February 20 2014.

[34] Francesca Iacopia, Neeraj Mishraa, Benjamin Vaughan Cunninga, Dayle Godinga, Sima Dimitrijeva, Ryan Brocka, Reinhold H. Dauskardta, Barry Wood, and John Boeck, "A catalytic alloy approach for highly uniform graphene on epitaxial SiC on silicon wafers", Journal of Materials Research, Volume 30, Issue 05, Pages 609-616, 2015.

[35] Jan Kunc, Yike Hu, James Palmer, Zelei Guo, John Hankinson, Salah H. Gamal, Claire Berger, and Walt A. de Heer, "Planar edge Schottky barrier-tunneling transistors using epitaxial graphene/SiC junctions" Nano Letters, Volume 14, Issue 9, Pages 5170-5, September 10 2014.

[36] Joseph Lyding, Joshua Wood, and Eric Pop, "Growing better graphene by finding the best copper surface", SPIE DOI: 10.1117/2.1201201.004110, March 12 2012.

[37] Youngbin Lee, Sukang Bae, Houk Jang, Sukjae Jang, Shou-En Zhu, Sung Hyun Sim, Young Il Song, Byung Hee Hong, and Jong-Hyun Ahn, "Wafer-Scale Synthesis and Transfer of Graphene Films", Nano Letters, Volume 10, Issue 2, Pages 490-493, 2010.

[38] Corey A. Joiner, Tania Roy, Zohreh Razavi Hesabi, Bhaswar Chakrabarti, and Eric M. Vogel, "Cleaning graphene with a titanium sacrificial layer", Applied Physics Letters, Volume 104, Page 223109, 2014

[39] Anindya Nath, Andrew D. Koehler, Glenn G. Jernigan, Virginia D. Wheeler, Jennifer K. Hite, Sandra C. Hernández, Zachary R. Robinson, Nelson Y. Garces, Rachael L. Myers-Ward4, Charles R. Eddy Jr., D. Kurt Gaskill4, and Mulpuri V. Rao, "Achieving clean epitaxial graphene surfaces suitable for device applications by improved lithographic process", Applied Physics Letters, Volume 104, Issue 22, Page 224102, 2014.

[40] Jeehwan Kim, Can Bayram, Hongsik Park, Cheng-Wei Cheng, Christos Dimitrakopoulos, John A. Ott, Kathleen B. Reuter, Stephen W. Bedell, and Devendra K. Sadana, "Principle of direct van der Waals epitaxy of single-crystalline films on epitaxial graphene", Nature Communications, Volume 5, Article number 4836, September 11 2014.

[41] Haomin Wang, Yihong Wu, Chunxiao Cong, Jingzhi Shang, and Ting Yu, "Hysteresis of electronic transport in graphene transistors", ACS Nano, Volume 4, Issue 12, Pages 7221-7228, 2010.

[42] Cory R Dean, Andrea F Young, Inanc Meric, Chris Lee, Lei Wang, Sebastian Sorgenfrei, Kenji Watanabe, Takashi Taniguchi, Phillip Kim, Kenneth L Shepard, and James Hon, "Boron nitride substrates for high-quality graphene electronics", Nature Nanotechnology, Volume 5, Issue 10, Pages 722-726, 2010.

[43] Heejun Yang, Jinseong Heo, Seongjun Park, Hyun Jae Song, David H. Seo, Kyung-Eun Byun, Philip Kim, InKyeong Yoo, Hyun-Jong Chung, and Kinam Kim, "Gra-

phene Barristor, a Triode Device with a Gate-Controlled Schottky Barrier", Science, Volume 336, Number 6085, Pages 1140-1143, June 1 2012.

[44] Thomas Mueller, Fengnian Xia, Phaedon Avouris, "Graphene photodetectors for high-speed optical communications", Nature Photonics, Volume 4, Issue 5, Pages 297-301, 2010.

[45] Xinghan Cai, Andrei B. Sushkov, Ryan J. Suess, Mohammad M. Jadidi, Gregory S. Jenkins, Luke O. Nyakiti, Rachael L. Myers-Ward, Shanshan Li, Jun Yan, D. Kurt Gaskill, Thomas E. Murphy, H. Dennis Drew, and Michael S. Fuhrer, "Sensitive room-temperature terahertz detection via the photothermoelectric effect in graphene", Nature Nanotechnology, Volume 9, Pages 814-20, 2014.

[46] Vasilios Georgakilas, Michal Otyepka, Athanasios B. Bourlinos, Vimlesh Chandra, Namdong Kim, K. Christian Kemp, Pavel Hobza, Radek Zboril, and Kwang S. Kim, "Functionalization of Graphene: Covalent and Non-Covalent Approaches, Derivatives and Applications", Chemical Reviews, Volume 112, Issue 11, pages 6156-6214, 2012.

[47] Andre K. Geim, and Irina V. Grigorieva, "Van der Waals heterostructures", Nature, Volume 499, Pages 419-425, July 25 2013.

[48] He Tian, Zhen Tan, Can Wu, Xiaomu Wang, Mohammad Ali Mohammad, Dan Xie, Yi Yang, Jing Wang, Lain-Jong Li, Jun Xu, and Tian-Ling Ren, "Novel Field-Effect Schottky Barrier Transistors Based on Graphene-MoS2 Heterojunctions", Scientific Reports, Issue 4, Article number 5951, 2014.

[49] Rafi Bistritzer and Allan H. MacDonald, "Moiré bands in twisted double-layer graphene", Proceedings of the National Academy of Sciences, Volume 108, Number 30, Pages 12233-12237, 2011.

[50] Wataru Norimatsu and Mau Kusunoki, "Epitaxial graphene on SiC{0001}: Advances and perspectives." Physical Chemistry Chemical Physics, Volume 16, Issue 8, Pages 3501-11, Feb 28 2014.

[51] Jeremy T. Robinson, James S. Burgess, Chad E. Junkermeier, Stefan C. Badescu, Thomas L. Reinecke, F. Keith Perkins, Maxim K. Zalalutdniov, Jeffrey W. Baldwin, James C. Culbertson, Paul E. Sheehan, and Eric S. Snow, "Properties of fluorinated graphene films", Nano Letters, Volume 10, Issue 8, Pages 3001-5, August 11 2010.

[52] Marc Dvorak, William Oswald, and Zhigang Wu, "Bandgap Opening by Patterning Graphene", Scientific Reports, Issue 3, Article Number 2289, 2013.

[53] Melinda Y. Han, Barbaros Özyilmaz, Yuanbo Zhang, and Philip Kim, "Energy Band-Gap Engineering of Graphene Nanoribbons", Physical Review Letters, Issue 98, Article Number 206805, May 16 2007.

[54] Anna Orlof, Julius Ruseckas, and Igor V. Zozoulenko, "Effect of zigzag and armchair edges on the electronic transport in single-layer and bilayer graphene nanoribbons with defects", Physics Review B Issue 88, Article Number 125409, September 4, 2013.

[55] Bart N. Szafranek, Gianluca Fiori, Daniel Schall, Daniel Neumaier, and Heinrich Kurz "Current Saturation and Voltage Gain in Bilayer Graphene Field Effect Transistors", Nano Letters, Volume 12, Issue 3, Pages 1324-1328, 2012.

[56] Xuetao Gan, Ren-Jye Shiu, Yuanda Gao, Inanc Meric, Tony F. Heinz, Kenneth Shepard, James Hone, Solomon Assefa, and Dirk Englund, "Chip-integrated ultrafast graphene photodetector with high responsivity", Nature Photonics, Volume 7, Pages 883-887 November 2013

[57] Christopher B. McKitterick, Heli Vora, Xu Du, Boris S. Karasik, and Daniel E. Prober, "Graphene Microbolometers with Superconducting Contacts for Terahertz Photon Detection", Journal of Low Temperature Physics, Volume 176, Issue 3-4, pp 291-298, August 2014.

[58] Shu-Jen Han, Alberto Valdes Garcia, Satoshi Oida, Keith A. Jenkins, and Wilfried Haensch, "Graphene radio frequency receiver integrated circuit", Nature Communications, Volume 5, Article number 3086, Pages 1-6, January 2014.

[59] Yanjie Wang, Bo-Chao Huang, Ming Zhang, Congqin Miao, Ya-Hong Xie, and Jason C. S. Woo, "High performance graphene FETs with self-aligned buried gates fabricated on scalable patterned Ni-Catalyzed Graphene", IEEE Symposium on VLSI Technology (VLSIT), Pages 116-117, Honolulu Hawaii, June 14-16 2011.

[60] Sujit S. Datta, Douglas R. Strachan, and A.T. Charlie Johnson Jr, "Gate coupling to nanoscale electronics." Physical Review B, Issue 79, Article Number 205404, 2009.

[61] Yanqing Wu, Keith A. Jenkins, Alberto Valdes-Garcia, Damon B. Farmer, Yu Zhu, Ageeth A. Bol, Christos Dimitrakopoulos, Wenjuan Zhu, Fengnian Xia, Phaedon Avouris, and Yu-Ming Lin, "State-of-the-Art Graphene High-Frequency Electronics", Nano Letters, Volume 12, Issue 6, Pages 3062-3067, 2012.

[62] Siyuranga O. Koswatta, Alberto Valdes-Garcia, Mathias B. Steiner, Yu-Ming Lin, and Phaedon Avouris, "Ultimate RF Performance Potential of Carbon Electronics", IEEE Microwave Theory and Techniques, IEEE Transactions on, Volume 59, Issue 10, Pages 2739-2750, 2011.

[63] Yuanda Gao, Ren-Jye Shiue, Xuetao Gan, Luozhou Li, Cheng Peng, Inanc Meric, Lei Wang, Attila Szep, Dennis Walker, Jr., James Hone, and Dirk Englund, "High-Speed Electro-Optic Modulator Integrated with Graphene-Boron Nitride Heterostructure and Photonic Crystal Nanocavity", Nano Letters, Volume 15, Pages 2001-2005, 2015.

[64] Xuetao Gan, Ren-Jye Shiue, Yuanda Gao, Solomon Assefa, Senior Member, IEEE, James Hone, and Dirk Englund, "Controlled Light–Matter Interaction in Graphene Electrooptic Devices Using Nanophotonic Cavities and Waveguides", IEEE Journal

of Selected Topics in Quantum Electronics, Volume 20, Number 1, January/February 2014.

[65] Kelvin J. A. Ooi, Hong Son Chu, Ping Bai, and Lay Kee Ang, "Electro-optical graphene plasmonic logic gates", Optics Letters, Volume 39, Issue 6, Pages 1629-1632, 2014.

[66] Maksym V. Strikha and Fedir T. Vasko, "Electro-optics of graphene: Field-modulated reflection and birefringence", Physical Review B, Issue 81, Article Number 115413, 2010.

[67] Kelvin J. A. Ooi, Lay Kee Ang, and Dawn T. H. Tan, "Waveguide Engineering of Graphene's Nonlinearity", Applied Physics Letters, Issue 105, Article Number 111110, 2014.

[68] Manijeh Razeghi, "Current status and future trends of infrared detectors", Opto-Electronics Review, Volume 6, Issue 3, Pages 155-194, 1998.

[69] Jun Yan, Mann Ho Kim, Jennifer A. Elle, Andrei B. Sushkov, Greg S. Jenkins, Howard M. Milchberg, Michael S. Fuhrer, and H. D. Drew, "Dual-gated bilayer graphene hot electron bolometer", Nature Nanotechnology, Volume 7, Pages 472-478, 2012.

[70] Victor Ryzhii, Taiichi Otsuji, Maxim Ryzhii, and Michael S. Shur, "Double graphene-layer plasma resonances terahertz detector", Journal of Physics D: Applied Physics, Volume 45, Number 30, Article Number 302001, 2012.

[71] Kiran Shankar Hazra, N. Sion, Anil Yadav, James McLauhglin, and Devi Shanker Misra, "Vertically aligned graphene based non-cryogenic bolometer" 2013, arXiv: 1301.1302.

[72] Thomas Mueller, Tony Low, Phaedon Avouris, and Marcus Freitag, "Increased Responsivity of Suspended Graphene Photodetectors" Nano Letters, Volume 13, Issue 4, Pages 1644-1648, 2013.

[73] Xu Du, Daniel E. Prober, Heli Vora, and Christopher B. Mckitterick, "Graphene-based Bolometers", Graphene and 2D Materials, Volume 1, Issue 1, Pages 2299-2321, April 2014.

[74] Alexander V. Klekachev, Amirhasan Nourbakhsh, Inge Asselberghs, Andre L. Stesmans, Marc M. Heyns, and S. De Gendl, "Graphene Transistors and Photodetectors", Electrochemical Society Interface, Volume 22, Issue 1, Pages 63-68, 2013.

[75] Jeong-Sun Moon, Kurt Gaskill and Paul Campbell, Graphene Transistors and RF Applications in the book "Physics and Applications of Graphene - Experiments" edited by Sergey Mikhailov, ISBN 978-953-307-217-3, InTech, April 4, 2011

Chapter 4

Photoactive Graphene — From Functionalization to Applications

Hang-Xing Wang and Hao-Li Zhang

Additional information is available at the end of the chapter

http://dx.doi.org/10.5772/61401

Abstract

Photoactive graphene fabricated by chemical functionalization of pristine graphene with different light-harvesting molecules has become one of the most exciting topics of graphene research in the last few years, which remarkably sustains and expands the graphene boom due to its great potentials in various applications. This chapter presents some important issues of photoactive graphene, covering material synthesis, electron/ energy- transfer interaction, organic photovoltaic and photocatalystic applications. Of particular interest is the utilization of graphene as a two- dimensional platform to anchor organic conjugated aromatic molecules and their applications in photo-energy conversion and photocatalysis. Challenges currently faced by researchers and future perspectives in this field are also discussed.

Keywords: Photoactive graphene, light-harvest, functionalization, photovoltaics, photocatalysis

1. Introduction

Graphene, a monolayer of graphite, has attracted intense interest in recent years because of its unique two-dimensional structure and excellent physical and chemical properties such as high electronic and thermal conductivity, great mechanical strength, and huge specific surface area. [1-4] However, due to its zero bandgap semiconductor characteristic and a highly transparent property in the visible spectrum, pristine graphene is severely limited in optoelectronic and photo-energy conversion applications.

Photoactive graphene in recent years bring up impressive phenomena that remarkably sustain and expand the graphene research interests. The rise of photoactive graphene in photonics and optoelectronics is evidenced by a spate of latest reports, typically photovoltaics and photocatalysts. Although there is some available literature covering various aspects of photoactive graphene,[1, 5, 6] a strategic update that reflects the newest progress, growing trends, and opening opportunities of photoactive graphene is required. This chapter pays particular attention to the development of photoactive graphene, including material preparation, the photophysical progress of excited organic molecules and graphene, as well as its applications in optoelectronics and photocatalysis.

1.1. What is photoactive graphene?

The term "photoactive graphene" generally refers to graphene that undergo a chemical or physical reaction when interacted with sunlight and/or ultraviolet light. Unlike the transparent pristine graphene, photoactive graphene shows optical response characteristics when light passes through it.

To date, two different approaches toward the preparation of photoactive graphene can be found in the literature. As shown in Figure 1 (a), the first is based on the bandgap engineering (opening and tuning) of pristine graphene,[7] and in the second strategy the photoactive graphene is obtained by chemical functionalization with photoactive moieties.[8] The most reported examples of the first route are roughly classified into four categories: (1) heteroatom doping;[9-11] (2) chemical modification;[12] (3) electrostatic field tuning;[13, 14] and (4) cutting graphene into nanoribbons.[15-19] As shown in Figure 1 (b), the second strategy for giving photoactive to graphene is chemical functionalization of graphene with photoactive units, including organic conjugated molecules and polymers, inorganic semiconductors particles and quantum dots, rare-earth metal complexes, and so on. This approach has been demonstrated to be a feasible route to achieve the photo-electron response of graphene to light.[8, 20] Herein, we will only discuss the progress and challenges related to the chemical functionalization approach to preparing photoactive graphene.

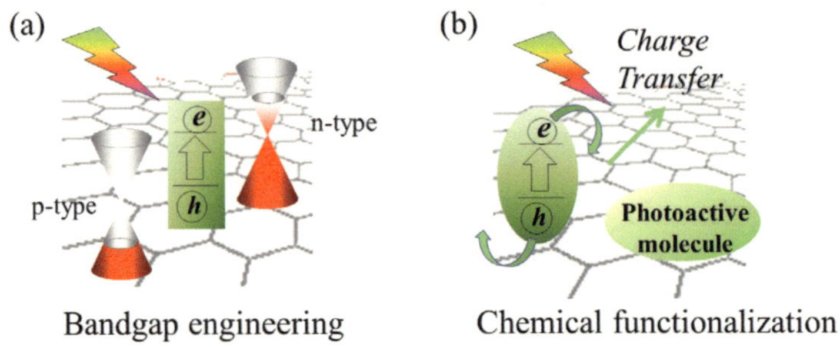

Figure 1. Two different approaches toward the photoactive graphene. One is based on the bandgap opening and tuning of pristine graphene (a), another strategy is chemical functionalization of graphene with photoactive moieties (b).

1.2. Why photoactive graphene?

Pristine graphene is made of sp^2 hybrid carbon atoms with the s, p_x, and p_y orbitals on each carbon atom forming three strong σ bonds with other three surrounding atoms.[21] The formed valence and conduction bands touch at the Brillouin zone corners (so-called Dirac or neutrality points) making graphene a zero bandgap semiconductor with poor photoactive characteristic.[22, 23] Experimentally, the transmittance of the mechanically exfoliated graphene is overwhelming (97.7%), and thus it absorbs only 2.3% of light that passes through it (Figure 2).[2, 24] So it is impossible for graphene to absorb light when used as a photoactive material (Geim *et al.*, 2008).

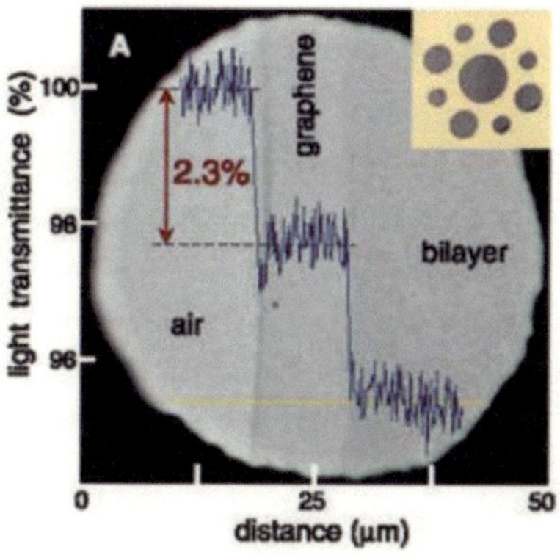

Figure 2. Optical adsorption property of graphene and its layer (Geim *et al.*, *Science* 2008, 320, 1308).

The poor photoactive characteristic of graphene leads to some major drawbacks in photo-energy conversion applications. For photovoltaic devices, using pristine graphene as active layer may significantly reduce the open circuit voltage, compared to conversional semiconductors.[25] Moreover, graphene is unfavorable for efficient photocurrent generation in photovoltaic devices. From a dynamics point of view, to output a photocurrent in an external circuit, the photo-generated carrier needs to separate from the photo-generation region before recombination.[26, 27] However, for graphene based transistor, the lifetime of photo-generated hot carriers was cooled down within several hundred femtoseconds, followed by electron–hole recombination on a picosecond timescale.[28, 29] Therefore, in order to improve the photo-energy conversion efficiency of graphene, it is necessary to extend the lifetime of photo-generated carriers of graphene.

Recently, endowing graphene photoactive property has been viewed as an effective approach to exploit the advantages of graphene in photovoltaics and photocatalytics.[30] The photoactive graphene can play a significant role in photovoltaics and photocatalysts resulting from the synergy effect of the charge transfer based in the photoactive graphene. For photoactive graphene, the propensity of graphene to interact with excited state photoactive moiety is often involve energy- and/or charge-transfer processes, where the excited fluorescent moieties is served as excellent probes by monitoring their emission evolution.[31] Meanwhile, the high charge mobility of graphene promotes it can be used as efficient acceptors to enhance photo-induced charge transfer for improving photocurrent conversion performance. The effective nonradiative deactivation of the excited photoactive moieties on graphene surface by interfacial charge transfer is necessary for the improvement of the conversion efficiency from optoelectronic and photo-energy to electricity.

In recent years, there has been a growing interest in graphene functionalized with photoactive units owing to their significance in both fundamental research and practical applications. Recent research results have demonstrated that chemical functionalization of graphene with photoactive moieties is a necessity to harvest its full potential.[8]

2. Rational design of photoactive graphene

Graphene can be functionalized at the basal plane and the edges.[32] On the basal plane, sp^2 hybridization of carbon leads to a strong covalent bonding, as well as delocalization of the π electrons. The sheet edges are considered the most reactive sites,[33] and the dangling bonds at edge sites of graphene are highly reactive to guest atoms or molecules. In addition, the functional molecules attached onto the basal plane of graphene lead to modification of the π–π conjugation and thus the physical and chemical properties and the electron density distribution. There are generally three major purposes for chemically functionalizing graphene: enabling its solution processing, tuning its energy level and gap, as well as providing photoactive functionalities.[8]

2.1. Processability

A major obstacle in the synthesis and processing of bulk-quantity graphene sheets is the preparation of monolayer graphene and its insolubility, with the latter being responsible for poor handling and manipulation during graphene processing.[34, 35] Graphene has a strong tendency to cluster together into aggregation, caused by the electrostatic forces and the strong π–π interaction between individual graphene flakes, which make further manipulation and device fabrication using graphene difficult.[36] For this reason, covalent and noncovalent manners for modification of graphene have been developed for exfoliation and dispersion of graphene.[37] Besides, by derivatizing graphene with different moieties, the solubility of graphene can be tuned to suit varied solvents needed for different applications. For example, chemically grafted CH_2OH-terminated regioregular poly(3-hexylthiophene) (P3HT) onto carboxylic groups of graphene oxide (GO) *via* amidation reaction; the resultant P3HT-grafted GO sheets are soluble in common organic solvents, which facilitates the structure/property characterization and the device fabrication by solution processing.[38]

2.2. Energy level tuning

Experiment and theory studies demonstrated that semiconductor graphene (p-type and n-type) can be obtained by modifying graphene with organic semiconductor molecules (electron-acceptor and electron-donor), which provides a simple and nondestructive way of tuning the Femi level and controlling the charge carriers concentration of graphene.[39] As shown in Figure 3, for the p-type graphene, the Fermi level shifts upward relative to the Dirac point when the electron-acceptor coverage increases. In contrast, for the n-type graphene, the Fermi level shifts downward relative to the Dirac point when the electron-donor coverage increases.[40]

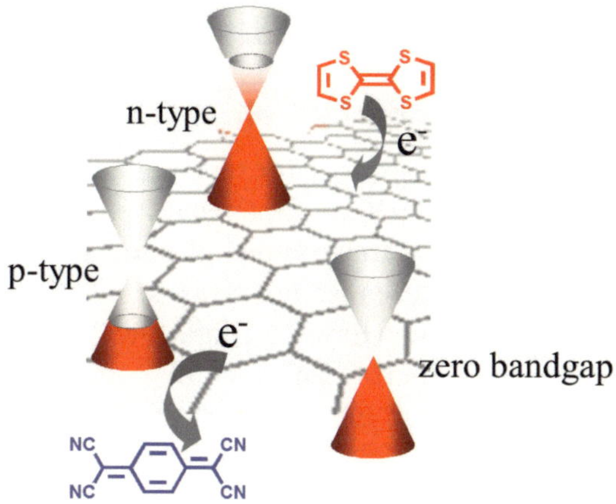

Figure 3. Doping graphene with semiconductor donor and acceptor molecules move the Fermi level up or down with respect to the Dirac point.

Wang *et al.* presented an effective route for preparation of both n-type and p-type graphene through adsorbing organic semiconductor molecules: one is the tetracyanoquinodimethane derivative (F4-TCNQ) and the other, the vanadyl-phthalocyanine (VOPc). The Kelvin probe force microscopy characterization results demonstrated that the F4-TCNQ molecules obtained electrons from graphene, but VOPc donated electrons to graphene. This phenomena indicated that chemical functionalization of graphene is a feasible approach for bandgap opening and tuning of graphene, which may have great implications for future large-scale applications of graphene-based nanoelectronics.[41] Pati *et al.* studied the modification in the electronic structure, as well as optical and transport properties of graphene induced by molecular charge transfer using ab initio density functional theory and Raman spectroscopic studies of modified graphene systems. They found that donor and acceptor molecules adsorbed onto the graphene surface exhibited effective molecular charge transfer, giving rise to mid-gap molecular levels with tuning of the band gap region near the Dirac point.[42]

2.3. Chemical functionalization

The chemistry of graphene is a powerful route to tailor its properties through introduction of various chemical functional groups to graphene.[32] For most applications, graphene need to be integrated with other functional materials, and the modification of graphene *via* chemical approach holds promise for tuning the electronic and optical properties of graphene, controlling interfaces with other materials, and tailoring surface chemical reactivity.[43] Therefore, functionalization of graphene with various functional components is considered to be crucial for graphene processing.[44, 45]

Similar to fullerenes and carbon nanotubes (CNTs), functionalization of graphene with different functional groups can open up new routes to hybrid materials that exhibit even more exciting features than graphene itself.[46] The design and feature tuning/altering of transparent pristine graphene by integrating a versatile electron-donor system has attracted more attention. However, modification of the flat and rigid structure of graphene is a more challenging work than that of the curvature structured fullerenes and carbon tubes because of the necessity to overcome a high-energy barrier.[47, 48]

Recently, rational design and efficient strategy for preparation of photoactive graphene have achieved considerable progresses, which are motivated by many potential applications of photoactive graphene. For example, a large number of photoactive graphene with different properties and structures have been designed and synthesized. As shown in Figure 4, photoactive moieties and graphene were linked through either covalent functionalization approach, such as amidation reaction, cycloaddition, Suzuki coupling, "click" chemistry, or noncovalent functionalization manner, including π–π interaction, electrostatic interaction, and electrostatic–π interaction.

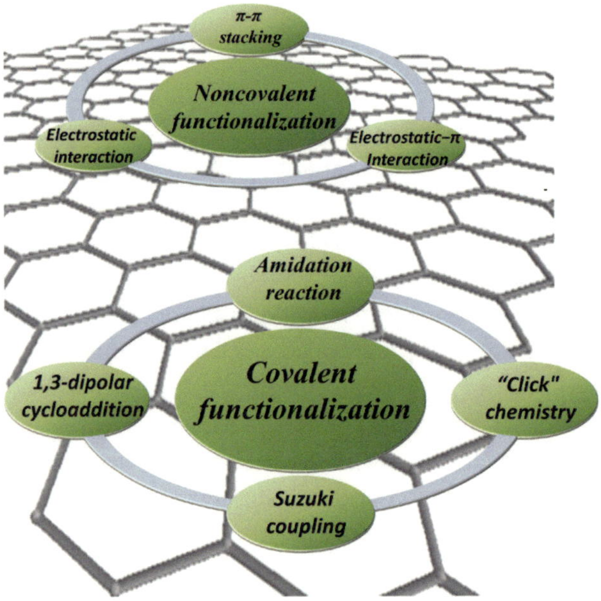

Figure 4. Chemical functionalization approach for preparation of photoactive graphene.

Photoactive organic moieties, including small molecules and conjugated polymers have been used to prepare photoactive graphene. The above-mentioned organic photoactive molecules are generally planar, electron-rich, and liable to photochemical electron-transfer process and show remarkably high extinction coefficients in the visible region. It is expected that by combining graphene with photoactive molecules, multifunctional graphene composites for optical and/or optoelectronic applications may be generated.[49]

http://dx.doi.org/10.5772/61401

2.3.1. *Covalent functionalization of graphene*

Recently, combining graphene with photoactive organic functional components, including small molecules and polymers, has attracted widespread attention, and a number of important photoactive moieties have been attached either to edge or the basal plane of graphene surface through covalent functionalization or noncovalent functionalization approaches.

Covalent functionalization of graphene with photoactive small molecules offers key advantages such as greater stability of the composite materials, and control over the degree of functionalization, and good reproducibility.[50] Meanwhile, conjugated polymers are one of the most successfully exploited classes of materials due to the incredible variety of chemical structures and their relatively low cost, facile processing, and their possible recyclability and applicability as sustainable materials.[51] Employing these organic photoactive molecules to functionalize graphene has grown to become a crucial branch in nanoscience and nanotechnology, which offers significant potential in the development of advanced photoactive graphene materials in numerous and diverse application areas.[52]

Amidation reaction: Chen *et al.* reported the first example of covalently attaching porphyrin units onto graphene, and explored the photophysical properties of the graphene composites.[53] The synthesized porphyrin–graphene composites consist of amino-containing porphyrin (TPP-NH_2) molecule and GO covalently bonded together *via* amidation reaction. Fourier transform infrared (FTIR) spectroscopic characterization confirmed that the TPP-NH_2 molecules had been covalently bonded to the edge of GO by the amide linkage. The linear relationship between the absorption and the concentrations of graphene moiety in the composites indicates good dispersibility of composite. Moreover, the fluorescence quenching of TPP-NH_2 was observed, indicating that there is a strong interaction between the excited state of TPP-NH_2 and graphene moieties in the composite. Kang *et al.* used arylamine-containing conjugated polymer TPAPAM to covalently modify GO.[54] The covalent attachment of TPAPAM onto the GO *via* amide linkage was confirmed by XPS and FTIR spectroscopy. In contrast to fluorescence quenching often observed in luminescence molecule–graphene systems, the steady-state fluorescence spectra showed that electronic interaction between TPAPAM and GO entities resulted in enhancement of the fluorescence intensity of the parent TPAPAM. In addition to the examples provided above, amidation reaction was also applied to link fullerene,[55] phthalocyanine zinc (PcZn),[56] and oligothiophene moieties to the surface of graphene.[57]

Cycloaddition reaction: Feringa *et al.* applied one-pot cycloaddition approach to prepare porphyrin derivative functionalized graphene composites.[37] The excited state energy/electron-transfer processes between graphene and the covalently attached porphyrin molecules was demonstrated from fluorescence quenching and reduced fluorescence lifetime phenomenon.[58] Guldi *et al.* presented their work on linking photoactive phthalocyanines (Pcs) to graphene surfaces.[50] Covalent functionalization of the fewer-layered graphene with Pcs was achieved through 1,3-dipolar cycloaddition and the esterification reaction yielded Pcs-graphene nanoconjugate.

Suzuki coupling reaction: The 2010 Nobel Prize for Chemistry rewarded a family of palladium-catalyzed coupling reactions for forging carbon–carbon bonds, which have already helped to create new graphene hybrid materials. Ma *et al.* reported covalent functionalization of graphene with polythiophene through Suzuki coupling reaction.[59] A donor–spacer–acceptor triad conjugated polymer containing fluorene, thiophene, and benzothiadazole moieties, which was covalently attached to r-GO *via* Suzuki polymerization procedure.[60] These polymer–graphene composites show excellent solubility in different type of solvents and exhibit superior optical-limiting performance. Moreover, Loh *et al.* applied Heck reaction to synthesize dye molecule functionalized graphene composite.[61] In their work, r-GO was covalently modified by diazonium, followed by the Heck reaction to give a 4-(2-(pyridin-4-yl)vinyl)phenyl group modified graphene. Considering the high efficiency of the palladium catalyzed C-C coupling reaction, we believe that more and more attention will be paid to the synthesis of photoactive-moieties–graphene hybrid materials.

"Click" chemistry: "Click" chemistry has emerged as a useful strategy for rapid and efficient attachment of functional groups to various materials since its reinvention in 2001.[62] In previous works, "click" chemistry has succeeded in linkage of various functional groups onto CNTs and fullerenes.[63, 64] Zhang *et al.* reported a facile approach for covalently attaching various photoactive organic molecules onto graphene surfaces *via* "click" chemistry.[65] Kaminska *et al.* presented a one-step protocol for simultaneous reduction and functionalization of GO with a dopamine derivative bearing an azide function. The chemical reactivity of the azide moieties was demonstrated by a post-functionalization with ethynylferrocene using "click" chemistry.[66] Salvio *et al.* treated GO suspension with sodium azide, and the obtained azido derivative can be used to functionalize the graphene oxide with long alkyl chains through a "click" chemistry approach. This functionalization results in the exfoliation of this material in organic solvent.[67] Salavagione *et al.* reported the preparation of polyfluorene-modified graphene by azide–alkyne "click" coupling.[68]

2.3.2. Noncovalent functionalization of graphene

Noncovalent functionalization strategy is advantageous in the preservation of the properties of the graphene, while weak forces between absorbed molecules and graphene may lower the load transfer in the composite, and as a result of free molecules and molecules adsorbed on graphene exist in equilibrium in the solution. Moreover, photoactive small molecules are commonly planar in structure and electron-rich; these advantages promote the interaction between the small molecules and graphene *via* π–π stacking, electrostatics interactions, and electrostatic–π interactions, as illustrated in Figure 4. Meanwhile, conjugated polyelectrolyte with highly electron-delocalized backbones and ionic side chains are water-soluble, fluorescent, rigid-rod polymers, which thereby combine the electronic properties of conjugated polymers with the electrostatic behavior of electrolytes. The conjugated polyelectrolytes and graphene hybrid materials generally have good solubility in polar solvents.

Via π–π stacking: Highly aromatic molecules may assemble themselves onto graphene surface *via* π–π stacking interaction. The π–π stacking interaction between aromatic skeleton of graphene and conjugated planar molecules afford synergistic binding interactions. Loh *et al.*

observed that the perylene wire could be coated on graphene surface to form a hybrid system. Such type of synergistic interaction between organic nanostructures and graphene affords a novel route to synthesis of hybrid materials with new properties and novel functions.[69] Mai *et al.* synthesized methyl blue functionalized graphene which exhibited excellent solubility and stability in water due to the photoactive molecules that were noncovalently stacked onto the basal plane of graphene.[70]

Via electrostatic interaction: GO and r-GO in aqueous dispersion is expected could act like a two-dimensional conjugated polyelectrolyte because of they are negatively charged, on which the cationic aromatic derivatives can be assembled through electrostatic and π–π stacking interactions. Shi *et al.* reported the supramolecular assembly and complexation of r-GO sheets with cationic porphyrin derivative in aqueous media.[71] In the UV-visible spectra, the Soret band of prophyrin showed a large bathochromic shift (37 nm) after linking on r-GO sheets. This phenomenon was attributed to the structure isomerization of porphyrin, which was caused by twist of their cationic methylpyridinium groups. For the structure of the unstrained porphyrin derivative, four cationic methylpyridinium moieties are nearly perpendicular to the plane of porphyrin because of the strong steric hindrance.[72] When the pyridinium groups rotate toward the coplanar conformation with respect to the flattened porphyrin ring, the π conjugation and electron-withdrawing effect of porphyrin will be enhanced.[73] Zhang and coworkers employed a specially designed anionic CPE to stabilize r-GO during the hydrazine-mediated reduction in aqueous solution.[74] The conjugated-polyelectrolyte (CPE) functionalized r-GO sheets show good solubility in a variety of polar solvents, because of the presence of double bonds along the backbone-endowed $PFVSO_3$ with a preferred coplanar backbone geometry, which matched the flat shape of r-GO and thus further enhanced the π–π interactions.

Via electrostatic–π interaction: The chemistry community now recognizes the electrostatic–π interaction as a major force for molecular recognition.[75] Jia *et al.* presented a novel molecular strategy based on cationic dye Rhodamine B (RhB) to functionalize graphene *via* in situ reduction.[76] RhB was designed to prevent the aggregation of graphene when it was modified to graphene surface through cation–π and π–π interaction. Zhang *et al.* reported the design and synthesis of a novel amphiphilic graphene composite by using amphiphilic coil-rod–coil conjugated triblock copolymer (PEG-OPE) as stabilizer to improve the dispersibility of r-GO. The rational designed PEG-OPE is composed of one lipophilic π-conjugated oligomer and two hydrophilic PEG coils. The conjugated rigid-rod backbone of PEG-OPE prefers to adsorb onto the basal plane of r-GO *via* strong π–π interaction; whilst the lipophilic side chains and two hydrophilic coils of the backbone would fly away from the surface of r-GO to form an amphiphilic outer layer, consequently facilitating the dispersion of modified r-GO common solvents.[77]

3. Energy transfer and charge transfer

It is believed that electron-transfer and energy-transfer processes between photoactive chromophores and graphene are the two fundamentally important processes responsible for the photophysical processes of photoactive graphene.

Generally, each decay step of an excited photoactive molecule is characterized by its own rate constant and each excited state is characterized by its lifetime. In solution, when the intramolecular deactivation processes are not too fast, that is, when the lifetime of the excited state is sufficiently long, an excited photoactive molecule may have a chance to encounter graphene. In such a case, some specific interaction can occur, leading to the deactivation of the excited state by second-order kinetic processes.[78]

Disentangling the detailed charge-transfer and energy-transfer dynamics in photoactive graphene composite is essential for a full understanding of their photophysical properties, and is expected to open new avenues for their unique and specific applications.[79] For this purpose, characterization of charge- and energy-transfer rates is essential. Although the investigations on the properties of the excited states and excitons are very challenging, this generally involves delicate fluorescence lifetime and transient absorption spectroscopic measurements.

3.1. Charge transfer

Excited-state photophysical processes between graphene and photoactive moieties have been of much importance because of their relevance to optoelectronic and photo-energy conversion applications.[80] In a considerable number of cases, the phenomena of graphene to quench fluorescence of aromatic molecules is shown to be associated with photo-induced electron-transfer process, and can be conveniently verified by the fluorescence decay and time-resolved transient absorption spectroscopic characterizations. These measurement results provide quantitative insights, both kinetically and spectroscopically, into the nature of the interactions of graphene and photoactive molecules.

Kamat *et al.* reported the excited electron-transfer interaction between the photo-excited porphyrin and graphene (Figure 5).[81] In their work, cationic 5,10,15,20-tetrakis (1-methyl-4-pyridinio)porphyrin tetra (p-toluenesulfonate), noted as TMPyP, was employed to noncovalent functionalization of graphene (Kamat *et al., 2010*). Upon complexation with graphene, the fluorescence lifetime of porphyrin was significantly reduced from 5 ns to 1 ns. Moreover, the femtosecond transient absorption measurements confirmed the formation of a short-lived singlet excited state of ^{1}(TMPyP)* and a subsequent longer-living porphyrin radical cation of (TMPyP)$^{\cdot+}$ with an absorption maximum around 515 nm, which clearly indicated the occurrence of electron-transfer process between TMPyP and graphene. Furthermore, it inferred that electron injection from the ^{1}(TMPyP)* to the graphene film is feasible because the oxidation potential of the ^{1}(TMPyP)* is -0.29 V *vs* normal hydrogen electrode (NHE), which is lower than the Fermi level of the graphene material (0 V *vs* NHE); the resulting energy gap hence provides sufficient driving force for the charge-transfer process.[7, 82]

Malig and coworkers reported the transient absorption characterization studies on the interactions of zinc phthalocyanines (ZnPc) oligomer–graphene composite both in the ground state and excited state. The experiment results confirmed that the nature of these interactions is electron transfer from ZnPc to graphene, both in the ground and in the excited state, affords an electron-transfer product that survives for several hundred picoseconds.[83] More interestingly, by combining with characterization results of the steady-state and femtosecond time-

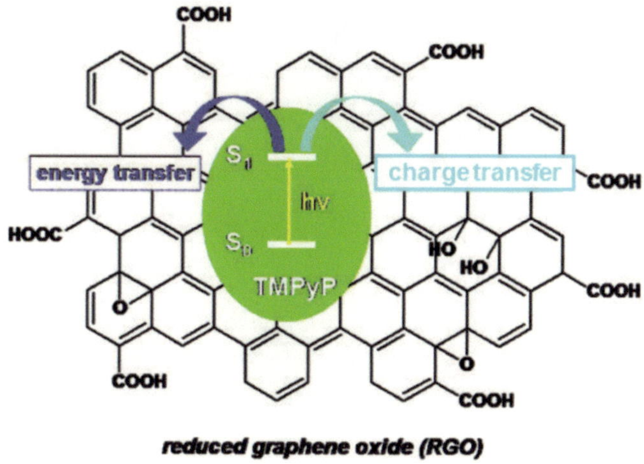

Figure 5. Photo-excited TMPyP molecules undergo charge-transfer interaction with r-GO (Kamat *et al.*, *ACS Nano* 2010, 4, 6697-6706).

resolved spectroscopy, Mohammed *et al.* found that the charge-transfer process at the porphyrin–graphene carboxylate interfaces could be tuned from zero to very sufficient and ultrafast by changing the electronic structure of the meso unit and the redox properties of the porphyrin cavity.[84]

3.2. Energy transfer

Graphene exhibits metallic behavior in many respects, in particular, graphene is shown to be a good exciton sink due to the highly efficient nonradiative energy transfer from the nearby fluorescent units through dipole–dipole coupling, which is also known as Forster-type resonant energy transfer (FRET).[85-89] Recently, FRET has been employed for interpreting the energy interaction of graphene combined with photoactive materials such as semiconductor nanoparticles and dyes (Figure 6).

Theoretical and experimental studies have disclosed efficient energy transfer to graphene and the process was found to be useful in identifying graphene sheets both on substrates and in solution.[90] Sebastian *et al.* studied the distance dependence of the rate of resonance energy transfer from the excited state of a dye (pyrene and nile blue) to the π skeleton of graphene.[91] Using the tight-binding model for the system and the Dirac cone approximation, the analytic expression for the rate of energy transfer from an electronically excited dye to graphene was obtained. It was found that graphene is a very efficient quencher of the electronically excited states and that the rate is proportional to d^{-4} (d is distance). Koppens *et al.* measured Rhodamine emitter lifetimes as a function of Rhodamine–graphene distance d, and found agreement with a universal scaling law, governed by the fine-structure constant. The observed emitter decay rate is enhanced 90 times (energy-transfer efficiency of ~99%) with respect to the decay in vacuum at distances d ≈ 5 nm.[88] Zhao *et al.* reported their study on employing ethidium bromide (EB) as a model for constructing an inexpensive and label-free biosensor to improve the sensitivity performance of GO–DNA-based sensors. Experiment results indicated that the fluorescence of EB was quenched by GO in the process of long-range resonance energy transfer.[92]

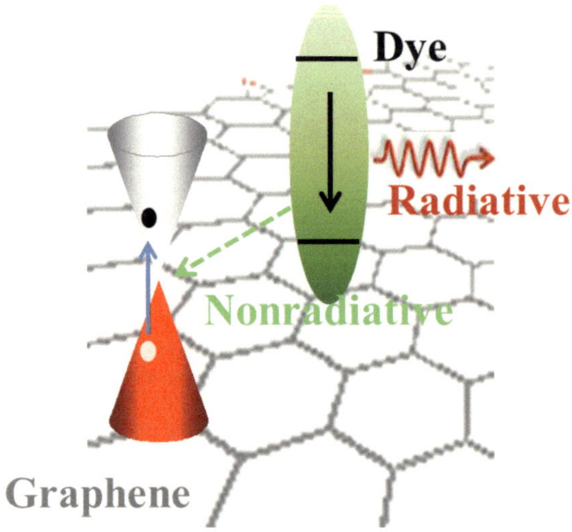

Figure 6. Nonradiative decay by dipolar coupling to electron–hole pair transition in the graphene surface and to a lower extent through the emission of radiation.

4. Applications of photoactive graphene

Covalent or noncovalent functionalization of graphene with various photoactive components has been considered to be crucial for graphene processing.[44, 45] The photoactive graphene *via* chemical approach holds promise for optimizing dispersibility in common solvents, tailoring surface chemical reactivity, and tuning the electronic properties of graphene.[43] Therefore, chemical functionalization strategy produced photoactive graphene has many advantages over pristine graphene and bare photoactive molecules, especially in the field of photo-energy conversion devices and as synergistic catalyst for organic synthesis reactions.[93, 94]

4.1. Organic Photovoltaic (OPV)

A typical OPV cell consists of a transparent conductor, a photoactive layer, and an electrode. OPV cells rely on organic small molecules or polymers for light absorption and charge transport. Different from the traditional inorganic semiconductors where free electrons and holes are easily generated under solar illumination, in OPV device, neither bilayer and planar heterojunction structure or an intermixed bulk heterojunction (BHJ) structure, a strongly

http://dx.doi.org/10.5772/61401

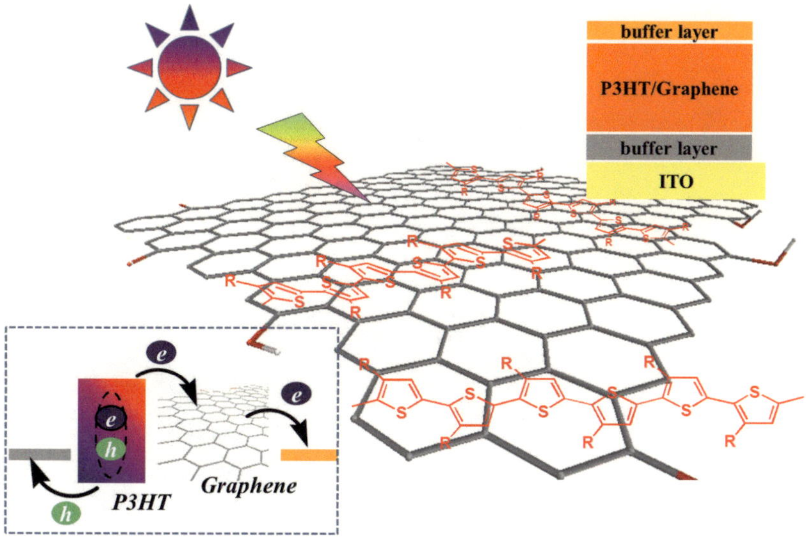

Figure 7. Organic solar cells prepared by using P3HT/graphene composite as photoactive layer.

bound electron–hole pair is generated, and the electron–hole pair needs to be separated by a acceptor material.[95, 96]

As shown in Figure 7, for P3HT-modified graphene-based solar cells, graphene is expected to be a good electron-acceptor candidate, because large donor/acceptor interfaces for charge generation and a continuous pathway for electron transfer will be formed owing to the large two-dimensional plane structure and the one-atom thickness property of graphene.

Dai *et al.* created a bilayer photovoltaic device with solution-cast P3HT-GO heterostructure as donor and electron acceptor of thermally evaporated C60 layer C_{60}.[38] The power conversion efficiency (PCE) of P3HT-GO solar cells is two times larger than that of the P3HT/C_{60} device. Liu *et al.* prepared a novel type OPV device with BHJ active layer of a solution-processable functionalized graphene (SPFGraphene) as the acceptor material and electron donor compound of P3HT.[97] In the ITO/PEDOT:PSS/P3HT:graphene/LiF/Al photovoltaic device, a PCE of 1.1% was achieved. The SPFGraphene acts as charge dissociation and provide percolation paths for electron transfer, which promote the active layer yields better carrier mobility, easy excitons splitting, and suppression of charge recombination, thereby improving photovoltaic action. Liu and coworkers introduced SPFGraphene in P3HT/PCBM photovoltaic devices. By taking advantage of the electron-accepting feature of fullerenes and the high electron transport capability of graphene, a PCE as high as 1.4% was obtained.[98]

4.2. Photocatalysts

Graphene-involved semiconductor photocatalysts have attracted extensive attention because of their usefulness in environmental and green chemical catalyst applications.[99] Recently, photoactive-graphene-based photocatalysis has been widely used to catalyze various organic

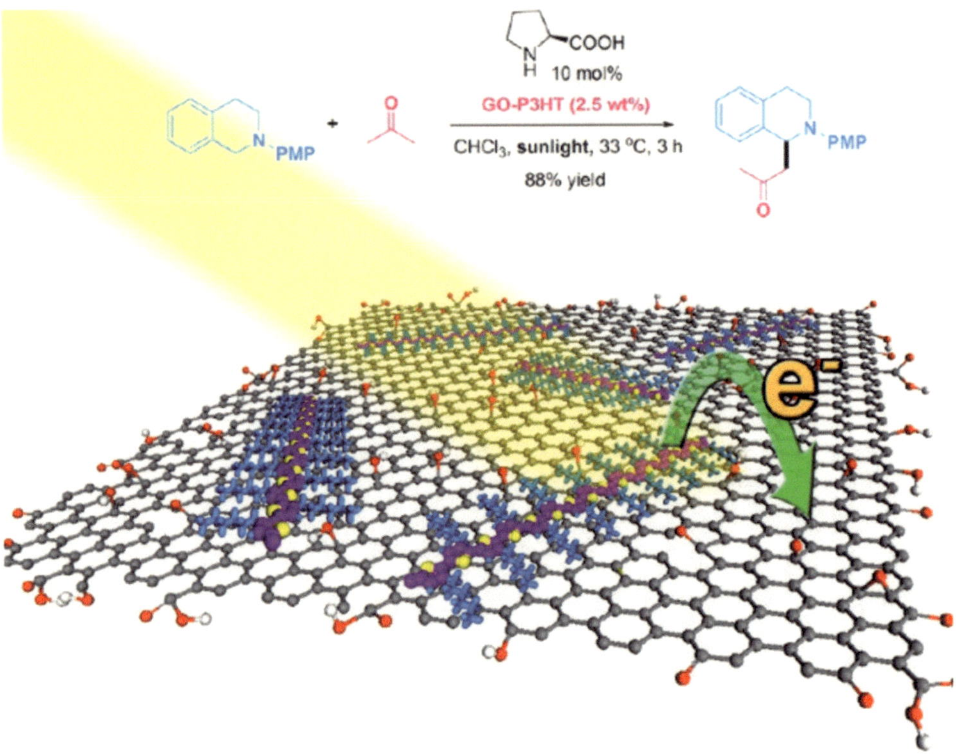

Figure 8. GO/P3HT composite as a synergistic photocatalyst (Loh *et al., J. Phys. Chem. Lett.* 2012, *3*, 2332-2336).

reactions. For example, in Figure 8, photoactive graphene of noncovalently bonded graphene–polymer (P3HT) composite shows significant advancement of photocatalystic performance in Mannich reaction over commercial photocatalyst P25 (Loh *et al.*, 2012).[100] Transient optical absorption studies have inferred that the tertiary amine is oxidized by the positive hole on the highest occupied molecular orbital (HOMO) of P3HT *via* single electron transfer to form the radical cation. At the same time, the excited electron is injected from the lowest unoccupied molecular orbital (LUMO) of P3HT into GO, which is then used to activate molecular oxygen to form the dioxygen radical anion; the latter can be stabilized by the aromatic scaffold in GO.[93] Pan *et al.* found that incorporation of Rose Bengal (RB) with GO sheet can provide higher catalyst actively of the visible light induced oxidative C-H functionalization of tertiary amines, even there was no direct physical interactions between RB and GO.[101]

5. Conclusions and outlooks

This chapter presented some important features of photoactive graphene, from material synthesis, electron/energy-transfer interaction to organic photovoltaic and photocatalystic applications. Of particular interest is the utilization of graphene as a two-dimensional platform

http://dx.doi.org/10.5772/61401

to anchor organic conjugated aromatic molecules and their applications in photo-energy conversion and photocatalysis.

However, despite many successful examples showing that photoactive graphene holds excellent properties and great potentials in various applications, there are still some problems that need to be overcome. One major drawback for the use of photoactive graphene in photo-energy conversion and photocatalystic application is light absorption without directly generating long-lived charge carriers.[21] Hence, one needs to take caution in optimizing the modification degree of photoactive moieties functionalized to graphene surface. Moreover, optimization of the charge carriers transfer channel of photoactive graphene, as well as precisely controlling any combination of direction, position, and distance of graphene and photoactive molecules are also necessary for high-efficiency photo-energy conversion application. More careful design of the chemical functionalization is necessary to exploit the electronic properties of graphene and photoactive groups, for example, multifunctional graphene materials with different moieties can provide new ways to design charge-transfer systems for light energy conversion.

From another point of view, chemists will be needed to develop strategies for rational design and facial synthesis of photoactive moiety; not only morphology and electronic structure, but also redox properties are also important for tuning the charge-transfer process of photoactive graphene, because the more precisely manipulated the charge-transfer efficiency of photoactive moiety–graphene interfaces, the more favorable it is to improve the photo-electron conversion efficiency and photocatalytic performance. With more progresses in the material optimization and novel device design, we believe that photoactive graphene will lead to a wide range of applications.

Finally, the photoactive graphene researches are still at their initial stages. With the multidisciplinary efforts from chemistry, physics, and materials science, we believe that much more progresses in the applications of photoactive graphene will become a reality in the near future.

Author details

Hang-Xing Wang[1,2] and Hao-Li Zhang[1]

1 Lanzhou University, Lanzhou, China

2 Hubei University, Wuhan, China

References

[1] Georgakilas, V., Otyepka, M., Bourlinos, A. B., Chandra, V., Kim, N., Kemp, K. C., HobzaR, P., Zboril, A., Kim, K. S. Chem. Rev. 2012, 112, 6156–6214.

[2] Bonaccorso, F., Sun, Z., Hasan, T., Ferrari, A. C. Nat. Photon. 2010, 4, 611-622.

[3] Geim, A. K., Novoselov, K. S. Nat.Mater. 2007, 6, 183-191.

[4] Fuhrer, M. S., Lau, C. N., MacDonald, A. H. MRS Bull. 2010, 35, 289-295.

[5] Novoselov, K. S., Falko, V. I., Colombo, L., Gellert, P. R., Schwab, M. G., Kim, K. Nature 2012, 490, 192-200.

[6] Dreyer, D. R., Ruoff, R. S., Bielawski, C. W. Angew. Chem. Int. Ed. 2010, 49, 9336-9344.

[7] Liu, L., Shen, Z. Appl. Phys. Lett. 2009, 95, 252104.

[8] Wang, H.-X., Wang, Q., Zhou, K.-G., Zhang, H.-L. Small 2013, 9, 1266-1283.

[9] Liu, H., Liu, Y., Zhu, D. J. Mater. Chem. 2011, 21, 3335-3345.

[10] Wehling, T. O., Novoselov, K. S., Morozov, S. V., Vdovin, E. E., Katsnelson, M. I., Geim, A. K., Lichtenstein, A. I. Nano Lett. 2008, 8, 173-177.

[11] Choi, J., Lee, H., Kim, K., Kim, B., Kim, S. J. Phys. Chem. Lett. 2010, 1, 505-509.

[12] Liu, J., Tang, J., Gooding, J. J. J. Mater. Chem. 2012, 22, 12435-12452.

[13] Zhang, Y., Tang, T.-T., Girit, C., Hao, Z., Martin, M. C., Zettl, A., Crommie, M. F., Shen, Y. R., Wang, F. Nature 2009, 459, 820-823.

[14] Ohta, T., Bostwick, A., Seyller, T., Horn, K., Rotenberg, E. Science 2006, 313, 951-954.

[15] Dutta, S., Pati, S. K. J. Mater. Chem. 2010, 20, 8207-8223.

[16] Jiao, L., Zhang, L., Wang, X., Diankov, G., Dai, H. Nature 2009, 458, 877-880.

[17] Kosynkin, D. V., Higginbotham, A. L., Sinitskii, A., Lomeda, J. R., Dimiev, A., Price, B. K., Tour, J. M. Nature 2009, 458, 872-875.

[18] Han, M. Y., Özyilmaz, B., Zhang, Y., Kim, P. Phys. Rev. Lett. 2007, 98, 206805.

[19] Saha, S. K., Baskey, M., Majumdar, D. Adv. Mater. 2010, 22, 5531-5536.

[20] Han, L., Wang, P., Dong, S. Nanoscale 2012, 4, 5814-5825.

[21] Novoselov, K. S., Geim, A. K., Morozov, S. V., Jiang, D., Zhang, Y., Dubonos, S. V., Gregorieva, I. V., Firsov, A. A. Science, 2004, 306, 666-669.

[22] Castro Neto, A. H., Guinea, F., Peres, N. M. R., Novoselov, K. S., Geim, A. K. Rev. Mod. Phys. 2009, 81, 109-162.

[23] Kim, K. S., Zhao, Y., Jang, H., Lee, S. Y., Kim, J. M., Ahn, J. H., Kim, P., Choi, J. Y., Hong, B. H. Nature 2009, 457, 706-710.

[24] Nair, R. R., Blake, P., Grigorenko, A. N., Novoselov, K. S., Booth, T. J., Stauber, T., Peres, N. M. R., Geim, A. K. Science 2008, 320, 1308.

[25] Shockley, W., Queisser, H. J. Appl. Phys. 1961, 32, 510-519.

[26] Ryuzaki, S., Onoe, J. Electr. Commun. JPN. 2013, 96, 1-8.

[27] Hernandez, W. Sensors 2008, 8, 35-50.

[28] Breusing, M., Ropers, C., Elsaesser, T. Phys. Rev. Lett. 2009, 102, 086809.

[29] Xia, F., Mueller, T., Lin, Y.-m., Valdes-Garcia, A., Avouris, P. Nat. Nanotechnol. 2009, 4, 839-843.

[30] Williams, G., Seger, B., Kamat, P. V. ACS Nano, 2008, 2, 1487-1491.

[31] Liu, Z., Liu, B., Ding, J., Liu, J. Anal. Bioanal. Chem. 2014, 406, 6885-6902.

[32] Yan, L., Zheng, Y. B., Zhao, F., Li, S., Gao, X., Xu, B., Weiss, P. S., Zhao, Y. Chem. Soc. Rev. 2012, 41, 97-114.

[33] Sharma, R., Baik, J. H., Perera, C. J., Strano, M. S. Nano Lett. 2010, 10, 398-405

[34] Huang, X., Qi, X., Boey, F., Zhang, H. Chem. Soc. Rev. 2012, 41, 666-686.

[35] Allen, M. J., Tung, V. C., Kaner, R. B. Chem. Rev. 2010, 110, 132-145.

[36] Chen, C., Zhai, W., Lu, D., Zhang, H., Zheng, W. Mater. Res. Bull. 2011, 46, 583-587.

[37] Zhang, X., Hou, L., Cnossen, A., Coleman, A. C., Ivashenko, O., Rudolf, P., van Wees, B. J., Browne, W. R., Feringa, B. L. Chem. Eur. J. 2011, 17, 8957-8964.

[38] Yu, D., Yang, Y., Durstock, M., Baek, J.-B., Dai, L. ACS Nano 2010, 4, 5633-5640.

[39] Medina, H., Lin, Y.-C., Obergfell, D., Chiu, P.-W. Adv. Funct. Mater. 2011, 21, 2687-2692.

[40] Lu, Y. H., Chen, W., Feng, Y. P., He, P. M. J. Phys. Chem. B 2009, 113, 2-5.

[41] Stankovich, S., Dikin, D. A., Dommett, G. H. B., Kohlhaas, K. M., Zimney, E. J., Stach, E. A., Piner, R. D., Nguyen, S. T., Ruoff, R. S. Nature 2006, 442, 282-286.

[42] Manna, A. K., Pati, S. K. Chem. Asian J.2009, 4, 855-860.

[43] Sun, Y. Q., Wu, Q., Shi, G. Q. Energy Environ. Sci. 2011, 4, 1113-1132.

[44] Yang, Z., Gao, R., Hu,N., Chai, J., Cheng, Y., Zhang, L., Wei, H., Kong, E. S., Zhang, Y. Nano-Micro Lett. 2012, 4, 1-9.

[45] Wang, Y., Li, Z., Wang, J., Li, J., Lin, Y. Trends Biotech. 2011, 29, 205-212.

[46] Kuila, T., Bose, S., Mishra, A. K., Khanra, P., Kim, N. H., Lee, J. H. Prog. Mater. Sci. 2012, 57, 1061-1105.

[47] Niyogi, S., Hamon, M. A., Hu, H., Zhao, B., Bhowmik, P., Sen, R., Itkis, M. E., Haddon, R. C. Acc. Chem. Res. 2002 , 35 , 1105-1113.

[48] Sun, Z., James, D. K., Tour, J. M. J. Phys. Chem. Lett. 2011, 2, 2425-2432.

[49] Malig, J., Jux, N., Guldi, D. Acc. Chem. Res. 2013, 46, 53-64.

[50] Ragoussi, M.-E., Malig, J., Katsukis, G., Butz, B., Spiecker, E., de la Torre, G., Torres, T., Guldi, D. M. Angew. Chem. Int. Ed. 2012, 51, 6421-6425.

[51] Potts, J. R., Dreyer, D. R., Bielawski, C. W., Ruoff, R. S. Polymer 2011, 52, 5-25.

[52] Schlierf, A., Samori, P., Palermo, V. J. Mater. Chem. C 2014, 2, 3129-3143.

[53] Xu, Y., Liu, Z., Zhang, X., Wang, Y., Tian, J., Huang, Y., Ma, Y., Zhang, X., Chen, Y. Adv. Mater. 2009, 21, 1275-1279.

[54] Zhuang, X.-D., Chen, Y., Liu, G., Li, P.-P., Zhu, C.-X., Kang, E.-T., Noeh, K.-G., Zhang, B., Zhu, J.-H., Li, Y.-X. Adv. Mater. 2010, 22, 1731-1735.

[55] Liu, Z. B., Xu, Y. F., Zhang, X. Y., Zhang, X. L., Chen, Y. S., Tian, J. G. J. Phys. Chem. B 2009, 113, 9681-9686.

[56] Zhu, J., Li, Y., Chen, Y., Wang, J., Zhang, B., Zhang, J., Blau, W. J. Carbon 2011, 49, 1900-1905.

[57] Zhang, X.-L., Zhao, X., Liu, Z.-B., Liu, Y.-S., Chen, Y.-S., Tian, J.-G. Opt. Express 2009, 17, 23959-23964.

[58] Choi, H., Santra, P. K., Kamat, P. V. ACS Nano 2012, 6, 5718-5726.

[59] Yao, Y., Gao, J., Bao, F., Jiang, S., Zhang, X., Ma, R. RSC Adv. 2015, 5, 42754-42761.

[60] Midya, A., Mamidala, V., Ang, P., Ang, P., Chen, Z.K., Ji, W. Small 2010, 6, 2292-2230.

[61] Bao, Q., Zhang, H., Yang, J.-x., Wang, S., Tang, D. Y., Jose, R., Ramakrishna, S., Lim, C. T., Loh, K. P. Adv. Funct. Mater. 2010, 20, 782-791.

[62] Kolb, H. C., Finn, M. G., Sharpless, K. B. Angew. Chem. Int. Ed. 2001, 40, 2004-2021.

[63] Campidelli, S. Curr. Org. Chem. 2011, 15, 1151-1159.

[64] Iehl, J., de Freitas, R. P., Nierengarten, J.-F. Tetra. Lett. 2008, 49, 4063-4066.

[65] Wang, H. X., Zhou, K. G., Xie, Y. L., Zeng, J., Chai, N. N., Li, J., Zhang, H. L. Chem. Commun. 2011, 47, 5747-5749.

[66] Kaminska, I., Das, M. R., Coffinier, Y., Niedziolka-Jonsson, J., Sobczak, J., Woisel, P., Lyskawa, J., Opallo, M., Boukherroub, R., Szunerits, S. ACS Appl. Mater. Inter. 2012, 4, 1016-1020.

[67] Salvio, R., Krabbenborg, S., Naber, W. J. M., Velders, A. H., Reinhoudt, D. N., van der Wiel, W. G. Chem. Eur. J. 2009, 15, 8235-8240.

[68] Castelaín, M., Martínez, G., Merino, P., Martín-Gago, J. Á., Segura, J. L., Ellis, G., Salavagione, H. J. Chem. Eur. J. 2012 , 18, 4965-4973.

[69] Wang, S., Goh, B. M., Manga, K. K., Bao, Q., Yang, P., Loh, K. P. ACS Nano 2010, 4, 6180-6186.

[70] Cai, X., Tan, S., Xie, A., Lin, M., Liu, Y., Zhang, X., Lin, Z., Wu, T., Mai, W. Mater. Res. Bull. 2011, 46, 2353-2358.

[71] Xu, Y., Zhao, L., Bai, H., Hong, W., Li, C., Shi, G. J. Am. Chem. Soc. 2009, 131, 13490-13497.

[72] Takagi, S., Shimada, T., Eguchi, M., Yui, T., Yoshida, H., Tryk, D. A., Inoue, H. Langmuir 2002, 18, 2265-2272.

[73] Heidner, E., Ladner, R.C., Perutz, M.F. J. Mol. Bio. 1976, 104, 707-722.

[74] Qi, X., Pu, K.-Y., Zhou, X., Li, H., Liu, B., Boey, F., Huang, W., Zhang, H. Small 2010, 6, 663-669.

[75] Dougherty, D. A. Acc. Chem. Res. 2013, 46, 885-893.

[76] Tang, Z., Lei, Y., Guo, B., Zhang, L., Jia, D. Polymer 53, 673-680.

[77] Qi, X., Pu, K.-Y., Li, H., Zhou, X., Wu, S., Fan, Q.-L., Liu, B., Boey, F., Huang, W., Zhang, H. Angew. Chem. Int. Ed. 2010, 49, 9426-9429.

[78] McClenaghan, N. D., Leydet, Y., Maubert, B., Indelli, M. T., Campagna, S. Coord. Chem. Rev. 2005, 249, 1336-1350.

[79] Hsu, C.-P. Acc. Chem. Res. 2009, 42, 509-518.

[80] Wu, J., Pisula, W., Muellen, K. ChemInform 2007, 38, 718-747.

[81] Wojcik, A., Kamat, P. V. ACS Nano 2010, 4, 6697-6706.

[82] Kalyanasundaram, K., Neumann-Spallart, M. J. Phys. Chem. 1982, 86, 5163-5169.

[83] Malig, J., Jux, N., Kiessling, D., Cid, J.-J., Vázquez, P., Torres, T., Guldi, D. M. Angew. Chem. Int. Ed. 2011, 50, 3561-3565.

[84] Aly, S. M., Parida, M. R., Alarousu, E., Mohammed, O. F. Chem. Commun. 2014, 50, 10452-10455.

[85] Persson, B. N. J., Lang, N. D. Phys. Rev. B 1982, 26, 5409-5414.

[86] Wang, Y., Kurunthu, D., Scott, G. W., Bardeen, C. J. J. Phys. Chem. C 2010, 114, 4153-4159.

[87] Chen, Z., Berciaud, S., Nuckolls, C., Heinz, T. F., Brus, L. E. ACS Nano 2010, 4, 2964-2968.

[88] Gaudreau, L., Tielrooij, K. J., Prawiroatmodjo, G. E. D. K., Osmond, J., de Abajo, F. J. G., Koppens, F. H. L. Nano Lett. 2013, 13, 2030-2035.

[89] Yeltik, A., Kucukayan-Dogu, G., Guzelturk, B., Fardindoost, S., Kelestemur, Y., Demir, H. V. J. Phys. Chem. C 2013, 117, 25298-25304.

[90] Chang, H., Tang, L., Wang, Y., Jiang, J., Li, J. Anal. Chem. 2010, 82, 2341-2346.

[91] Swathi, R. S., Sebastian, K. L. J. Chem. Phys. 2008, 129, 054073.

[92] Jiang, Y., Tian, J., Chen, S., Zhao, Y., Wang, Y., Zhao, S. J. Fluoresc. 2013, 23, 697-703.

[93] Su, C. L., Loh, K. P. Acc. Chem. Res. 2012, 10, 2275-2285.

[94] Wan, X., Huang, Y., Chen, Y. Acc. Chem. Res. 2012, 45, 598-607.

[95] Kalman, G. J., Hartmann, P., Donkó, Z., Golden, K. I. Phys. Rev. Lett. 2007, 236801.

[96] Bakulin, A. A., Dimitrov, S. D., Rao, A., Chow, P. C. Y., Nielsen, C. B., Schroeder, B. C., McCulloch, I., Bakker, H. J., Durrant, J. R., Friend, R. H. J. Phys. Chem. Lett. 2013, 4, 209-215.

[97] Liu, Q., Liu, Z., Zhang, X., Yang, L., Zhang, N., Pan, G., Yin, S., Chen, Y., Wei, J. Adv. Funct. Mater. 2009, 19, 894-904.

[98] Liu, Z., He, D., Wang, Y., Wu, H., Wang, J., Wang, H. Solar Energ. Mater. Solar Cells 2010, 94, 2148-2153.

[99] Xiang, Q., Yu, J., Jaroniec, M. Chem. Soc. Rev. 2011, 41, 782-796.

[100] Wang, S., Nai, C. T., Jiang, X.-F., Pan, Y., Tan, C.-H., Nesladek, M., Xu, Q.-H., Loh, K. P. J. Phys. Chem. Lett. 2012, 3, 2332-2336.

[101] Pan, Y., Wang, S., Kee, C. W., Dubuisson, E., Yang, Y., Loh, K. P., Tan, C.-H. Green Chem. 2011, 13, 3341-3344.

Chapter 5

Compton Effect in Graphene and in the Graphene-Like Dielectric Medium

Miroslav Pardy

Additional information is available at the end of the chapter

Abstract

In the introduction and the second part of the chapter, we discuss the Compton effect in general, and the modern viewpoint on the 2-dimensional carbon crystals called graphene, where graphene unique properties arise from he collective behavior of electrons governed by the Dirac equation. The Dirac equation in graphene physics is used for so called pseudospin of pseudoelectron formed by the hexagonal lattice composed of the systems of two equilateral triangles with the corresponding particular wave functions. The total wave function of an electron moving in the hexagonal system is superposition of the particular wave functions. The crucial step in the graphene physics is the definition of the new spinor function where spinor function is solution of the Pauli equation in the nonrelativistic situation and Dirac equation of the generalized case. The corresponding mass of such effective electron is proved to be approximately zero.

In the third part of the chapter, we deal with the Dirac equation and its Volkov solution and in the fourth part of the chapter, we discuss the Volkov solution in a dielectric medium.

The fifth part of the chapter, deals with the Compton effect derived from Volkov solution of the Dirac equation while the sixth part of the chapter, deals with the calculation of the Compton effect with ultrashort laser pulse, where the pulse is of the Dirac delta-function form.

The seven part of the chapter, deals with Compton effect initiated by two orthogonal plane waves. We solve the Dirac equation for two different four-potentials of the plane electromagnetic waves and we specify the solutions of the Dirac equation for two orthogonal plane waves. The modified Compton formula for the scattering of two photons on an electron is determined.

The conclusion eighth part involves possible perspectives of the Compton effect with regard to the scientific and technological meaning of the results derived in our contribution.

Keywords: Compton effect, Dirac equation, Volkov solution, dielectric, laser pulse, graphene, silicene.

1. Introduction

The Compton effect is the light-particle interaction where the wavelength of scattered photon is changed. The difference between the Compton effect and the Einstein photoeffect consists in the fact that during the photoeffect, the energy of photon is transmitted to electron totally.

Compton used in his original experiment [1] the energy of the X-ray photon (~20 keV) which was very much larger than the binding energy of the atomic electron, so the electrons could be treated as being free. Compton scattering usually refers to the interaction involving only the electrons of an atom. However, the nuclear Compton effect was confirmed too. The effect is important because it demonstrates that light cannot be explained purely as a wave phenomenon. The Compton experiment proved that light is composed of particle-like objects with energy $E = \hbar\omega$. The interaction between electrons and high-energy photons is such that the electron takes part of the photon initial energy, and a photon containing the remaining energy is emitted in a different direction from the original. If the scattered photon still has enough energy left, the process may be repeated.

Compton, in his paper [1], derived a simple formula relating the shift of wavelength to the scattering angle of the X-rays by postulating that each scattered X-ray photon interacted with only one electron. His paper involves the information on experiments for verification of the equation:

$$\lambda' - \lambda = \frac{h}{mc}(1 - \cos\theta), \tag{1}$$

where λ' is wavelength of the scattered X-ray and θ is the angle between the incident and scattered X-ray. The scattering was considered in the laboratory frame where electron was at rest. Let us remark immediately that eq. (1) has also a limit for $m \to 0$, if and only if $\theta \to 0$. It corresponds to the situation of graphene sheet, where the mass of the so-called pseudoelectron can be considered zero. This limit can be verified immediately by the visible light and not by the X-rays. In case of using the X-rays in graphene, we get still the original Compton process with the real electrons (ionization process in graphene) and not the process with the so-called pseudoelectrons. The considered process was the so-called one-photon process with the symbolic equation

$$\gamma + e \to \gamma + e. \tag{2}$$

The differential cross section corresponding to eq. (2) was derived by Klein and Nishina in the form

$$\frac{d\sigma}{d\Omega} = \frac{1}{2}\frac{e^4}{m^2c^4}\left(\frac{\omega'}{\omega}\right)^2\left(\frac{\omega'}{\omega} + \frac{\omega}{\omega'} - \sin^2\theta\right). \tag{3}$$

It is the ratio of the number of scattered photons into the unit solid angle Ω over the number of incident photons. At the present time with the high-power lasers, there is a possibility to realize so-called multiphoton scattering according to equation $N\gamma + e \to M\gamma + e$, where N and M are numbers of photons participating in the scattering. N photons are absorbed at a single point and, after some time, M photons are emitted at the distant point. Let us remark that in case of the Raman effect, the equation describing the Raman process is $\gamma + A \to \gamma + A^*$ where A denotes atom, or molecule, and A^* denotes excited atom, or excited molecule. This Raman process involves also the interaction of the fullerene C_{60} with photons.

Equation (3) has also limit for $m \to 0$, if and only if the angle θ is solution of eq. $(\omega'/\omega + \omega/\omega' - \sin^2\theta) = 0$. Such limit corresponds to the Compton effect in graphene with zero mass of pseudoelectron.

We calculate the Compton process as a result of the Volkov solution [2] of the 3D and 2D Dirac equation. The Volkov solution involves not only the one-photon scattering but also the multiphoton scattering of photons on electron. In the time of the Compton experiment in 1922, the Volkov solution was not known, because the Dirac equation was published in 1928 and the Volkov solution in 1935 [2].

In the next part of the chapter, we discuss the modern viewpoint on the two-dimensional carbon crystals – graphene. Then we discuss the Dirac equation with plane wave and its Volkov solution including dielectric medium. Then we deal with the calculation of the Compton effect with ultrashort laser pulse in graphene and graphene-like sheet as silicene. We consider the case that a charged particle moves in the parallel direction to the silicene sheet. We include also the classical situation of a charged particle accelerated by the δ-function form of impulsive force. Then, we discuss the corresponding quantum theory based on the Volkov solution of the Dirac equation. The modified Compton formula for frequency of photons generated by the laser pulse is derived. The discussion is extended to the Dirac equation for two different waves, and the Volkov solution is then determined for the orthogonal two plane waves. The last part of the chapter is the conclusion, where we consider the scientific and technological perspectives of the results derived in our contribution.

2. Graphene

Graphene is a two-dimensional carbon sheet which is very consistent (100 times stronger than steel). It is a very good conductor of heat and electricity. Graphene was investigated by theorists for decades; however, it was first generated in the laboratory in 2003. Being two-dimensional, it interacts in a special way with light and with other materials. Researchers have discovered the bipolar transistor effect, the charge ballistic transport, and the large quantum oscillations.

Technically, graphene is a crystalline allotrope of carbon with two-dimensional geometrical properties. More than 70 years ago, Peierls [3] and Landau [4] proved that the two-dimensional crystal is not stable from the viewpoint of thermodynamics and cannot exist. The thermodynamic displacements of atoms in such a crystal are of the same size as the distances between atoms at any finite temperature. Mermin [5] accepted the theoretical arguments in 1968, and it seemed that many experimental observations were in accord with the Landau-Peierls-Mermin theory.

However, in 2004, Geim and Novoselov [6], [7] with co-workers at the University of Manchester in England produced a crystalline sheet of carbon just one-atom thick. Then, the Geim group was able to isolate graphene and was able to visualize the new crystal medium using a simple optical microscope. The Landau-Peierls-Mermin proof remained as the historical document.

After some time, the new sophisticated methods generating graphene sheets were invented. The graphene sheets were, e.g., synthesized by passing liquid ethanol droplets into an argon plasma. The authors of this method are Dato et al. [8].

Graphene is composed of the benzene rings (C_6H_6) without their H-atoms. Graphene is only one of the crystalline forms of carbon which crystallize as diamond, graphite, fullerene (C_{60}), carbon nanotube, and glassy carbon.

Unique physical properties of graphene are caused by the collective behavior of the quasiparticles called pseudoelectrons having pseudospins, which move according to the Dirac equation in the hexagonal lattice.

The Dirac fermions in graphene carry unit electric charge. Strong interactions between the electrons and the carbon atoms result in linear dispersion relation $E = v_F|\mathbf{p}|$, where v_F is the so-called Fermi-Dirac velocity, $\mathbf{p}$ being the momentum of a pseudoelectron. The Fermi velocity is approximately only about 300 times less than the speed of light.

The pseudospin of the pseudoelectron follows from the hexagonal form of graphene. Every hexagonal cell system is composed of the systems of two equilateral triangles. The fermions in the triangle sub-lattice systems can be described by the wave functions φ_1 and φ_2. Then the adequate wave function of the fermion moving in the hexagonal structure is their superposition, or $\psi = c_1\varphi_1 + c_2\varphi_2$, where c_1 and c_2 are functions of coordinate $\mathbf{x}$ and functions φ_1, φ_2 are functions of the wave vector $\mathbf{k}$ and coordinate $\mathbf{x}$. The crucial step in graphene theory is the definition of the bispinor function with components φ_1, φ_2 [9].

The relativistic generalization of nonrelativistic equation $E = v_F|\mathbf{p}|$ is evidently the Dirac-Weyl equation for the description of neutrino which can be transcribed in four-component spinor form as

$$p_\mu \gamma^\mu = 0 \tag{4}$$

and it is possible to prove that this spinor function is solution of the Pauli equation in the nonrelativistic situation. The corresponding mass of such effective electron is proved approximately to be zero. So, it follows from this formalism that to describe the Compton effect on graphene is to solve the Compton effect with quasielectron with zero mass.

The introduction of the Dirac relativistic Hamiltonian in graphene physics has the physical meaning that we describe the graphene physics by means of electron-hole medium. It is the analogue of the Dirac theory of the electron-positron vacuum in quantum electrodynamics. However, the pseudoelectron and pseudospin in graphene physics are not an electron and the spin of quantum electrodynamics (QED), because QED is the relativistic quantum theory of the interaction of real electrons and photons where mass of an electron is defined by classical mechanics and not by collective behavior in the hexagonal sheet called graphene.

The graphene sheet can be considered as the special form of the more general 2D-graphene-like sheets, where, for instance, silicene has the similar structure as graphene [10] .

On the other hand, amorphous solids – glasses – lack long-range translational periodicity in the atomic structure. However, due to chemical bonds, glasses do possess a high degree of short-range order with respect to local atomic polyhedra. It means that such structures can be considered as the graphene-like structures with the appropriate index of refraction, being necessary for the existence of the Čherenkov effect and the Compton effect in dielectric medium.

The last but not least graphene-like structure can be represented by graphene-based polaritonic crystal sheet [11], which can be used for the Čherenkov effect and the Compton effect in the graphene-like dielectric medium with the index of refraction n.

3. Volkov solution of the Dirac equation in vacuum

Volkov solution of the Dirac equation is the mathematical solution of the Dirac equation with the plane-wave potential. The derivation of the Volkov solution of the Dirac equation in vacuum is described in the textbook of Berestetzkii et al. [12]. The four-potential in the Dirac equation

$$(\gamma(p - eA) - m)\psi = 0, \tag{5}$$

is

$$A^\mu = A^\mu(\varphi); \quad \varphi = kx. \tag{6}$$

We suppose that the four-potential satisfies the Lorentz gauge condition

$$\partial_\mu A^\mu = k_\mu (A^\mu)' = (k_\mu A^\mu)' = 0, \tag{7}$$

where the prime denotes derivation with regard to φ. From the last equation follows

$$kA = const = 0, \tag{8}$$

because we can put the constant to zero. The tensor of electromagnetic field is

$$F_{\mu\nu} = k_\mu A'_\nu - k_\nu A'_\mu. \tag{9}$$

Instead of the linear Dirac equation (5), we consider the quadratic equation, which we get by multiplying the linear equation by operator $(\gamma(p - eA) + m)$, [12]. We get

$$\left[(p - eA)^2 - m^2 - \frac{i}{2}eF_{\mu\nu}\sigma^{\mu\nu}\right]\psi = 0. \tag{10}$$

Using $\partial_\mu(A^\mu\psi) = A^\mu\partial_\mu\psi$, which follows from eq. (7), and $\partial_\mu\partial^\mu = \partial^2 = -p^2$, with $p_\mu = i(\partial/\partial x^\mu) = i\partial_\mu$, we get the quadratic Dirac equation for the four-potential of the plane wave:

$$[-\partial^2 - 2i(A\partial) + e^2A^2 - m^2 - ie(\gamma k)(\gamma A')]\psi = 0. \tag{11}$$

We are looking for the solution of the last equation in the form

$$\psi = e^{-ipx}F(\varphi). \tag{12}$$

After insertion of this relation into (11), we get with ($k^2 = 0$)

$$\partial^\mu F = k^\mu F', \quad \partial_\mu\partial^\mu F = k^2F'' = 0, \tag{13}$$

the following equation for $F(\varphi)$:

$$2i(kp)F' + [-2e(pA) + e^2A^2 - ie(\gamma k)(\gamma A')]F = 0. \tag{14}$$

The integral of the last equation is of the form

$$F = \exp\left\{-i\int_0^{kx}\left[\frac{e(pA)}{(kp)} - \frac{e^2}{2(kp)}A^2\right]d\varphi + \frac{e(\gamma k)(\gamma A)}{2(kp)}\right\}\frac{u}{\sqrt{2p_0}}, \tag{15}$$

where $u/\sqrt{2p_0}$ is the arbitrary constant bispinor.

Al powers of $(\gamma k)(\gamma A)$ above the first are equal to zero, since

$$(\gamma k)(\gamma A)(\gamma k)(\gamma A) = -(\gamma k)(\gamma k)(\gamma A)(\gamma A) + 2(kA)(\gamma k)(\gamma A) = -k^2A^2 = 0. \tag{16}$$

Then we can write

$$\exp\left\{e\frac{(\gamma k)(\gamma A)}{2(kp)}\right\} = 1 + \frac{e(\gamma k)(\gamma A)}{2(kp)}. \tag{17}$$

So, the solution is of the form

$$\psi_p = R\frac{u}{\sqrt{2p_0}}e^{iS} = \left[1 + \frac{e}{2kp}(\gamma k)(\gamma A)\right]\frac{u}{\sqrt{2p_0}}e^{iS}, \tag{18}$$

where u is an electron bispinor of the corresponding Dirac equation

$$(\gamma p - m)u = 0, \tag{19}$$

with the normalization condition $\bar{u}u = 2m$.

The mathematical object S is the classical Hamilton-Jacobi function, which was determined in the form

$$S = -px - \int_0^{kx} \frac{e}{(kp)} \left[(pA) - \frac{e}{2}(A)^2\right] d\varphi. \tag{20}$$

The current density is

$$j^\mu = \bar{\psi}_p \gamma^\mu \psi_p, \tag{21}$$

where $\bar{\psi}_p$ is defined as the transposition of (18), or

$$\bar{\psi}_p = \frac{\bar{u}}{\sqrt{2p_0}} \left[1 + \frac{e}{2kp}(\gamma A)(\gamma k)\right] e^{-iS}. \tag{22}$$

After insertion of ψ_p and $\bar{\psi}_p$ into the current density, we have

$$j^\mu = \frac{1}{p_0} \left\{p^\mu - eA^\mu + k^\mu \left(\frac{e(pA)}{(kp)} - \frac{e^2 A^2}{2(kp)}\right)\right\}, \tag{23}$$

which is in agreement with the formula in the Meyer article [13].

If $A^\mu(\varphi)$ are periodic functions, and their time-average value is zero, then the mean value of the current density is

$$\overline{j^\mu} = \frac{1}{p_0} \left(p^\mu - \frac{e^2}{2(kp)} \overline{A^2} k^\mu\right) = \frac{q^\mu}{p_0}. \tag{24}$$

4. Volkov solution in a dielectric medium

The mathematical approach to the situation where we consider plane wave solution in a medium is the same, only with the difference that the Lorentz condition must be replaced according to Schwinger et al. by the following one [14]:

$$\partial_\mu A^\mu = kA' = (\mu\varepsilon - 1)(\eta\partial)(\eta A) = (\mu\varepsilon - 1)(\eta k)(\eta A') \tag{25}$$

with the specification $\eta^{\mu} = (1, \mathbf{0})$ as the unit time-like vector in the rest frame of the medium [14].

For periodic potential A^{μ}, we then get from eq. (25) instead of $kA = 0$ the following equation:

$$kA = (\mu\varepsilon - 1)(\eta k)(\eta A). \tag{26}$$

Then, we get instead of eq. (14) the following equation for function $F(\varphi)$:

$$2i(kp)F' + [-2e(pA) + e^2A^2 - ie(\gamma k)(\gamma A') - ie(\mu\varepsilon - 1)(\eta k)(\eta A')]F = 0. \tag{27}$$

The solution of the last equation is the solution of the linear equation of the form $y' + Py = 0$, and it means it is of the form $y = C\exp(-\int Pdx)$, where C is some constant. So, we can write the solution as follows:

$$F = \exp\left\{-i\int_0^{kx}\left[\frac{e}{(kp)}(pA) - \frac{e^2}{2(kp)}A^2\right]d\varphi + \frac{e(\gamma k)(\gamma A)}{2(kp)} + \frac{e}{2(kp)}\alpha\right\}\frac{u}{\sqrt{2p_0}}, \tag{28}$$

where

$$\alpha = (\mu\varepsilon - 1)(\eta k)(\eta A) \tag{29}$$

The wave function ψ is then the modified wave function (18), which we can write in the form

$$\psi_p = \left[1 + \sum_{n=1}^{\infty}\left(\frac{e}{2(kp)}\right)^n (2\alpha)^{n-1}(\gamma k)(\gamma A)\right]\frac{u}{\sqrt{2p_0}}e^{iS}e^{T}, \tag{30}$$

where

$$T = \frac{e}{2(kp)}(\mu\varepsilon - 1)(\eta k)(\eta A), \tag{31}$$

and where we used in the last formula the following relation:

$$[(\gamma k)(\gamma A)]^n = (2\alpha)^{n-1}(\gamma k)(\gamma A). \tag{32}$$

So, we see that the influence of the medium on the Volkov solution is involved in $\exp(T)$, where T is given by eq. (31) and in the new term which involves the sum of the infinite number of coefficients.

5. Compton effect from Volkov solution

We determine the Compton process in vacuum as a result of the Volkov solution of the Dirac equation. The Volkov solution involves not only the one-photon scattering but also the multiphoton scattering of photons on electron. In the time of the Compton experiment in 1922, the Volkov solution was not known, because the Dirac equation was published in 1928 and the Volkov solution in 1935.

Now, let us consider electromagnetic monochromatic plane wave which is polarized in circle. We write the four-potential in the form

$$A = a_1 \cos\varphi + a_2 \sin\varphi, \tag{33}$$

where the amplitudes a_i are the same and mutually perpendicular, or

$$a_1^2 = a_2^2 = a^2, \quad a_1 a_2 = 0. \tag{34}$$

The Volkov solution for the standard vacuum situation is of the form

$$\psi_p = \left\{1 + \left(\frac{e}{2(kp)}\right)[(\gamma k)(\gamma a_1)\cos\varphi + (\gamma k)(\gamma a_2)\sin\varphi]\right\}\frac{u(p)}{\sqrt{2q_0}} \quad \times$$
$$\exp\left\{-ie\frac{(a_1 p)}{(kp)}\sin\varphi + ie\frac{(a_2 p)}{(kp)}\cos\varphi - iqx\right\}, \tag{35}$$

where

$$q^\mu - p^\mu - e^2\frac{a^2}{2(kp)}k^\mu. \tag{36}$$

because it follows from eq. (24).

We know that the matrix element M corresponding to the emission of photon by electron in the electromagnetic field is as follows [12]:

$$S_{fi} = -ie^2\int d^4x\bar{\psi}_{p'}(\gamma e'^*)\psi_p\frac{e^{ik'x}}{\sqrt{2\omega'}}, \tag{37}$$

where ψ_p is the electron wave function before the interaction of electron with the laser pulse and $\psi_{p'}$ is the electron wave function after emission of photon with components $k'^\mu = (\omega', \mathbf{k}')$. Symbol e'^* is the four polarization vector of emitted photon.

Then, we get the following linear combination in the matrix element:

$$e^{-i\alpha_1 \sin\varphi + i\alpha_2 \cos\varphi} \tag{38}$$

$$e^{-i\alpha_1 \sin\varphi + i\alpha_2 \cos\varphi} \cos\varphi \tag{39}$$

$$e^{-i\alpha_1 \sin\varphi + i\alpha_2 \cos\varphi} \sin\varphi. \tag{40}$$

where

$$\alpha_1 = e\left(\frac{a_1 p}{kp} - \frac{a_1 p'}{kp'}\right), \tag{41}$$

and

$$\alpha_2 = e\left(\frac{a_2 p}{kp} - \frac{a_2 p'}{kp'}\right). \tag{42}$$

Now, we can expand the exponential function into the Fourier transformation where the coefficients of the expansion will be B_s, B_{1s}, B_{2s}. So we write

$$e^{-i\alpha_1 \sin\varphi + i\alpha_2 \cos\varphi} = e^{-i\sqrt{\alpha_1^2+\alpha_2^2}\sin(\varphi-\varphi_0)} = \sum_{s=-\infty}^{\infty} B_s e^{-is\varphi} \tag{43}$$

$$e^{-i\alpha_1 \sin\varphi + i\alpha_2 \cos\varphi} \cos\varphi = e^{-i\sqrt{\alpha_1^2+\alpha_2^2}\sin(\varphi-\varphi_0)} \cos\varphi = \sum_{s=-\infty}^{\infty} B_{1s} e^{-is\varphi} \tag{44}$$

$$e^{-i\alpha_1 \sin\varphi + i\alpha_2 \cos\varphi} \sin\varphi = e^{-i\sqrt{\alpha_1^2+\alpha_2^2}\sin(\varphi-\varphi_0)} \sin\varphi = \sum_{s=-\infty}^{\infty} B_{2s} e^{-is\varphi}. \tag{45}$$

The coefficients B_s, B_{1s}, B_{2s} can be expressed by means of the Bessel function as follows [12]:

$$B_s = J_s(z) e^{is\varphi_0} \tag{46}$$

$$B_{1s} = \frac{1}{2}\left[J_{s+1}(z) e^{i(s+1)\varphi_0} + J_{s-1}(z) e^{i(s-1)\varphi_0}\right] \tag{47}$$

$$B_{2s} = \frac{1}{2i}\left[J_{s+1}(z)e^{i(s+1)\varphi_0} - J_{s-1}(z)e^{i(s-1)\varphi_0}\right], \tag{48}$$

where the quantity z is defined in [12] through α components as follows:

$$z = \sqrt{\alpha_1^2 + \alpha_2^2} \tag{49}$$

and

$$\cos\varphi_0 = \frac{\alpha_1}{z}; \quad \sin\varphi_0 = \frac{\alpha_2}{z} \tag{50}$$

Functions B_s, B_{1s}, B_{2s} are related to one another as follows:

$$\alpha_1 B_{1s} + \alpha_2 B_{2s} = sB_s, \tag{51}$$

which follows from the well-known relation for Bessel functions:

$$J_{s-1}(z) + J_{s+1}(z) = \frac{2s}{z}J_s(z) \tag{52}$$

The matrix element (37) can be written in the form [12]

$$S_{fi} = \frac{1}{\sqrt{2\omega' 2q_0 2q_0'}}\sum_s M_{fi}^{(s)}(2\pi)^4 i\delta^{(4)}(sk + q - q' - k'), \tag{53}$$

where the δ-function involves the law of conservation in the form

$$sk + q = q' + k', \tag{54}$$

where, respecting eq. (24),

$$q^\mu = p^\mu - \frac{e^2}{2(kp)}\overline{A^2}k^\mu. \tag{55}$$

Using the last equation, we can introduce the so-called effective mass of electron immersed in the periodic wave potential as follows:

$$q^2 = m_*^2; \quad m_* = m\sqrt{\left(1 - \frac{e^2}{m^2}\overline{A^2}\right)} \tag{56}$$

Formula (56) represents the mass renormalization of an electron mass in the field A. In other words, the mass renormalization is defined by the equation

$$m_{\text{phys}} = m_{\text{bare}} + \delta m \tag{57}$$

where δm follows from eq. (56). The quantity m_{phys} is the physical mass that an experimenter would measure if the particle were subject to Newton's law $\mathbf{F} = m_{\text{phys}}\mathbf{a}$. In case of the periodic field of laser, the quantity δm has the finite value. The renormalization is not introduced here "by hands," but it follows immediately from the formulation of the problem of electron in the wave field.

We can write

$$q^2 = q'^2 = m^2(1+\xi^2) \equiv m_*^2, \tag{58}$$

where for plane wave (35) with relations (36)

$$\xi = \frac{e}{m}\sqrt{-a^2}. \tag{59}$$

It may be easy to see that eq. (58) has very simple limit for $m = 0$ which is the graphenic case. Or, in other words, with the help of eq. (59) $m_*^2(m=0) = -e^2a^2$.

According to [12], the matrix element in (53) is of the form

$$M_{fi}^{(s)} = -e\sqrt{4\pi}\bar{u}(p')\left\{\left((\gamma e') - e^2a^2\frac{(ke')}{2(kp)}\frac{(\gamma k)}{(kp')}\right)B_s + \right.$$
$$e\left(\frac{(\gamma a_1)(\gamma k)(\gamma e')}{2(kp')} + \frac{(\gamma e')(\gamma k)(\gamma a_1)}{2(kp)}\right)B_{1s} +$$
$$\left. e\left(\frac{(\gamma a_2)(\gamma k)(\gamma e')}{2(kp')} + \frac{(\gamma e')(\gamma k)(\gamma a_2)}{2(kp)}\right)B_{2s}\right\}u(p) \tag{60}$$

It is possible to show that the total probability of the emission of photons from unit volume in unit time is [12]

$$W = \frac{e^2m^2}{4q_0}\sum_{s=1}^{\infty}\int_0^{u_s}\frac{du}{(1+u)^2} \times$$
$$\left\{-4J_s^2(z) + \xi^2\left(2+\frac{u^2}{1+u}\right)\left(J_{s+1}^2(z) + J_{s-1}^2(z) - 2J_s^2(z)\right)\right\}, \tag{61}$$

where

$$u = \frac{(kk')}{(kp')}, \quad u_s = \frac{2s(kp)}{m_*^2}, \quad z = 2sm^2 \frac{\xi}{\sqrt{1+\xi^2}} \sqrt{\frac{u}{u_s}\left(1 - \frac{u}{u_s}\right)}. \tag{62}$$

Variables $\alpha_{1,2}$ are to be expressed in terms of variables u and u_s from eq. (62).

When $\xi \ll 1$ (the condition for perturbation theory to be valid), the integrand (61) can be expanded in powers of ξ. For the first term of the sum W_1, we get

$$W_1 \approx \frac{e^2 m^2}{4p_0}\xi^2 \int_0^{u_1} \left[2 + \frac{u^2}{1+u} - 4\frac{u}{u_1}\left(1 - \frac{u}{u_1}\right)\right] \frac{1}{(1+u)^2} du =$$
$$\frac{e^2 m^2}{4p_0}\xi^2 \left[\left(1 - \frac{4}{u_1} - \frac{8}{u_1^2}\right) \ln(1+u_1) + \frac{1}{2} + \frac{8}{u_1} - \frac{1}{2(1+u_1)^2}\right] \tag{63}$$

with $u_1 \approx 2(kp)/m^2$. It is possible to determine the second and the next harmonics as an analogy with the Berestetzkii approach; however, the aim of this article was only to illustrate the influence of the dielectric medium on the Compton effect.

Let us consider eq. (54) in the form

$$sk + q - k' = q'. \tag{64}$$

Equation (64) has physical meaning for $s = 1, 2, \ldots N$, s, N being positive integers. $s = 1$ means the conservation of energy momentum of the one-photon Compton process and $s = 2$ of the two-photon Compton process and $s = N$ means the multiphoton interaction with N photons of laser beam with an electron. The multiphoton interaction is nonlinear and differs from the situation where electron scatters twice or more as it traverses the laser focus.

By analogy, the original Einstein photoelectric equation must be replaced by the more general multiphoton photoelectric equation in the form

$$s\hbar\omega = \frac{1}{2}mv^2 + E_i, \tag{65}$$

where E_i is the binding energy of the outermost electron in the atomic system. It means that the ionization effect occurs also in the case that $\hbar\omega < E_i$ in case that the number of participating photons is $s > E_i/\hbar\omega$. We will not solve furthermore this specific problem.

We introduce the scattering angle θ between $\mathbf{k}$ and $\mathbf{k}'$. In other words, The scattering angle θ is measured with respect to the incident photon direction. Then, with $|\mathbf{k}| = n\omega$ and $|\mathbf{k}'| = n\omega'$, where n is index of refraction of the dielectric, we get from the squared eq. (64) in the rest system of electron, where $q = (m_*, 0)$, the following equation:

$$s\frac{1}{\omega'} - \frac{1}{\omega} = \frac{s}{m_*}(1 - n^2 \cos\theta), \tag{66}$$

which is a modification of the original equation for the Compton process

$$\frac{1}{\omega'} - \frac{1}{\omega} = \frac{1}{m}(1 - \cos\theta). \tag{67}$$

Using relation $m_*^2(m = 0) = -e^2a^2$, we get eq. (66) for the situation of the Compton effect in the graphene sheet:

$$s\frac{1}{\omega'} - \frac{1}{\omega} = \frac{s}{m_*(m = 0)}(1 - n^2 \cos\theta) = \frac{s}{e\sqrt{-a^2}}(1 - n^2 \cos\theta), \tag{68}$$

So, we see that the last Compton formula differs from the original one only by the existence of the renormalized mass and the occurrence of index of refraction.

We know that the original Compton formula can be written in the form suitable for the experimental verification, namely:

$$\Delta\lambda = 4\pi\frac{\hbar}{mc}\sin^2\frac{\theta}{2}, \tag{69}$$

which was used by Compton for the verification of the quantum nature of light [1]. The limiting case with $m \to 0$ has the appropriate angle limit $\theta \to 0$.

If we consider the Compton process in dielectric, then the last formula goes to the following form:

$$\Delta\lambda = 2\pi\frac{\hbar}{mc}(1 - n^2 \cos\theta). \tag{70}$$

It is evident that relation $\lambda' - \lambda \geq 0$ follows from eq. (1). However, if we put

$$1 - n^2 \cos\theta \leq 0, \tag{71}$$

or equivalently

$$\frac{1}{n^2} \leq \cos\theta \leq 1, \tag{72}$$

then we see that for some angles determined by eq. (72), the relation $\lambda' - \lambda \leq 0$ follows. This surprising result is the anomalous Compton effect which is caused by the index of refraction of the medium. To our knowledge, it was not published in the optical or particle journals.

The limiting case with $m \to 0$ has the appropriate angle limit $n^2 \cos\theta \to 1$. It means that there is of the angle shift in this case in comparison with the Compton effect in vacuum.

The equation $sk + q = q' + k'$ is the symbolic expression of the nonlinear Compton effect in which several photons are absorbed at a single point, but only a single high-energy photon is emitted. The second process where electron scatters twice or more as it traverses the laser focus is not considered here. The nonlinear Compton process was experimentally confirmed, for instance, by Bulla et al. [15].

The formula (66) can be also expressed in terms of λ as follows:

$$s\lambda' - \lambda = \frac{2\pi s}{m_*}(1 - n^2 \cos\theta) \tag{73}$$

where we have put $\hbar = c = 1$. In the case of the graphene two-dimensional carbon sheet with zero mass of the pseudoelectron, we replace the renormalized mass by $m_*(m = 0) = e\sqrt{-a^2}$, which is the new renormalized mass in graphene.

Formula (73) can be used for the verification of the Compton effect in a dielectric medium, and on the other hand, the index refraction follows from it in the following form:

$$n^2 = \frac{1}{\cos\theta}\left[1 - \frac{m_*}{2\pi s}(s\lambda' - \lambda)\right]. \tag{74}$$

It means, if we know the $\theta, \lambda, \lambda', s, m_*$, we are able to determine the index of refraction of some dielectric medium from the Compton effect. To our knowledge, this method was not published in the optical journals. In the graphene case, we write as usual $m_*(m = 0) = e\sqrt{-a^2}$ in the last formula.

6. Compton effect initiated by a laser pulse

Let us start with the classical theory of interaction of particle with an impulsive force. We idealize the impulsive force by the dirac δ-function. Newton's second law for the interaction of a massive particle with mass m with an impulsive force $P\delta(t)$ is as follows:

$$m\frac{d^2x}{dt^2} = P\delta(t), \tag{75}$$

where P is some constant.

Using the Laplace transform on the last equation, with

$$\int_0^\infty e^{-st}x(t)dt = X(s), \tag{76}$$

$$\int_0^\infty e^{-st}\ddot{x}(t)dt = s^2X(s) - sx(0) - s\dot{x}(0), \tag{77}$$

$$\int_0^\infty e^{-st}\delta(t)dt = 1, \tag{78}$$

we obtain

$$ms^2X(s) - msx(0) - m\dot{x}(0) = P. \tag{79}$$

For a particle starting from rest with $\dot{x}(0) = 0, x(0) = 0$, we get

$$X(s) = \frac{P}{ms^2}, \tag{80}$$

and using the inverse Laplace transform, we obtain

$$x(t) = \frac{P}{m}t \tag{81}$$

and

$$\dot{x}(t) = \frac{P}{m}. \tag{82}$$

Let us remark that if we express δ-function by the relation $\delta(t) = \dot{H}(t)$, H being defined as a step function, then from eq. (75) follows $\dot{x}(t) = P/m$, immediately. The physical meaning of the quantity P can be deduced from equation $F = P\delta(t)$. After t-integration, we have $\int Fdt = mv = P$, where m is mass of a body and v its final velocity (with $v(0) = 0$). It means that the value of P can be determined a posteriori and then this value can be used in more complex equations than eq. (75). Of course it is necessary to suppose that δ-form of the impulsive force is adequate approximation of the experimental situation.

If we consider the δ-form electromagnetic pulse, then we can write

$$F_{\mu\nu} = a_{\mu\nu}\delta(\varphi). \tag{83}$$

where $\varphi = kx = \omega t - \mathbf{kx}$. In order to obtain the electromagnetic impulsive force in this form, it is necessary to define the four-potential in the following form:

$$A_\mu = a_\mu H(\varphi), \tag{84}$$

where function H is the Heaviside unit step function defined by the relation

$$H(\varphi) = \begin{cases} 0, & \varphi < 0 \\ 1, & \varphi \geq 0 \end{cases}. \tag{85}$$

If we define the four-potential by eq. (85), then the electromagnetic tensor with impulsive force is of the form

$$F_{\mu\nu} = \partial_\mu a_\nu - \partial_\nu a_\mu = (k_\mu a_\nu - k_\nu a_\mu)\delta(\varphi) = a_{\mu\nu}\delta(\varphi). \tag{86}$$

To find motion of an electron in the δ-form electromagnetic force, we must solve immediately the Lorentz equation, or, to solve Lorentz equation in general with four-potential $A_\mu = a_\mu A(\varphi)$ and then replace the four-potential by the eta-function. Following Meyer [13], we apply his method and then replace $A_\mu(\varphi)$ by $a_\mu H(\varphi)$ in the final result.

The Lorentz equation reads:

$$\frac{dp_\mu}{d\tau} = \frac{e}{m} F_{\mu\nu} p^\nu = \frac{e}{m}(k_\mu a \cdot p - a_\mu k \cdot p) A'(\varphi), \tag{87}$$

where the prime denotes derivation with regard to φ, τ is proper time, and $p_\mu = m(dx_\mu/d\tau)$. After multiplication of the last equation by k^μ, we get with regard to the Lorentz condition $0 = \partial_\mu A^\mu = a^\mu \partial_\mu A(\varphi) = k_\mu a^\mu A'$ or $k \cdot a = 0$ and $k^2 = 0$ the following equation:

$$\frac{d(k \cdot p)}{d\tau} = 0 \tag{88}$$

and it means that $k \cdot p$ is a constant of the motion and it can be defined by the initial conditions, for instance, at time $\tau = 0$. If we put $p_\mu(\tau = 0) = p^0_\mu$, then we can write $k \cdot p = k \cdot p^0$. At this moment, we have

$$k \cdot p = \frac{mk \cdot dx}{d\tau} = m\frac{d\varphi}{d\tau}, \tag{89}$$

or

$$\frac{d\varphi}{d\tau} = \frac{k \cdot p^0}{m}. \tag{90}$$

So, using the last equation and relation $d/d\tau = (d/\varphi)d\varphi/d\tau$, we can write eq. (87) in the form

$$\frac{dp_\mu}{d\varphi} = \frac{e}{k \cdot p^0}(k_\mu a \cdot p - a_\mu k \cdot p^0)A'(\varphi) \tag{91}$$

giving (after multiplication by a^μ)

$$\frac{d(a \cdot p)}{d\varphi} = -ea^2 A' \tag{92}$$

or

$$a \cdot p = a \cdot p^0 - ea^2 A. \tag{93}$$

Substituting the last formula into (91), we get

$$\frac{dp_\mu}{d\varphi} = e\left(a_\mu - \frac{k_\mu a \cdot p^0}{k \cdot p^0}\right)\frac{dA}{d\varphi} - \frac{e^2 a^2}{2k \cdot p^0}\frac{d(A^2)}{d\varphi}k_\mu. \tag{94}$$

This equation can be immediately integrated to give the resulting momentum in the form

$$p_\mu = p_\mu^0 - e\left(A_\mu - \frac{A^\nu p_\nu^0 k_\mu}{k \cdot p^0}\right) - \frac{e^2 A^\nu A_\nu k_\mu}{2k \cdot p^0}. \tag{95}$$

Now, if we put into this formula the four-potential (84) of the impulsive force, then for $\varphi > 0$ when $H > 1$, we get

$$p_\mu = p_\mu^0 - e\left(a_\mu - \frac{a^\nu p_\nu^0 k_\mu}{k \cdot p^0}\right) - \frac{e^2 a^\nu a_\nu k_\mu}{2k \cdot p^0}. \tag{96}$$

The last equation can be used to determine the magnitude of a_μ. It can be evidently expressed as the number of k-photons in electromagnetic momentum. For $\varphi < 0$, it is $H = 0$ and, therefore, $p_\mu = p_\mu^0$

It is still necessary to say what is the practical realization of the δ-form potential. We know from the Fourier analysis that the Dirac δ-function can be expressed by integral in the following form:

$$\delta(\varphi) = \frac{1}{\pi}\int_0^\infty \cos(s\varphi)ds. \tag{97}$$

So, the δ-potential can be realized as the continual superposition of the harmonic waves. In case it will not be possible to realize it experimentally, we can approximate the integral formula by the summation formula as follows:

$$\delta(\varphi) \approx \frac{1}{\pi}\sum_0^\infty \cos(s\varphi). \tag{98}$$

Now, the quantum mechanical problem is to find solution of the Dirac equation with the δ-form four-potential (84) and, from this solution, determine the quantum motion of the charged particle under this potential. After insertion of Ψ_p and $\bar{\Psi}_p$ into the current density $j^\mu = \Psi_p^*\gamma^\mu)\Psi_p)$, we have

$$j^\mu = \frac{1}{p_0}\left\{p^\mu - eA^\mu + k^\mu\left(\frac{e(pA)}{(kp)} - \frac{e^2A^2}{2(kp)}\right)\right\}. \tag{99}$$

which is evidently related to eq. (23).

The so-called kinetic momentum corresponding to j^μ is as follows:

$$J^\mu = \Psi_p^*(p^\mu - eA^\mu)\Psi_p) = \bar{\Psi}_p\gamma^0(p^\mu - eA^\mu)\Psi_p) =$$
$$\left\{p^\mu - eA^\mu + k^\mu\left(\frac{e(pA)}{(kp)} - \frac{e^2A^2}{2(kp)}\right)\right\} + k^\mu\frac{ie}{8(kp)p_0}F_{\alpha\beta}(u^*\sigma^{\alpha\beta}u), \tag{100}$$

where

$$\sigma^{\alpha\beta} = \frac{1}{2}(\gamma^\alpha\gamma^\beta - \gamma^\beta\gamma^\alpha). \tag{101}$$

Now, we express the four-potential by the step function. In this case, the kinetic momentum contains the tensor $F_{\mu\nu}$ involving δ-function. It means that there is a singularity at point $\varphi = 0$. This singularity plays no role in the situation for $\varphi > 0$ because, in this case, the δ-function is zero. Then, the kinetic momentum is the same as j^μ. The emission of photons by electron in the delta-pulse force follows from the matrix element M corresponding to the emission of photon by electron in the electromagnetic field [12]:

$$M = -ie^2\int d^4x\bar{\Psi}_{p'}(\gamma e'^*)\Psi_p\frac{e^{ik'x}}{\sqrt{2\omega'}}, \tag{102}$$

where Ψ_p is the wave function of an electron before interaction with a pulse and $\Psi_{p'}$ is the wave function of an electron after interaction and emission of photon with components $k'^{\mu} = (\omega', \mathbf{k}')$. The symbol e'^* is the four-polarization vector of emitted photon.

If we write Volkov wave function Ψ_p in the form (18), then, for the impulsive vector potential (84), we have

$$S = -px - \left[e\frac{ap}{kp} - \frac{e^2}{2kp}a^2\right]\varphi, \quad R = \left[1 + \frac{e}{2kp}(\gamma k)(\gamma a)H(\varphi)\right]. \tag{103}$$

So, we get the matrix element in the form

$$M = g\int d^4x\bar{\Psi}_{p'}O\Psi_p\frac{e^{ik'x}}{\sqrt{2\omega'}}, \tag{104}$$

where $O = \gamma e'^*$, $g = -ie^2$ in case of the electromagnetic interaction and

$$\bar{\Psi}_{p'} = \frac{\bar{u}}{\sqrt{2p'_0}}\bar{R}(p')e^{-iS(p')}. \tag{105}$$

In such a way, using the above definitions, we write the matrix element in the form

$$M = \frac{g}{\sqrt{2\omega'}}\frac{1}{2p'_0 2p_0}\int d^4x\bar{R}(p')OR(p)e^{-iS(p')+iS(p)}e^{ik'x}. \tag{106}$$

The quantity $\bar{R}(p')$ follows immediately from eq. (103), namely:

$$\bar{R}' = \overline{\left[1 + \frac{e}{2kp'}(\gamma k)(\gamma a)H(\varphi)\right]} = \left[1 + \frac{e}{2kp'}(\gamma a)(\gamma k)H(\varphi)\right]. \tag{107}$$

Using

$$-iS(p') + iS(p) = i(p' - p) + i(\alpha' - \alpha)\varphi, \tag{108}$$

where

$$\alpha = \left(e\frac{ap}{kp} - \frac{e^2}{2}\frac{a^2}{kp}\right), \quad \alpha' = \left(e\frac{ap'}{kp'} - \frac{e^2}{2}\frac{a^2}{kp'}\right), \tag{109}$$

we get

$$M = \frac{g}{\sqrt{2\omega'}} \frac{1}{2p_0' 2p_0} \int d^4x \bar{u} \bar{R}(p') O R(p) u e^{i(p'-p)x} e^{i(\alpha'-\alpha)\varphi} e^{ik'x}. \quad (110)$$

Putting

$$\bar{R}(p') O R(p) e^{i(\alpha'-\alpha)\varphi} = \frac{1}{2\pi} \int_{-\infty}^{\infty} D(s) e^{-is\varphi} ds \quad (111)$$

with the inverse transform

$$D(s) = \int_{-\infty}^{\infty} d\varphi e^{is\varphi} \bar{R}(p') O R(p) e^{i(\alpha'-\alpha)}, \quad (112)$$

we get after x-integration

$$M = \frac{g}{\sqrt{2\omega'}} \frac{1}{\sqrt{2p_0' 2p_0}} \int ds \bar{u}(p') D(s) u(p) \delta^{(4)}(ks + p - k' - p'). \quad (113)$$

We see from the presence of the δ-function in eq. (113) that during the process of the interaction of electron with the laser pulse, the energy-momentum conservation law holds good

$$sk + p = k' + p'. \quad (114)$$

The last equation has physical meaning for $s = 1, 2, \ldots N$, N being a positive integer. $s = 1$ means the conservation of energy momentum of the one-photon Compton process and $s = 2$ of the two-photon Compton process and $s = N$ means the multiphoton interaction with N photons of laser pulse with electron. The multiphoton interaction is nonlinear and differs from the situation where electron scatters twice or more as it traverses the laser focus. The analogical situation is valid for the photoelectric equation.

Now, let us determine $D(s)$. With regard to the mathematical relation $H^2(\varphi) = H(\varphi)$, we can put

$$\bar{R}(p') O R(p) = A + BH(\varphi) \quad (115)$$

where

$$A = \gamma e'^{*} \tag{116a}$$

and

$$B = \frac{e}{2(kp)}(\gamma e'^{*})(\gamma k)(\gamma a) + \frac{e}{2(kp')}(\gamma a)(\gamma k)(\gamma e'^{*}) + \frac{e^2}{4(kp)(kp')}(\gamma a)(\gamma k)(\gamma e'^{*})(\gamma k)(\gamma a). \tag{116b}$$

In such a way,

$$D(s) = A\int_{-\infty}^{\infty} e^{i(\alpha'-\alpha+s)\varphi}d\varphi + B\int_{-\infty}^{\infty} e^{i(\alpha'-\alpha+s)\varphi}H(\varphi)d\varphi = (2\pi)A\delta(\alpha'-\alpha+s) + (2\pi i)B\frac{1}{\alpha'-\alpha+s} \tag{117}$$

as a consequence of the integral

$$\int_0^{\infty} e^{-\varepsilon x}\sin mx dx = \frac{m}{\varepsilon^2 + m^2} \tag{118}$$

for $m \to 0$.

The total probability of the emission of photons during the interaction of the laser pulse with electron is as follows:

$$W = \int \frac{1}{2}\sum_{spin.polar.} \frac{|M|^2}{VT}\frac{d^3p'd^3k'}{(2\pi)^6}. \tag{119}$$

There are two substantial mathematical operations for the evaluation of W. The one step is to use, after performing $|M|^2$, the following mathematical identity:

$$(2\pi)^8\delta^{(4)}(sk + p - p' - k')\delta^{(4)}(p + s'k - p' - k') = (2\pi)^4 VT\frac{\delta(s-s')}{\delta(0)}\delta^{(4)}(p + sk - p' - k'), \tag{120}$$

where V is the space volume and T is time interval, and the second step is the determination of Trace, because according to the quantum electrodynamics of a spin, it is possible to show that [12]

$$\frac{1}{2}\sum_{spin.polar.}|M|^2 = \frac{1}{2}\mathrm{Tr}\left\{(\gamma p' + m)D(\gamma p + m)\gamma^0 D^+ \gamma^0\right\}, \tag{121}$$

where in our case, quantity D is given by eq. (117).

In order to determine *Tr* or spur of the combinations of γ-matrix, it is suitable to know some relations. For instance:

$$\mathrm{Tr}(a\gamma)(b\gamma) = 4ab, \quad \mathrm{Tr}(a\gamma)(b\gamma)(c\gamma) = 0, \tag{122}$$

$$\mathrm{Tr}(a\gamma)(b\gamma)(c\gamma)(d\gamma) = 4\left[(ab)(cd) - (ac)(bd) + (ad)(bc)\right]. \tag{123}$$

Then,

$$\mathrm{Tr}\left[(\gamma p' + m)D(\gamma p + m)\bar{D}\right] = S_1 + S_2 + S_3 + S_4; \quad \bar{D} = \gamma^0 D^+ \gamma^0 \tag{124}$$

with

$$S_1 = \mathrm{Tr}[\gamma p' D \gamma p \bar{D}] \tag{125}$$

$$S_2 = \mathrm{Tr}[m D \gamma p \bar{D}] \tag{126}$$

$$S_3 = \mathrm{Tr}[m \gamma p' D \bar{D}] \tag{127}$$

$$S_4 = \mathrm{Tr}[m^2 D \bar{D}], \tag{128}$$

where D is given by eq. (117) and A and B are given by eqs. (116b) and (116b).

Now, it is evident that the total calculation is complex and involves many algebraic operations with γ-matrices and δ-functions. At this moment, we restrict the calculations to the most simple approximation where we replace the term in brackets in eq. (18) by unit, and so we write instead of eq. (18)

$$\Psi_p \sim \frac{u}{\sqrt{2p_0}} e^{iS} \tag{129}$$

which is usually used in similar form for the nonrelativistic calculations as it is discussed in [12]. Then,

$$M = \frac{g}{\sqrt{2\omega'}} \frac{1}{2p_0' 2p_0} \int dx^4 \bar{u}(\gamma e'^*) u e^{i(p'-p)x} e^{i(\alpha'-\alpha)\varphi} e^{ik'x} =$$
$$\frac{g}{\sqrt{2\omega'}} \frac{1}{2p_0' 2p_0} \bar{u}(\gamma e'^*) u \delta^{(4)}(lk + p - p' - k'), \tag{130}$$

where

$$l = \alpha - \alpha'. \tag{131}$$

In this simplified situation, $\bar{R}OR$ reduces to $A = \gamma e'^*$. Then, using relation $\bar{\gamma^\mu} = \gamma^\mu$ with a consequence $\bar{A} = A$ and relations (122) and (123), we get

$$S_1 = \mathrm{Tr}[\gamma p' A \gamma p \bar{A}] = 4\left[(p'e'^*)(pe'^*) - (pp')(e'^*e'^*) + (p'e'^*)(pe'^*)\right] \tag{132}$$

$$S_2 = \mathrm{Tr}[mA\gamma p\bar{A}] = 0 \tag{133}$$

$$S_3 = \mathrm{Tr}[m\gamma p' A\bar{A}] = 0 \tag{134}$$

$$S_4 = \mathrm{Tr}[m^2 A\bar{A}] = 4m^2(e'^*e'^*). \tag{135}$$

At this moment, we can write probability of the process W in the form:

$$W = \int \frac{1}{2} \sum_{spin.\ polar.} \frac{|M|^2}{VT} \frac{d^3p' d^3k'}{(2\pi)^6} =$$
$$\int \frac{d^3p' d^3k'}{(2\pi)^6} \frac{1}{2} (S_1 + S_2 + S_3 + S_4) \frac{1}{(2\pi)^2} \delta^{(4)}(lk + p - p' - k') =$$
$$\int \frac{d^3p' d^3k'}{(2\pi)^6} \frac{1}{2} \delta^{(4)}(lk + p - p' - k') 4\left\{(p'e'^*)(pe'^*) + (e'^*e'^*)(m^2 - (pp'))\right\}. \tag{136}$$

The presence of the δ-function in the last formula is expression of the conservation law $lk + p = k' + p'$, which we write in the form

$$lk + p - k' = p'. \tag{137}$$

If we introduce the angle Θ between $\mathbf{k}$ and $\mathbf{k}'$, then, with $|\mathbf{k}| = \omega$ and $|\mathbf{k}'| = \omega'$, we get from the squared eq. (137) in the rest system of electron, where $p = (m, 0)$, the following equation:

$$l\frac{1}{\omega'} - \frac{1}{\omega} = \frac{l}{m}(1 - \cos\Theta); \quad l = \alpha - \alpha', \tag{138}$$

which is a modification of the original equation for the Compton process

$$\frac{1}{\omega'} - \frac{1}{\omega} = \frac{1}{m}(1 - \cos\Theta). \tag{139}$$

We observe that the basic difference between single-photon interaction and δ-pulse interaction is the factor $l = \alpha - \alpha'$.

We know that the last formula of the original Compton effect can be written in the form suitable for the experimental verification, namely:

$$\Delta\lambda = 4\pi\frac{\hbar}{mc}\sin^2\frac{\Theta}{2}, \tag{140}$$

which was applied Compton for the verification of the quantum nature of light.

We can express eq. (138) in a new form. From equation $lk + p = k' + p'$, we get after multiplying it by k in the rest frame of electron

$$kp' = \omega m - \omega\omega'(1 - \cos\Theta) \tag{141}$$

Then, l in eq. (138) is given by the formula ($a \equiv (v, \mathbf{w})$):

$$l = \frac{2evm - e^2a^2}{2\omega m} - \frac{2eap' - e^2a^2}{2\omega[m - \omega'(1 - \cos\Theta)]}. \tag{142}$$

This equation (138) can be experimentally verified by the similar methods which were used by Compton for the verification of his formula. However, it seems that the interaction of the photonic pulse substantially differs from the interaction of a single photon with electron.

The equation $lk + p = k' + p'$ is the symbolic expression of the nonlinear Compton effect in which several photons are absorbed at a single point, but only single high-energy photon is emitted. The second process where electron scatters twice or more as it traverses the laser focus is not considered here. The nonlinear Compton process was experimentally confirmed, for instance, by Bulla et al. [15].

The present text is a continuation of the author discussion on laser acceleration [16, 17, 18], where the Compton model of laser acceleration was proposed.

The δ-form laser pulses are here considered as an idealization of the experimental situation in laser physics. Nevertheless, it was demonstrated theoretically that at the present time, the zeptosecond and sub-zeptosecond laser pulses of duration $10^{-21} - 10^{-22}$ s can be realized by the petawatt lasers [19].

7. Compton effect initiated by two orthogonal plane waves

The project with the two laser waves is the next goal and the future direction of the laser physics of elementary particles. The two laser beams can replace many laser beams in the thermonuclear reactor such as ITER in Cadarache near Aix-en-Provence in France. At the same time, the two-laser system can be considered in chemistry as a catalyzer which was not known before the laser physics.

We solve the Dirac equation for two different four-potentials of the plane electromagnetic waves. We specify the solutions of the Dirac equation for two orthogonal plane waves.

The modified Compton formula for the scattering of two photons on an electron is determined. The solution of the Dirac equation for two waves was found by Sen Gupta [20, 21] in the form of the Fourier series, however without immediate application. The solution of the Dirac equation for two waves with the perpendicular polarization was given by Lyulka [22-25] who described the decay of particles in two laser fields. The derivation of two-wave solution is not presented in his articles. So, we investigated the situation and presented new results.

The total four-potential V_μ is a superposition of the potentials A_μ and B_μ as follows:

$$V_\mu = A_\mu(\varphi) + B_\mu(\chi), \tag{143}$$

where $\varphi = kx$ and $\chi = \kappa x$ and $k \neq \kappa$.

The Lorentz condition gives

$$\partial_\mu V^\mu = 0 = k_\mu \frac{\partial A^\mu}{\partial \varphi} + \kappa_\mu \frac{\partial B^\mu}{\partial \chi} = k_\mu A^\mu_\varphi + \kappa_\mu B^\mu_\chi, \tag{144}$$

where the subscripts φ, χ denote partial derivatives. Equation (144) takes a more simple form if we notice that partial differentiation with respect to φ concerns only A_μ and partial differentiation with respect to χ concerns only B_μ. So we write instead eq. (144).

$$\partial_\mu V^\mu = 0 = k_\mu (A^\mu)' + \kappa_\mu (B^\mu)' = kA' + \kappa B'. \tag{145}$$

Without loss of generality, we can write instead of eq. (145) the following one:

$$k_\mu (A^\mu)' = 0; \quad \kappa_\mu (B^\mu)' = 0, \tag{146}$$

or,

$$kA = const = 0; \quad \kappa B = const = 0, \tag{147}$$

putting integrating constant to zero.

The electromagnetic tensor $F_{\mu\nu}$ is expressed in the new variables as in eq. (5)

$$F_{\mu\nu} = k_\mu A'_\nu - k_\nu A'_\mu + \kappa_\mu B'_\nu - \kappa_\nu B'_\mu. \tag{148}$$

Now, we write Dirac equation for the two potentials in the form

$$[-\partial^2 - 2ie(V\partial) + e^2 V^2 - m^2 - \frac{i}{2} e F_{\mu\nu}\sigma^{\mu\nu}]\psi = 0, \tag{149}$$

where $V = A + B$, $F_{\mu\nu}$ is given by eq. (148) and the σ-term is defined as follows:

$$\frac{i}{2} e F_{\mu\nu}\sigma^{\mu\nu} = ie(\gamma k)(\gamma A') + ie(\gamma\kappa)(\gamma B') \tag{150}$$

We assume the solution of eq. (149) in the Volkov form, or

$$\psi = e^{-ipx} F(\varphi, \chi). \tag{151}$$

After performing all operations in eq. (149), we get the partial differential equation for function $F(\varphi, \chi)$:

$$\begin{aligned} -2k\kappa F_{\varphi\chi} + (2ipk - 2ikB)F_\varphi + (2ip\kappa - 2ieA\kappa)F_\chi \quad + \\ (e^2(A+B)^2 - 2e(A+B)p - ie(\gamma k)(\gamma A_\varphi) - ie(\gamma\kappa)(\gamma B_\chi))F = 0. \end{aligned} \tag{152}$$

Equation (152) was simplified by the author [20], putting $k\kappa = 0$. However, to get the Compton effect-initiated two orthogonal waves, we ignore this simplification and write eq. (152) in the following form:

$$aF_\varphi + bF_\chi + cF = 2k\kappa F_{\varphi\chi}, \tag{153}$$

where

$$a = 2ipk - 2iekB; \quad b = 2ip\kappa - 2ie\kappa A \tag{154}$$

and

$$c = e^2(A+B)^2 - 2e(A+B)p - ie(\gamma k)(\gamma A') - ie(\gamma\kappa)(\gamma B') \tag{155}$$

and the term of the two partial derivations is not present because of $k\kappa = 0$.

For the field which we specify by the conditions

$$kB = 0; \quad \kappa A = 0; \quad AB = 0, \tag{156}$$

we have

$$\begin{aligned} 2ipkF_\varphi + 2ip\kappa F_\chi + (e^2A^2 + e^2B^2 - 2epA - 2epB - ie(\gamma k)(\gamma A') \quad - \\ ie(\gamma\kappa)(\gamma B'))F = 2k\kappa F_{\varphi\chi}. \end{aligned} \tag{157}$$

Now, let us put

$$F(\varphi, \chi) = X(\varphi)Y(\chi). \tag{158}$$

After insertion of eq. (158) into eq. (157) and division of the new equation by XY, we get the terms depending only on φ and on χ. Or we get

$$\begin{aligned} \left(2i(pk + ik\kappa)\frac{X'}{X} + e^2A^2 - 2epA - ie(\gamma k)(\gamma A')\right) \quad + \\ \left(2i(p\kappa + ik\kappa)\frac{Y'}{Y} + e^2B^2 - 2epB - ie(\gamma\kappa)(\gamma B')\right) = 0 \end{aligned} \tag{159}$$

So, there are terms dependent on φ and terms dependent on χ only in eq. (159). The only possibility is that they are equal to some constant λ and $-\lambda$. Then,

$$2i(pk + ik\kappa)X' + (e^2A^2 - 2epA - ie(\gamma k)(\gamma A'))X = \lambda X \tag{160}$$

and

$$2i(p\kappa + ik\kappa)Y' + (e^2B^2 - 2epB - ie(\gamma\kappa)(\gamma B'))Y = -\lambda Y \tag{161}$$

We put $\lambda = 0$ without loss of generality. In such a way, the solution of eq. (159) is the solution of two equations only. Because the form of every equation is similar to the form of eq. (14), we can write the solution of these equations as follows:

$$X = \left[1 + \frac{e}{2(kp + ik\kappa)}(\gamma k)(\gamma A)\right] \frac{u}{\sqrt{2p_0}} e^{iS_1}, \tag{162}$$

with

$$S_1 = \int_0^{kx} \frac{e}{(kp + ik\kappa)} \left[(pA) - \frac{e}{2}(A)^2\right] d\varphi. \tag{163}$$

and

$$Y = \left[1 + \frac{e}{2(\kappa p + ik\kappa)}(\gamma\kappa)(\gamma B)\right] \frac{u}{\sqrt{2p_0}} e^{iS_2}, \tag{164}$$

with

$$S_2 = -\int_0^{\kappa x} \frac{e}{(\kappa p + ik\kappa)} \left[(pB) - \frac{e}{2}(B)^2\right] d\chi. \tag{165}$$

The total solution is then of the form

$$\psi_p = \left[1 + \frac{e}{2(kp + ik\kappa)}(\gamma k)(\gamma A)\right] \left[1 + \frac{e}{2(\kappa p + ik\kappa)}(\gamma\kappa)(\gamma B)\right] \frac{u}{\sqrt{2p_0}} e^{i(S_1(A) + S_2(B))}. \tag{166}$$

In the case of the two non-collinear laser beams, the problem was solved by Lyulka in 1974 for two linearly polarized waves [22]:

$$A = a_1 \cos\varphi; \quad B = a_2 \cos(\chi + \delta) \tag{167}$$

with the standard conditions for φ, χ, k, κ, and δ being the phase shift.

The two-wave Volkov solution is given by eq. (166), and the matrix elements with corresponding calculation ingredients are given by the standard method as it was shown by Lyulka [22].

It was shown in [22] that

$$q^\mu = p^\mu - e^2 \frac{a_1^2}{2(kp)} k^\mu - e^2 \frac{a_2^2}{2(\kappa p)} \kappa^\mu \tag{168}$$

and

$$m_*^2 = m^2 \left(1 - \frac{e^2 a_1^2}{m^2} - \frac{e^2 a_2^2}{m^2} \right), \tag{169}$$

which is for the massless graphene limit, the following one $m_*^2(m = 0) = -e^2(a_1^2 + a_2^2)$.

The matrix element involves the extended law of conservation. Namely:

$$sk + t\kappa + q = q' + k' + \kappa', \tag{170}$$

where the s and t are natural numbers. The last equation has natural interpretations. The photon object with momenta sk and $t\kappa$ interacts with electron with momentum q. After interaction, the electron has a momentum q' and two photons are emitted with momenta k' and κ'.

We can write eq. (170) in the equivalent form:

$$sk + q - k' = q' + \kappa' - t\kappa. \tag{171}$$

From the squared form of the last equation and after some modification, we get the generalized equation of the double Compton process for $s = t = 1$:

$$\frac{1}{\omega'} - \frac{1}{\omega} = \frac{1}{m_*}(1 - \cos\Theta) + \frac{\Omega' - \Omega}{\omega\omega'} - \frac{\Omega\Omega'}{\omega\omega' m_*}(1 - \cos\Xi), \tag{172}$$

where Ξ is the angle between the 3-momentum of the κ-photon and the 3-momentum of the κ'-photons with frequency Ω and Ω', respectively. In the situation of graphene with the massless limit, we replace the renormalized mass in eq. (172) by $m_*(m = 0) = e\sqrt{-a_1^2 - a_2^2}$.

Let us remark that if the frequencies of the photons of the first wave substantially differ from the frequencies of the photons of the second wave, then eq. (172) can be experimentally verified by the same method as the original Compton formula. To our knowledge, formula (172) is not involved in the standard textbooks on quantum electrodynamics.

8. Perspectives

We have considered the Compton effect in the framework of the Volkov solution of the Dirac equation assuming that the process occurred in vacuum and in medium with the index of refraction n. The determination of the index of refraction follows from the Compton effect. Mass renormalization of electron is involved in the Volkov solution.

The harmonic oscillator with frequency ω_0 and the dispersion theory leads to the known formula for the index of refraction of matter [26]. The index of refraction derived in the dispersion theory based on the damped oscillator is given by the formula [27]

$$n = 1 + 2\pi N \frac{e^2}{m} \frac{\omega_0^2 - \omega^2}{(\omega_0^2 - \omega^2)^2 + \gamma^2\omega^2}, \tag{173}$$

where N is number of electrons in the unit of volume. The modern aspects can be found in the Crenshaw article [28]. So, to consider the Compton effect in dielectric medium is the perspective problem.

The interesting result of our article is the derivation that for some scattering angles given by eq. (70), there exists the so-called anomalous Compton effect, where the wavelengths of scattered photons are shorter than the wavelengths of the original photons. To our knowledge, information of this effect was not published in the physical journals.

The Compton scattering is, at the present time, the elementary laboratory problem because for the monochromatic X-rays for $\lambda = 1$, the shift of wavelength is several percent. This is quantity which can be easily measured. On the other hand, the Compton wavelength shift for the visible light is only 0,01 percent. It means that the measurement of the Compton effect for the visible light in the dielectric medium involves the subtle approach.

We have also discussed the problem of the Dirac equation with the two-wave potentials of the electromagnetic fields. While the Volkov solution for one potential is well known for a long time, the Compton process with two beams was not investigated experimentally by any laboratory.

It is possible to consider the situation with the sum of N waves, or

$$V = \sum_{i=1}^{N} A_i(\varphi_i) \quad \varphi_i = k_i x. \tag{174}$$

The problem of the laser compression of target by many beams, involving the Compton effect, is one of the actual problems of the contemporary laser physics. The goal of the experiments is to generate the physical implosion in the spherical target. The light energy is absorbed by the target and generates a high-temperature plasma with high pressure of a few hundred megabars. For the process sufficiently spherically symmetric, the central area is heated up to 5–10 keV and fusion reaction starts [29] . The solution of that problem in the general form is a difficult one, and it can be solved only by the special laser institutions such as the Lebedev Institute of Physics, the Lawrence Livermore National Laboratory, and so on.

New experiments can be realized and new measurements performed by means of the laser pulses, giving new results and discoveries.

Author details

Miroslav Pardy

*Address all correspondence to: pamir@physics.muni.cz

Department of Physical Electronics, Masaryk University, Brno, Czech Republic

References

[1] Compton, A. A. (1923). A quantum theory of the scattering of X-rays by light elements, *Phys. Rev.*, 21(5), 483-502.

[2] Volkov, D. M. (1935). Über eine Klasse von Lösungen der Diracschen Gleichung, *Z. Physik*, 94, 250-260.

[3] Peierls, R. E. (1934). Bemerkungen über Umwandlungstemperaturen, *Helvetica Physica Acta* 7, 81; ibid: (1935). Quelques proprietes typiques des corpses solides, *Ann. Inst. Henri Poincaré*, 5, 177.

[4] Landau, L. D. (1937). Zur Teorie der Phasenumwandlungen II, *Phys. Z. Sowjetunion*, 11, 26.

[5] Mermin, N. D. (1968). Crystalline order in two dimensions, *Phys. Rev.*, 176, 250.

[6] Novoselov, K. S., Geim, A. K., Morozov, S. V., Jiang, D., Katsnelson, M. I., Grigorieva, I. V., Dubonos, S. V. and Firsov, A. A. (2005). Two-dimensional gas of massless Dirac fermions in graphene, *Nature*, 438/10.

[7] Kane, Ch. L. (2005). Erasing electron mass, *Nature*, 438, 168.

[8] Dato, A., Radmilovic, V., Lee, Z., Philips, J. and Frenklach, M. (2008). Substrate-free gas-phase synthesis of graphene sheets, *Nano Lett.*, 8(7), 2012.

[9] Lozovi, Yu. E.; Merkulov, S. P. and Sokolik, A. A. (2008). Collective electron phenomena in graphene, *Uspekhi Fiz. Nauk*, 178(7), 758.

[10] Ezawa, M. (2013). Photo-induced topological phase transition and single Dirac-cone state in silicene, *Phys. Rev. Lett.*, 110, 026603.

[11] Bludov, Yu. V., Peres, N. M. R. and Vasilevskiy, M. I. (2012). Graphene-based polaritonic crystal, *Phys. Rev. B* 85, 245409.

[12] Berestetzkii, V. B, Lifshitz E. M. and Pitaevskii, L. P. (1989). Quantum Electrodynamics, (Nauka, Moscow).

[13] Meyer, J. W. (1971). Covariant classical motion of electron in a laser beam, *Phys. Rev. D: Part. Fields*, 3(2), 621.

[14] Schwinger, J., Tsai, W. Y. and Erber, T. (1976). Classical and quantum theory of synergic synchrotron-Čerenkov radiation, *Ann. Phys. (NY)*, 96(2), 303.

[15] Bulla, C., *et al.*, (1996). Observation of nonlinear effects in Compton scattering, *Phys. Rev. Lett.*, 76, 3116.

[16] Pardy, M. (1998). The quantum field theory of laser acceleration, *Phys. Lett. A*, 243, 223.

[17] Pardy, M. (2001). The quantum electrodynamics of laser acceleration, *Radiat. Phys. Chem.*, 61, 321.

[18] Pardy, M. (2003). Electron in the ultrashort laser pulse, *Int. J. Theor. Phys.*, 42(1), 99.

[19] Pardy, M. (2004). Compton effect in dielectric medium, arXiv:hep-ph/0406008v1, 1 Jun.

[20] Sen Gupta, N. D. (1967). On the solution of the Dirac equation in the field of two beams of electromagnetic radiation, *Zeitschrift für Physik*, 200, 13.

[21] Pardy, M. (2006). Volkov solution for an electron in two wave fields, *Int. J. Theor. Phys.*, 45(3), 663.

[22] Lyulka, V. A. (1974). Quantum effects in intense electromagnetic fields, *Sov. Phys. JETP*, 40, 815.

[23] Lyulka, V. A. (1975). Decays of elementary particles in the field of of the intense electromagnetic wave, *Zh. Exp. Teor. Fiz.*, 69, 3(9), 800.

[24] Lyulka, V. A. (1977). Quantum effects in the field of the non-chromatic electromagnetic wave, *Zh. Exp. Teor. Fiz.*, 72(3), 865.

[25] Lyulka, V. A. (1985). Weak processes in the intense electron in the ultrashort laser pulse electromagnetic fields, *J. Nucl. Phys.*, 5(11), 1211.

[26] Sokolov, A. A., Loskutov, Yu. M. and Ternov, I. M. (1962). Quantum Mechanics, (GIMP, RSFSR, Moscow). (in Russian).

[27] Garbuny, M. (1965). Optical Physics, (Academic Press, New York, London).

[28] Crenshaw, M. E. (2004). Quantum optics of dielectrics in momentum-space, *Opt. Commun.*, 235, 153.

[29] Nakai, S. and Mima, K. (2004). Laser driven inertial fusion energy: present and perspective, *Rep. Prog. Phys.*, 67, 321.

Chapter 6

Nanoplatform Based on Vertical Nanographene

Mineo Hiramatsu, Hiroki Kondo and Masaru Hori

Additional information is available at the end of the chapter

http://dx.doi.org/10.5772/61318

Abstract

Self-organized graphite sheet nanostructures composed of graphene have been studied intensively. Carbon nanowalls and related sheet nanostructures are layered graphenes with open boundaries. The sheets form a self-supported network of wall structures with thicknesses in the range from a few nanometers to a few tens of nanometers, and with a high aspect ratio. The large surface area and sharp edges of carbon nanowalls could prove useful for a number of different applications. Fabrication techniques of carbon nanowalls and possible applications using carbon nanowalls as nanoplatform in the area of electrochemistry and tissue engineering have been described. Radical injection technique was successfully applied to fabricate straight and large-size monolithic carbon nanosheet. The structure of carbon nanowalls was controlled by changing the total pressure and input power. In addition, the structure of carbon nanowalls was modified by O_2 plasma etching and H_2O_2 treatment. Using carbon nanowalls as platform would be the most promising and important application. Carbon nanowalls were used as electrode to detect several biomolecules. In addition, carbon nanowalls were oxidized by the surface treatment using atmospheric pressure plasma, and proteins such as bovine serum albumin were immobilized on these surface. Moreover, carbon nanowalls were used as scaffold for cell culturing. The dependence of the cell culturing rates and morphological changes of HeLa cells on carbon nanowall scaffolds with different densities and wettability were systematically investigated. Nanoplatform based on vertical nanographene offers great promise for providing a new class of nanostructured electrodes for electrochemical sensing, biosensing and energy conversion applications.

Keywords: carbon nanowalls, vertical nanographene, nanoplatform, electrochemistry, biosensing

1. Introduction

It is well known that sp^2 carbon can lead to various kinds of layered structures. Among these structures, graphene (monolayer and few layers) is an actual two-dimensional material with the large anisotropy between the in-plane and out-of-plane directions. Planar graphene films with respect to the substrate have been synthesized by thermal decomposition of carbon-terminated silicon carbide and chemical vapor deposition (CVD) on metals such as nickel (Ni) and copper (Cu) substrates [1-3]. On the other hand, plasma-enhanced CVD (PECVD) is among the early methods to synthesize vertically standing carbon sheet structures [4-17]. These structures are called as carbon nanowalls (CNWs), carbon nanoflakes, and carbon nanosheets. CNWs and related nanocarbon structures consist of nanographene sheets standing vertically on a substrate. Figure 1 shows a schematic illustration of CNWs, where few-layer graphenes composed of nanographite domains form a self-supported network of wall structures. The mazelike architecture of CNWs with large-surface-area graphene planes and a high density of graphene edges would be useful as platform for electrochemical applications as well as tissue engineering such as scaffold for cell culturing [18-25].

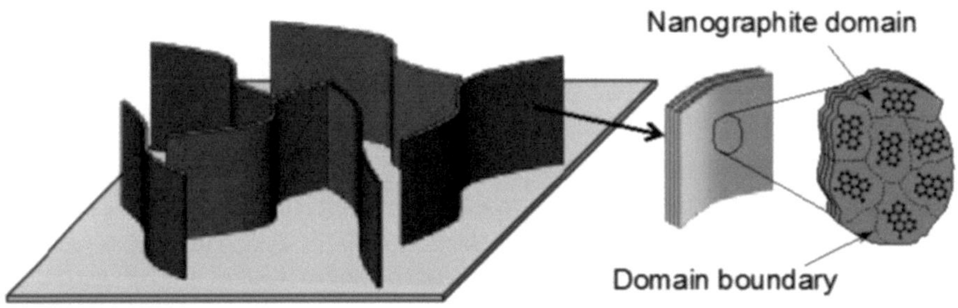

Figure 1. Schematic illustration of CNWs.

CNWs and related sheet nanostructures have been synthesized using several PECVD techniques, which are similar to those utilized for growing carbon nanotubes (CNTs) and diamond thin films. For the growth of CNWs, typically, a mixture of methane (CH_4) and hydrogen (H_2) is employed as source gases. A certain amount of hydrogen (H) atoms are required for growing CNWs. In general, microwave plasma and inductively coupled plasma (ICP) have been used for the growth of CNWs. These are high-density plasmas and are suitable for decomposing H_2 molecules efficiently. Or more specifically, radio frequency (rf) capacitively coupled plasma (CCP) with H radical injection and very high frequency (VHF) plasma with H radical injection have been applied to synthesize of CNWs. Pressures are ranging from 1 Pa to atmospheric pressure. Preparation of metal catalysts such as iron (Fe) and cobalt (Co) on the substrate is essential for the growth of CNTs. Unlike the CNT growth, CNWs do not require such catalysts for their nucleation. CNW growth has been conducted on several substrates including Si, SiO_2, and Al_2O_3 without the use of catalysts at substrate temperatures of 500-700°C [5]. In view of the practical use of CNWs for device applications such as biosensors or electrochemical sensors in micrototal analysis system, further investigations should be performed to enable control of structures and surface properties of CNWs.

In this chapter, fabrication techniques of CNWs and possible applications using CNWs as nanoplatform in the area of electrochemistry and tissue engineering are described. In the beginning, characterizations of CNWs are outlined. Then synthesis method for CNWs using VHF CCP with H radical injection is presented. Radical injection technique was successfully applied to fabricate straight and large-size monolithic carbon nanosheet. The VHF CCP with H radical injection was developed with the aim of achieving large-area growth of CNWs with a reasonable growth rate. The structure of CNWs was controlled by changing the total pressure and VHF power. In addition, the structure of CNWs was modified by O_2 plasma etching and hydrogen peroxide (H_2O_2) treatment.

In the latter half of this chapter, the electrochemical application of CNWs is described. Biosensing with CNWs is a promising application. Dopamine, ascorbic acid, and uric acid are compounds of great biomedical interest, which all are essential biomolecules in our body fluids. CNWs were used as electrode to detect these biomolecules. In addition, CNWs were oxidized by the surface treatment using atmospheric pressure plasma, and proteins such as bovine serum albumin were immobilized on these surface. Electrochemical properties of surface-decorated electrodes were investigated. Moreover, CNWs were used as scaffold for cell culturing. The dependence of the cell-culturing rates and morphological changes of HeLa cells on CNW scaffolds with different densities and wettability were systematically investigated.

2. Brief description of carbon nanowalls

CNWs are mazelike architecture consisting of few-layer graphenes standing vertically on a substrate, as was illustrated in Figure 1. The CNW sheet itself is composed of nanodomains of a few tens of nanometers in size. Scanning electron microscopy (SEM) images of CNWs with different morphology are shown in Figures 2(a) -2(d). The morphology of CNWs depends on the synthesis conditions, including pressure, substrate temperature, source gas mixtures, and the type of plasma used for the synthesis. Typical mazelike architecture (Figure 2(a)), isolated vertical nanosheets (Figure 2(b)), and highly branched type (Figure 2(c)) have been fabricated. Moreover, straight and aligned CNWs with regular spacing (Figure 2(d)) was obtained on the substrate set perpendicular to the electrode plate in the case of growth using rf CCP with H radical injection [9].

Figures 3(a) and 3(b) show typical transmission electron microscopy (TEM) images of CNW with a micrometer-high planar nanosheet structure, which was synthesized using electron beam excited plasma-enhanced CVD [12]. Despite the relatively smooth surface, each sheet in CNWs is actually composed of nanographite domains of a few tens of nanometers distinguished by domain boundaries as shown in Figure 3(a). Graphene layers are clearly observed in the high-resolution TEM image of the CNW shown in Figure 3(b). The spacing between neighboring graphene layers was approximately 0.34 nm.

Figure 4 shows a typical Raman spectrum of CNW film formed on Si substrate, which was measured at room temperature using a 514.5-nm line of an argon laser. Typical Raman spectrum for the CNWs has two strong peaks at 1590 cm^{-1} (G band), indicating the formation of graphitized structure and at 1350 cm^{-1} (D band) corresponding to the disorder-induced

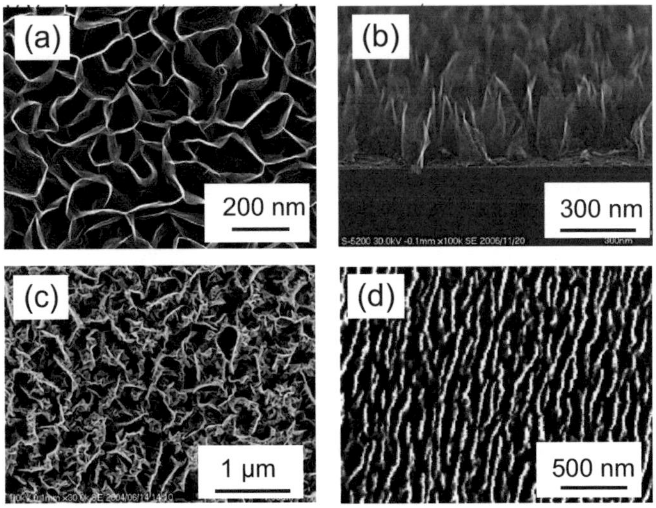

Figure 2. SEM images of CNWs with different morphologies.

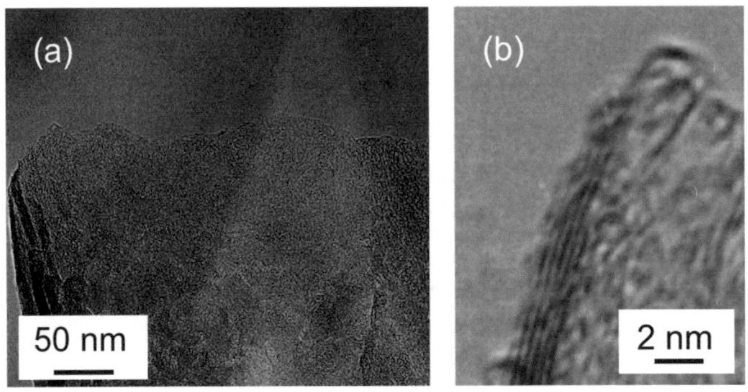

Figure 3. (a) TEM image of CNW grown using electron beam excited plasma-enhanced CVD and (b) high-resolution TEM image of CNW showing graphene layers [12].

phonon mode. The peak intensity of D band is comparable or twice as high as that of G band. The sharp and strong D band peak suggests a more nanocrystalline structure, and the presence of graphene edges and small graphite domains. It is noted that the G band peak is accompanied by a shoulder peak at 1620 cm^{-1}. This shoulder peak is often designated as D' band and associated with finite-size graphite crystals and graphene edges [26,27]. The strong and sharp D band peak and D' band peak are prevalent features of CNWs [8,11,13]. The 2D band peak at 2690 cm^{-1} is used to confirm the presence of few-layer graphene. It originates from a double resonance process that links phonons to the electronic band structure [28,29].

Carbon materials such as grassy carbon and conductive doped diamond have been widely used in both analytical and industrial electrochemistry due to their low cost, wide potential window, relatively inert electrochemistry, and electrocatalytic activity for a variety of redox

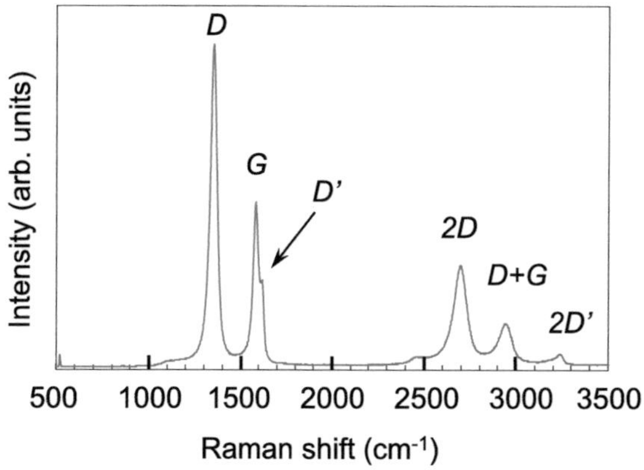

Figure 4. Typical Raman spectrum of CNWs.

reactions. For the electrochemical applications, these carbon-based electrodes are often decorated with catalyst nanoparticles such as platinum (Pt). As was illustrated in Figure 1, CNW film has many graphene edges, and the CNW sheet itself is composed of nanographite domains of a few tens of nanometers in size. These graphene edges and domain boundaries are chemically reactive and are modified easily with several types of surface termination, e.g., $C\text{-}NH_2$, C-OH, and C-COOH. Furthermore, Pt nanoparticles were preferably deposited on the defects such as grain boundaries on the surface of graphite [24]. Therefore, the structure of CNWs can be suitable for the platform of the electrochemical and biosensing applications. This kind of vertical-nanographene-based electrochemical platform with the high surface area and electrocatalytic activity offers great promise for creating revolutionary nanostructured electrodes for electrochemical sensing and biosensing, fuel cells and energy-conversion applications.

The morphology and electrical properties of CNW film depend on the synthesis conditions, including source gases, pressure, process temperature, and the type of plasma used for the growth of CNWs. We can expect a wide variety of applications based on their structure or electrical properties. For the electron emitters, CNWs with atomically thin edges, moderate spacing, and excellent height uniformity are required. For the membrane filter using honeycomb structure of CNWs, the spacing between adjacent nanowalls should be controlled. Moreover, CNWs should be detached from the substrate to obtain freestanding membrane, and the CNW membrane should be attached to the different materials. On the other hand, less aligned, dense CNW film with large surface area can be used as gas storage application, while vertical alignment and crystallinity of CNWs can be less crucial.

In view of the practical applications using CNWs, desirable structures and electrical and chemical properties of CNWs depend on the area of their applications. Therefore, structures, electrical properties, surface chemical properties of CNWs and related sheet nanostructures should be controlled according to their applications. Although the nucleation mechanism of CNWs is still unclear, ion bombardment on the substrate would have some effect on the nucleation of nanographene at the very early growth stage [30]. The growth of CNWs was enhanced occasionally by using metal substrates such as Ni and iron (Fe) [4]. The spacing between adjacent nanowalls and thickness of nanowalls would be affected by the density ratio of C_xH_y radicals to H atoms [31]. The addition of Ar into the source gas would induce the secondary nucleation at the wall surface, resulting in the formation of highly branched CNWs with high surface to volume ratio as shown in Figure 2(c). On one hand, branching could be suppressed and straight and large-size monolithic carbon nanosheet could be obtained by the addition of oxygen into the source gas [32]. In view of the practical use of CNWs for device applications such as biosensors or electrochemical sensors in the form of micrototal analysis system, postprocesses such as integration techniques, including etching and coating of CNWs and surface functionalization should be established. Figure 5 shows schematic illustration of CNW structures that should be controlled in the nucleation and growth stages and modified by the postprocesses, including etching and surface functionalization. Hereafter, we describe the recent development of CNW fabrication with emphasis on the structure control for realizing carbon nanoplatform working in the area of electrochemical and bio applications.

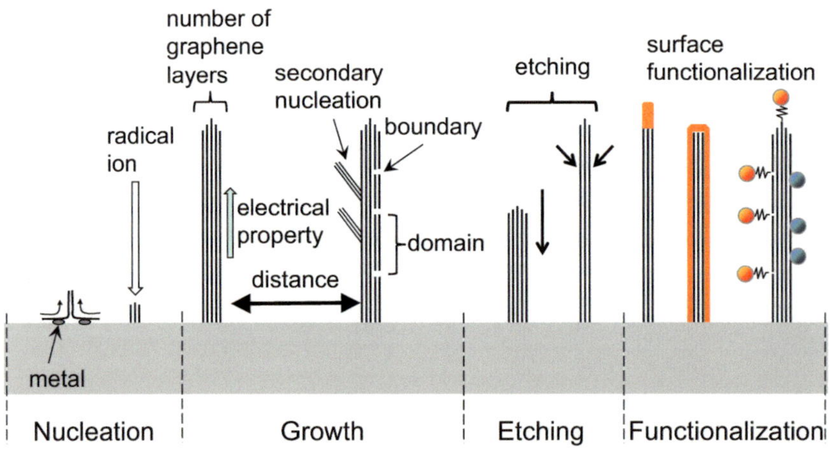

Figure 5. Schematic illustration of structures of CNWs to be controlled in the nucleation and growth stages and modified by the postprocesses, including etching and surface functionalization.

3. Fabrication of carbon nanowalls using Radical Injection Plasma Enhance Chemical Vapor Deposition (RI-PECVD)

3.1. Growth of carbon nanowalls from fluorocarbon/hydrogen mixture

In the case of PECVD with hydrocarbon/hydrogen system, for example, both CH_3 radical and H atoms are thought to play important roles in the formation of several carbon structures. The parallel-plate CCP might be useful to produce plenty of hydrocarbon radicals such as CH_3

radicals effectively [33] and also useful for the large-area deposition of the film. However, the CCP itself is not suitable for the growth of diamond and nanodiamond films because of the shortage of H atoms [34]. In contrast, high-density plasmas such as microwave plasma and ICP are suitable for dissociating H_2 molecules efficiently.

Although we have managed to control radical densities in the plasma by changing the mixing ratio of source gases, it is not easy to produce multiple species with different roles effectively at the same time using single plasma. As a solution, hydrocarbon or fluorocarbon gases were excited by a parallel-plate CCP, while the H atom density around the growing surface was actively increased by the injection from the external high-density H_2 plasma. This is the idea of radical injection. We have previously developed a radical-injection plasma-enhanced chemical vapor deposition (RI-PECVD) system that has allowed superior control of the properties of CNWs [5,9,10,14,15,32,35,36]. Figure 6 shows a schematic of the RI-PECVD system using very high frequency (VHF: 100MHz) plasma mainly, which was developed with the aim of achieving large-area growth of CNWs with a reasonable growth rate [10,11,14,15,32,35]. The RI-PECVD system used here is composed of a parallel plate VHF CCP and a surface wave-excited microwave (2.45 GHz) H_2 plasma (H_2 SWP) in tandem structure, as shown in Figure 6 [14,35]. In the case of CNW growth using fluorocarbon source gas, C_2F_6 was introduced into the VHF CCP region with a flow rate of 50 sccm. H_2 with a flow rate of 100 sccm was introduced into the microwave SWP region, and H radicals were injected into the VHF CCP. The total gas pressure ranged from 0.6 to 1.2 torr (80 to 160 Pa). During the CNW growth, the substrate was heated at 600°C. By using this system, the heated substrate was showered with both fluorocarbon (or hydrocarbon) radicals and plenty of H atoms in a controlled manner.

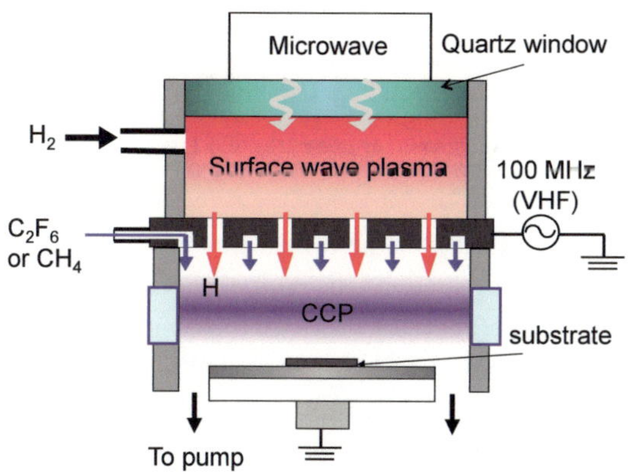

Figure 6. Schematic of VHF plasma assisted by H_2 microwave surface wave plasma [14,35].

Figure 7(a) shows the H and C atom densities in the VHF plasma of RI-PECVD employing C_2F_6/H_2 mixture, together with the height variation of CNW films. The measurement of atom densities was conducted using the vacuum ultraviolet absorption spectroscopy (VUVAS) system as a function of the total pressure during the CNW growth at microwave and VHF powers of 250 and 270 W, and C_2F_6 and H_2 flow rates of 50 and 100 sccm, respectively [35]. As the total pressure increased in the range from 0.1 to 0.6 torr (13.3 to 80 Pa), the H atom density increased drastically from 10^{12} to 10^{14} cm^{-3}, while the C atom density was almost constant at 5 × 10^{12} cm^{-3}, as shown in Figure 7(a). In contrast, the height of the CNW films decreased with an increase in the total pressure. SEM images of CNWs grown at different total pressures are shown in Figures 7(b) -7(d) [35]. CNW film with narrow interspaces was obtained at a low total pressure of 0.1 torr (13.3 Pa), while CNW film with wider interspaces of 30-40 nm was grown at a high pressure of 0.6 torr (80 Pa). As the H/C atom density ratio increased, the growth rate of the CNWs decreased and the average interspaces between the walls became wider.

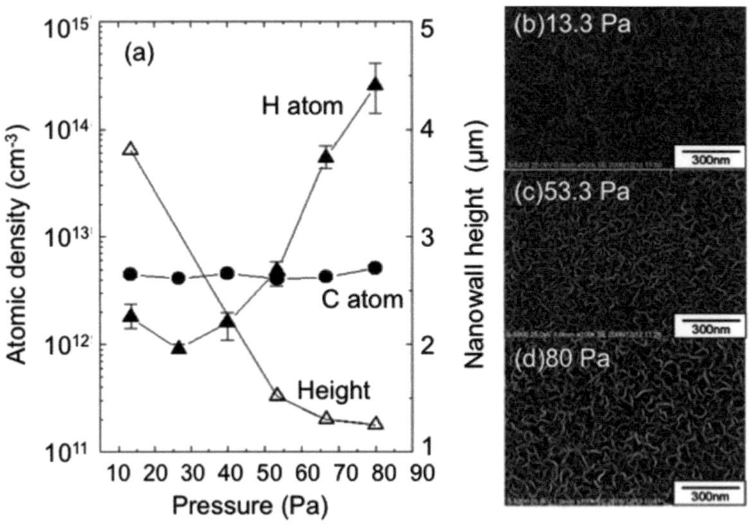

Figure 7. (a) H and C atom densities in VHF plasma employing C_2F_6/H_2 mixtures, together with the height variation of CNW films, as a function of total pressure during the CNW formation at a microwave power of 250 W and a VHF power of 270 W. (b-d) SEM images of CNWs grown at pressures of 0.1, 0.4, and 0.6 torr (13.3, 53.3 and 80 Pa), respectively [35].

In addition, O_2 and N_2 gases were introduced into the VHF-CCP region. The crystallinity of vertically standing CNWs is improved by introducing O_2 into the plasma used for CNW growth [32], while N addition is expected to control the electrical properties of the CNWs [35]. Figure 8(a) shows the cross-sectional SEM image of the typical CNW film grown on a Si substrate, and the inset shows the top view SEM image of the same CNW film. CNW growth was conducted on a Si substrate using a C_2F_6/H_2 mixture, resulting in the formation of slightly branching carbon sheets standing almost vertically on the substrate, as shown in Figure 8(a). The thickness of CNW film grown for 30 min was approximately 1 μm. Figure 8(b) shows TEM image of CNWs grown using C_2F_6/H_2, where randomly oriented, small overlapping multilayered graphene domains were observed. The inset shows magnified image of the area enclosed by square, where a bent multilayered graphene structure with a thickness of approximately 9 nm was observed. In contrast, Figure 8(c) shows a cross-sectional SEM image of the CNW film grown for 40 min using C_2F_6/H_2 with O_2 addition, and the inset shows the top view SEM image of the same CNW film. CNW films grown with O_2 had larger plane sheets with wider interspaces than those grown without O_2. By the addition of O_2 at a flow rate of 5 sccm into source gas mixture, less branching, monolithic graphene sheets were obtained, as shown in Figure 8(c). However, as a result of O_2 addition, the growth rate was reduced by approximately 33%. Figure 8(d) shows TEM image of CNWs grown with additional O_2, together with magnified image of square area as an inset. Monolithic self-sustaining graphene sheets larger than 200 nm in size were clearly observed in the CNWs grown with O_2. A highly orientated, smooth multilayered graphene structure was clearly observed in the inset of Figure 8(d) [32].

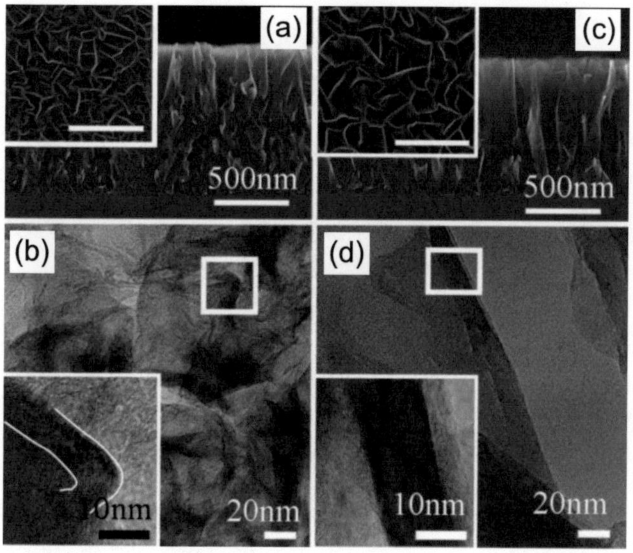

Figure 8. (a) Cross-sectional SEM image of CNW film grown on a Si substrate for 30 min using C_2F_6/H_2, together with SEM top view image of identical CNW film as an inset. (b) TEM image of bent CNWs grown using C_2F_6/H_2, together with magnified image of square area as an inset. (c) Cross-sectional SEM image of CNW film grown for 40 min using C_2F_6/H_2 with additional O_2 gas, together with SEM top view image of identical CNW film as an inset. (d) TEM image of straight CNWs grown with additional O_2, together with magnified image of square area as an inset [32].

Here, the morphology and structure of deposits formed using C_2F_6/H_2 in the early growth stage were investigated in detail. Figures 9(a)-9(d) show tilted SEM images of the deposits formed during the nucleation of CNW growth. At the very early stage of nucleation, as shown in Figure 9(a), nanoislands were formed on the Si substrate in 30 s. The density of nanoislands (number of nanoislands per area) increased with growth period. In 1 min, most of the surface of Si was covered with nanoislands (Figure 9(b)). The thickness of this layer composed of nanoislands was approximately 10 nm. At this moment, some nanoflakes have started to form at the aggregations of nanoislands forming the first layer. Subsequently, randomly oriented nanoflakes were formed on the first layer (Figure 9(c)). In 3 min, these sheet structures grow preferentially in a vertical direction to form vertical CNWs, while the number density of these nanoflakes was less than that observed at 2 min (Figure 9(d)). Figure 9(e) shows a cross-sectional TEM image of CNWs grown for 30 min, indicating that the interfacial layer exists between the CNWs and the Si surface. The thickness of the interface layer is approximately 10 nm, which is identical to the thickness of the first layer formed during the nucleation stage. Similar interface layer was also observed in the CNW films grown on Si and SiO_2 substrates using inductively coupled plasma (ICP) with $CH_4/H_2/Ar$ mixtures [13].

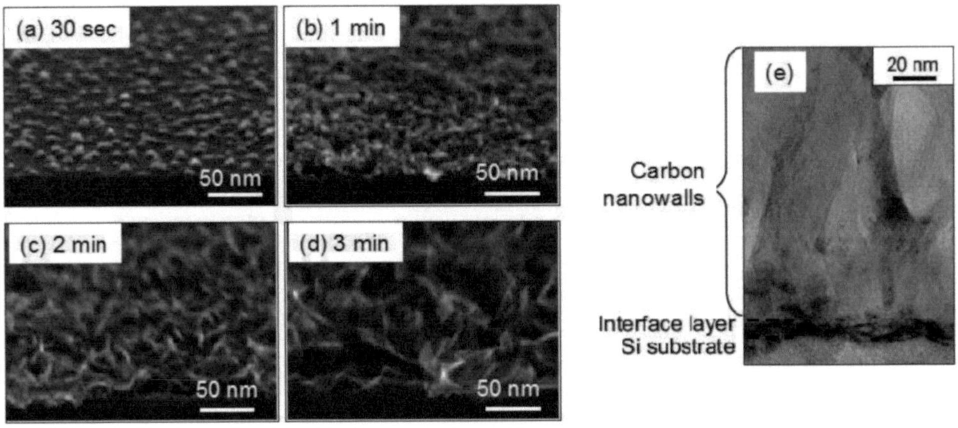

Figure 9. SEM images of the deposits formed on a Si substrate using a C_2F_6/H_2 system for (a) 30 s, (b) 1 min, (c) 2 min, and (d) 3 min. (e) Cross-sectional TEM image of CNWs and an interface layer synthesized for 30 min [14].

So far, several papers have been published on the observation of CNW growth in the early growth stage and the nucleation mechanism for the formation of vertical layered-graphenes on Si and SiO_2 substrates using various CVD methods [5,7,8,13-17,30]. It is common in previous studies that there is an induction period of 1-5 min for the nucleation of vertical nanographene. In addition, there exists an interface layer between the vertical nanographenes and substrate surface. Raman spectra were recorded for the deposits in the initial growth stages. D- and G-bands were not observed in the Raman spectra of nanoislands formed on the substrate for 1 min or less despite the fact that carbon was detected in these samples by X-ray photoelectron spectroscopy (XPS) analysis [14]. The nanoislands and the interface layer underlying two-dimensional nanographene are considered to be amorphous carbon. In most cases using several PECVD methods, the interface layer under the CNWs is considered to be an amorphous carbon [5,7,15,17,30,37]. Due to the existence of amorphous carbon interface layer, it is possible to grow CNWs and similar structures on a variety of substrates without catalyst. In contrast, Zhu et al. suggested that graphenes parallel to the substrate surface would grow at first. In their model, at the grain boundaries of these horizontal few-layer graphenes, spreading edge of the top layers of few-layer graphenes would curl upward, resulting in the vertical orientation of these sheets [8].

Figures 10(a)-10(d) show tilted SEM images of the deposits on the Si surface during the nucleation of CNW growth with O_2 addition. At 30 s, no deposits were observed on the surface [Figure 10(a)]. In 1 min, nanoislands were formed on the Si substrate [Figure 10(b)]. It took longer time to nucleate nanoislands in the case of the growth with O_2 gas addition, compared with the case without O_2 shown in Figure 9(a). In 2 min, some small two-dimensional nanoflakes have started to grow at isolated nanoislands, while the fractional surface coverage was low, as shown in Figure 10(c). A distinct interface layer was not formed. As shown in Figure 10(d), isolated wall structures had grown in 3 min, while the number density of start-up CNWs was lower than that without O_2,

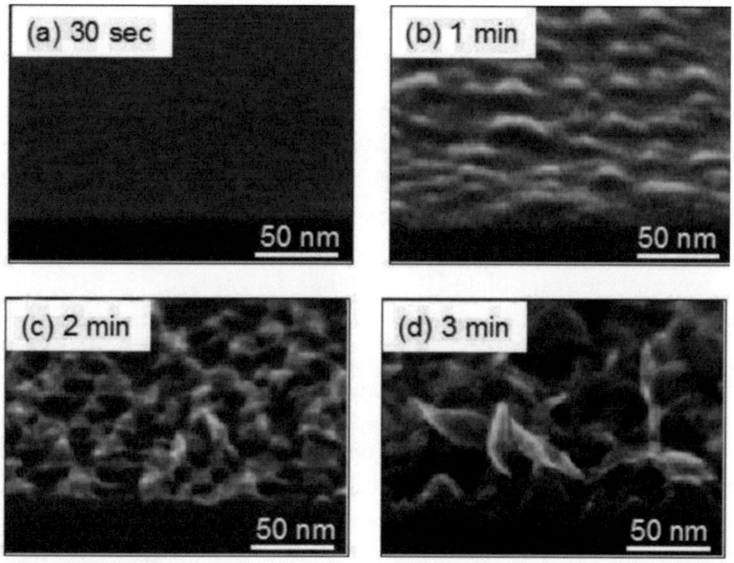

Figure 10. SEM images of the deposits formed on Si substrate using C_2F_6/H_2 with O_2 addition for (a) 30 s, (b) 1 min, (c) 2 min, and (d) 3 min [14].

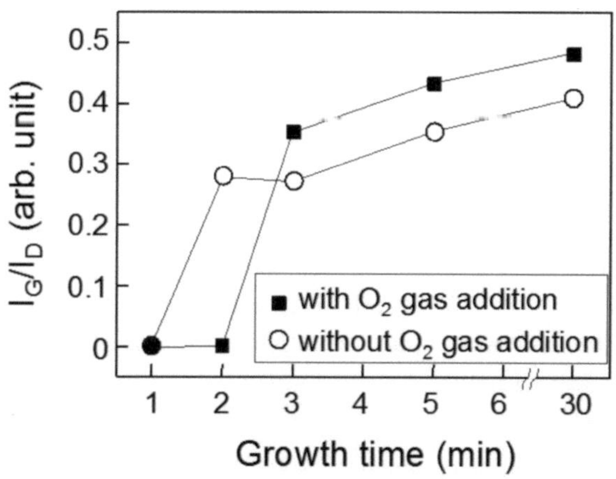

Figure 11. Temporal behaviors of I_G/I_D ratios of CNWs formed with and without O_2 gas addition [14].

Raman spectra were recorded for the deposits formed without and with O_2 addition in the initial growth stages. The intensity ratios of the G-band (I_G) to the D-band (I_D) of the deposits formed with and without O_2 gas addition as functions of growth period are shown in Figure 11. In the case of the CNW growth without O_2 addition, distinct G-band peak was observed in 2 min, and the I_G/I_D ratio increased gradually with increasing growth period, indicating that vertical nanographene formation started after the 2-min growth. In the case of the CNW growth with O_2 addition, on the other hand, distinct G-band peak was observed at 3 min, indicating that vertical nanographene formation started after 3 min. Moreover, the I_G/I_D ratio was higher than that for the CNWs synthesized without O_2, revealing that the O_2 gas addition is effective for obtaining highly graphitized CNWs.

In the process without O_2 gas addition, amorphous nanoislands were formed on the Si substrate, and the Si surface was completely covered with these nanoislands in the initial stage, resulting in the formation of a 10-nm-thick amorphous carbon interface layer. Vertical nanographene started to grow at nuclei on the surface of the interface layer. On the other hand, distinct interface layer was not formed in the process with O_2 gas addition, and vertical nanographenes were formed on isolated nanoislands. O_2 gas addition to C_2F_6/H_2 is effective in suppressing the formation of carbon nanoislands and thereby in controlling CNW nucleation.

3.2. Growth of carbon nanowalls from methane/hydrogen mixture

As is obvious, CNWs can also be fabricated employing the CH_4/H_2 mixture using RI-PECVD. In terms of controlling the wall density (or interspaces between adjacent nanowalls), total pressure and VHF power were changed. In these experiments, the heights of the CNWs were adjusted to 800 ± 50 nm, by varying the growth period. For all growth conditions, the films were uniform and exhibited a similar morphology. The thickness of individual CNWs in the films was approximately 10 nm. To consider what chemical species would affect the determination of wall density in these experiments, a plasma diagnosis was carried out using optical emission spectroscopy (OES). By introducing Ar gas into plasma region with a flow rate of 3 sccm, the actinometric measurements were carried out. Here, for determining the relative densities of H atoms and CH radicals, the emission intensity ratios ([CH]/[Ar] and [Hα]/[Ar]) were monitored by detecting the spectral lines associated with Hα 656.1 nm (excitation threshold energy, E = 12.1 eV), CH 431.2 nm (E = 14.6 eV), and Ar 751.4 nm (E = 13.27 eV).

Figure 12 shows top view and cross-sectional SEM images of CNWs grown on SiO_2 substrates at total pressures of (a) 1 Pa, (b) 3 Pa, and (c) 5 Pa under a constant VHF power of 300 W. As the total pressure increased, the wall density decreased or interspaces between adjacent nanowalls increased. Figure 12(d) shows the intensities of the CH and Hα emissions relative to Ar as a function of total pressure at a constant VHF power of 300 W. As the total pressure increased, [CH]/[Ar] decreased and [H]/[Ar] increased. Figure 13 shows top view and cross-sectional SEM images of CNWs grown on SiO_2 substrates at VHF powers of (a) 200 W, (b) 300 W, and (c) 400 W under a constant total pressure of 1 Pa. As the VHF power increased, the wall density increased or interspaces between adjacent nanowalls decreased. As shown in Figure 13(d), [CH]/[Ar] increased and [H]/[Ar] decreased with the increase of VHF power. It was found from the results shown in Figures 12 and 13 that the wall density could be controlled using the total pressure and the VHF power. The above results suggest that H and CH radicals are the important chemical species and the density ratio [CH]/[H] can be useful and simple index for controlling the wall density. Obviously, the OES provides information about only radicals of which optical emission transitions are permitted by selection rules. Other important carbon-containing species, including CH_3, CH_2, and C_2H_2 should be measured by other diagnostics such as absorption spectroscopy and mass spectrometry for further investigation on the growth mechanism.

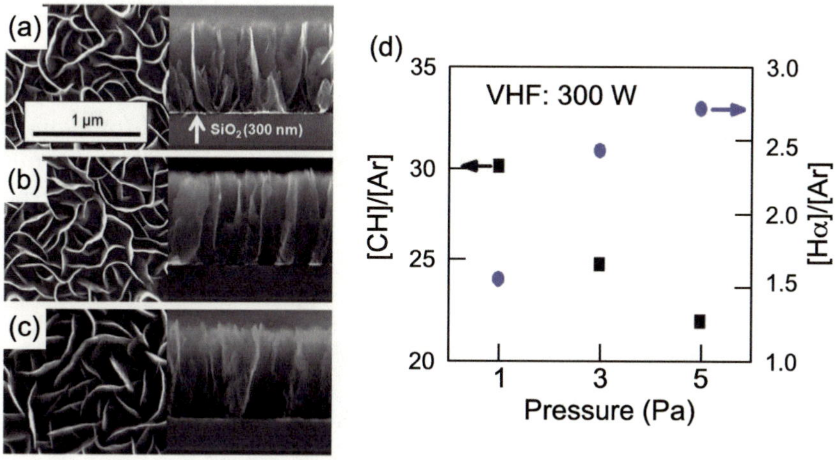

Figure 12. Top view and cross-sectional SEM images of CNWs grown on SiO_2 substrates at total pressures of (a) 1 Pa, (b) 3 Pa, and (c) 5 Pa at constant VHF power of 300 W. (d) [CH]/[Ar] and [H]/[Ar] ratios as a function of total pressure [31].

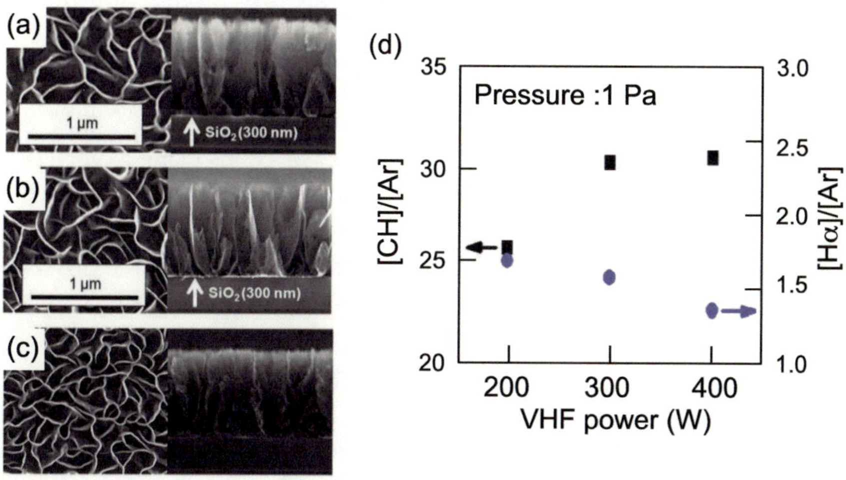

Figure 13. Top view and cross-sectional SEM images of CNWs grown on SiO_2 substrates at VHF powers of (a) 200 W, (b) 300 W, and (c) 400 W at constant total pressure of 1 Pa. (d) [CH]/[Ar] and [H]/[Ar] ratios as a function of VHF power [31].

3.3. Etching of carbon nanowalls

The selective oxidation techniques of the edges without modification of the graphene planes are essential for the fabrication of novel carbon nanoelectronic devices. A selective etching from the top edges of CNWs using remote oxygen plasma has been demonstrated [38]. CNW film samples were prepared on Si substrates for 30 min using the RI-PECVD system employing C_2F_6/H_2 as described in Section 3.1. The CNW film sample was exposed to oxygen atoms provided using remote ICP source, where two electrically grounded metal meshes were installed at the exit in order to remove irradiation of electrons and ions [38]. The CNW sample was placed on the heater stage 20 cm from the exit of the remote ICP.

Figures 14(a) and 14(b) show SEM images of CNW films before and after atomic oxygen etching, respectively. As a result of the atomic oxygen etching at 700°C for 5 min, the height of CNW film was reduced by approximately 160 nm, without change of wall thickness. In contrast, CNWs were not etched in the O_2 atmosphere without plasma at 700°C. These results indicate that atomic oxygen would react with the top edges of CNWs preferably without ion irradiations.

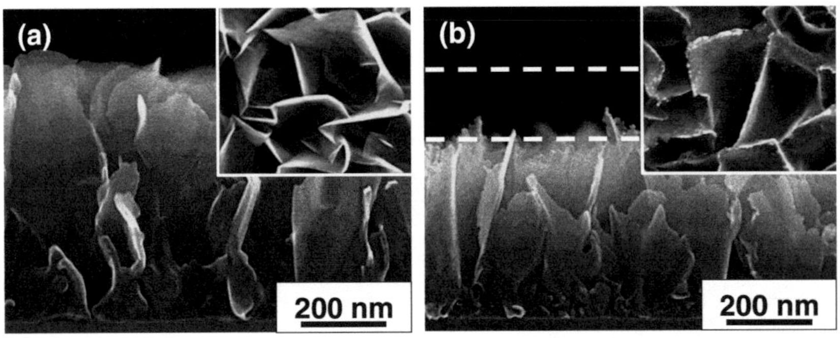

Figure 14. Cross-sectional SEM images of the CNW films (a) before and (b) after atomic oxygen etching at 700°C for 5 min. Insets show views from the top [38].

For comparison, we examined reactive ion etching (RIE) reactions using oxygen plasma. The RIE was carried out for 1 min using dual frequency (60 and 2 MHz) CCP system [38]. Figure 15 shows cross-sectional SEM images of CNW samples (a) before and (b) after oxygen RIE at 20°C for 1 min. As a result of the oxygen RIE, the height of the CNWs decreased drastically, and the top edges of the CNWs were sharpened and spearlike structures were formed.

Moreover, CNWs were subjected to hydrogen peroxide (H_2O_2) treatment [39]. It has been reported that H_2O_2 treatment can induce oxidative functional groups, such as hydroxyl groups on CNT surfaces, and can selectively oxidize disordered parts on the graphene surface [40]. Accordingly, H_2O_2 treatment has potential for modifying the surfaces of CNWs composed of nanographene domains. CNW film samples were prepared on Si substrates for 45 min using the RI-PECVD system employing C_2F_6/H_2 as described in Section 3.1. The CNW film samples were treated with 30% H_2O_2 solution for 6 and 12 h at 90°C. Then these samples were dried in air at 110°C on a hot plate. Figures 16(a) -16(c) show cross-sectional SEM images of CNWs

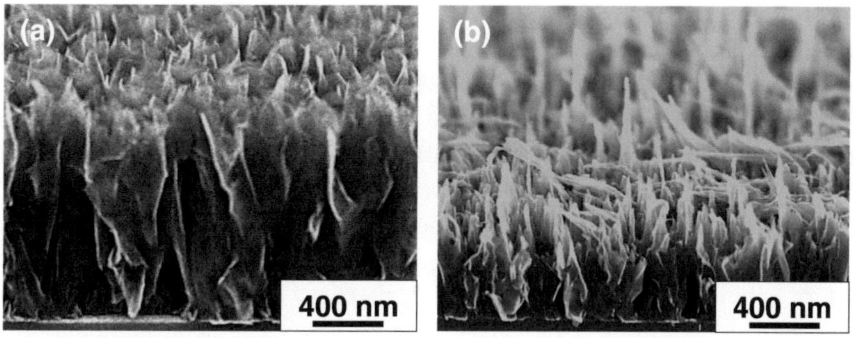

Figure 15. Cross-sectional SEM images of the CNW films (a) before and (b) after oxygen RIE for 1 min [38].

before and after the H_2O_2 treatment for 6 and 12 h. The magnified views of CNW sheets before and after the H_2O_2 treatment were shown in Figures 16(d) -16(f). As a result of H_2O_2 treatment, characteristic nanometer-scale asperities were formed on the wall surfaces of the CNWs as shown in Figures 16(d) -16(f), while the height of CNWs hardly changed. The size of the dents observed in Figures 16(e) and 16(f) was 20-30 nm. The morphology with nanometer-sized asperities on the surfaces of CNWs was stable after the H_2O_2 treatment. This result indicates that the radicals in H_2O_2 solution, such as hydroxyl radicals, react preferentially with the surfaces of CNWs at domain boundaries and induce the characteristic changes in their morphology.

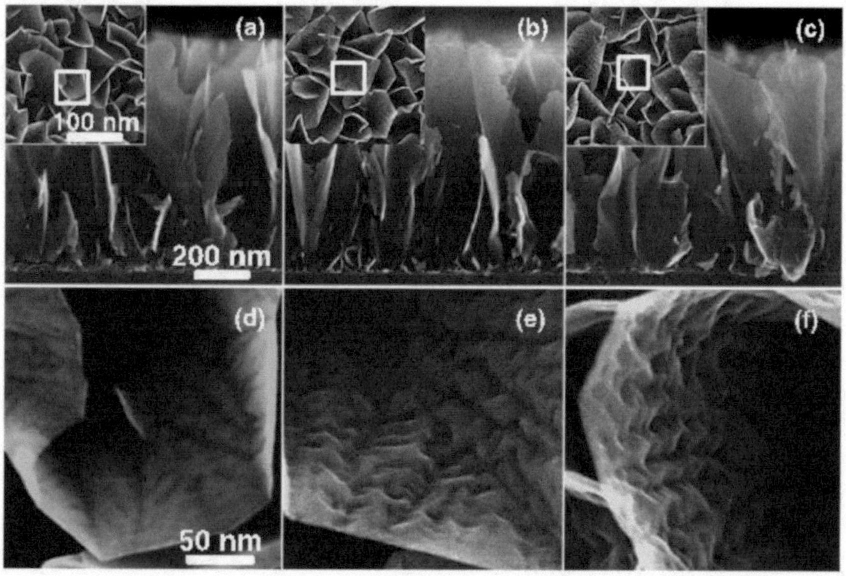

Figure 16. Cross-sectional SEM images of the CNWs (a) before and after H_2O_2 treatment for (b) 6 and (c) 12 h. Insets show top views. (d-f) Magnified views of the areas denoted by white squares in the insets [39].

In the case of atomic oxygen etching, CNWs are selectively etched from the top edges with almost no change in wall surface morphology, as shown in Figure 14(b). On the other hand, the H_2O_2 treatment induces the characteristic changes in their morphology with keeping the size of CNWs constant. The nanometer-scale asperities on the CNW surface increase the surface area, which would be useful as a platform for supporting metal nanoparticles and organopollutant degradation devices [24,41]. It is noted that such asperities could be reduced by O radical exposure after H_2O_2 treatment, resulting in the reduction of the thickness of CNW sheets. These results, including atomic oxygen etching, oxygen RIE, and H_2O_2 treatment suggest the possibility of realizing etching and thickness control of walls in CNWs, which should be essential for controlling the electrical properties of graphene materials and realizing their applications to electronic devices.

4. Applications of nanoplatform based on vertical nanographene

4.1. Platform for biosensing

Carbon materials have been widely used in both analytical and industrial electrochemistry due to their low cost, wide potential window, relatively inert electrochemistry, and electrocatalytic activity for a variety of redox reactions. Recently, graphene has proved to be an excellent nanomaterial for applications in electrochemistry. Graphene-based materials with large surface area are useful as electrodes for electrochemical sensors and biosensors [42-44]. Electrochemical activity of CNW electrode has been investigated by cyclic voltammetry measurements in an aqueous solution of ferrocyanide and a faster electron transfer between the electrolyte and the nanosheet surface has been demonstrated [21-23]. Dopamine (DA) is a hormone and neurotransmitter that plays a very important role in the human brain and body. Since the changes in the concentration of DA are closely linked to a human's health status, its detection has gained significant attention. Ascorbic acid (AA) and uric acid (UA) are also compounds of great biomedical interest, which all are essential biomolecules in our body fluids. Chemically reduced graphene oxide modified glassy carbon electrode was used to detect these neurotransmitters and biological molecules [42]. In these days, researches on the sensing of biological molecules became popular. Figure 17 shows examples of cyclic voltammogram responses of CNW electrode in the phosphate buffer solution (PBS) with UA, AA, and their mixture at 100 mV/s scan rate. Shang and coworkers demonstrated the excellent electrocatalytic activity of multilayer graphene nanoflakes in simultaneous determination of DA, AA, and UA in PBS [20].

Very recently, electrochemical glutamate biosensor for bioelectronic applications has been demonstrated using platinum (Pt)-functionalized graphene nanoplatelet prepared from graphene oxides [45]. Among the neurotransmitters detected by biosensors, L-glutamate is one of the most important in the mammalian central nervous system, playing a vital role in many physiological processes. The glutamate biosensor is based on the oxidation of glutamate in the presence of glutamate oxidase.

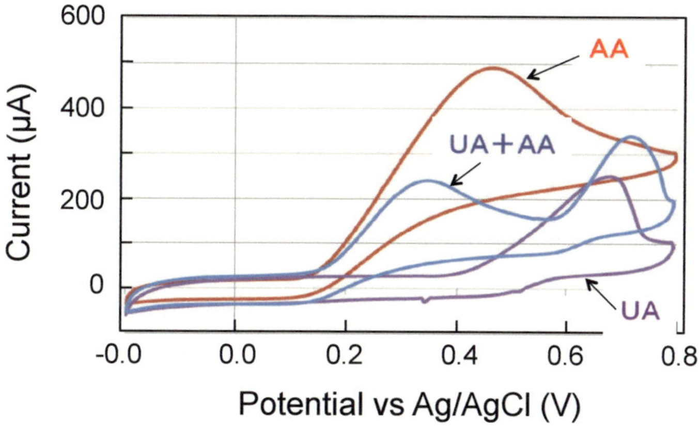

Figure 17. Cyclic voltammogram responses of CNW electrode in PBS with AA, UA, and their mixture at 100 mV/s scan rate.

$$\text{Glutamate} + O_2 + H_2O \rightarrow \alpha-\text{ketoglutarate} + NH_3 + H_2O_2 \quad (1)$$

The H_2O_2 produced in this reaction is electroactive at electrodes such as Pt, although it is inactive at many typical carbon-based electrodes. Therefore, it is necessary to add various electrocatalytic materials such as Pt nanoparticles, hydrous iridium oxide, Prussian blue, or peroxidase enzymes on the surface of carbon-based electrodes. As was illustrated in Figure 1, CNW film has many graphene edges on the top, and the CNW sheet itself is composed of nanodomains of a few tens of nanometers in size [46]. Pt nanoparticles were preferably deposited on the defects such as grain boundaries on the surface of graphite [24]. Therefore, the structure of CNWs can be suitable for the platform of the electrochemical and biosensing applications. In many cases, Pt nanoparticles have been prepared by the reduction of Pt salt precursors such as H_2PtCl_6 in solution. As an alternative approach to support the metal nanoparticles on the surface of carbon nanostructures, including dense CNTs and CNWs with narrow interspaces, we developed a new method of deposition using the supercritical carbon dioxide (sc-CO_2) as a solvent of metal-organic compounds. We demonstrated the preparation of dispersed Pt nanoparticles using the metal-organic chemical fluid deposition (MOCFD) employing supercritical fluid (SCF) [47-49]. TEM image of the surface of the CNW supporting Pt nanoparticles is shown in Figure 18(a). Pt nanoparticles were prepared on the CNWs by the SCF-MOCFD method for 30 min. The pressure and temperature of sc-CO_2 were 10 MPa and 130°C, respectively, and the temperature of CNWs was maintained at 180°C. Pt nanoparticles were prepared preferentially at the domain boundaries of CNW surface, as shown in Figure 18(a).

Furthermore, nanoparticles of the anatase phase of TiO_2 were prepared on the entire surface of CNWs by SCF-MOCFD method at a substrate temperature of 180°C. Figure 19 shows TEM image of the CNWs supporting TiO_2 nanoparticles after SCF-MOCFD for 30 min and the statistical distribution of the observed nanoparticle size [41]. For decomposing methylene blue

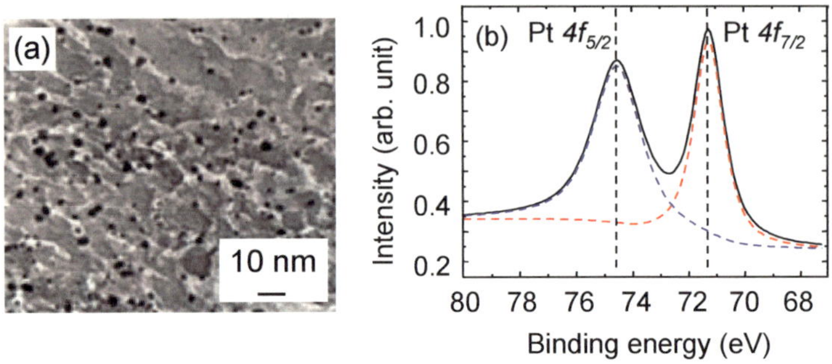

Figure 18. (a) TEM image of the surface of the CNW supporting Pt nanoparticles after the SCF-MOCFD for 30 min, and (b) XPS profile of the Pt 4f region of the Pt-supported CNW after SCF-MOCFD [50].

under ultraviolet irradiation, a high photocatalytic decomposition rate of 6 mg/h was obtained for 1 mg of TiO_2 supported on CNWs [41]. CNW-based electrochemical platform, which possesses large surface area with edges and electrochemical activity, offers great promise for providing a new class of nanostructured electrodes for electrochemical sensing and biosensing.

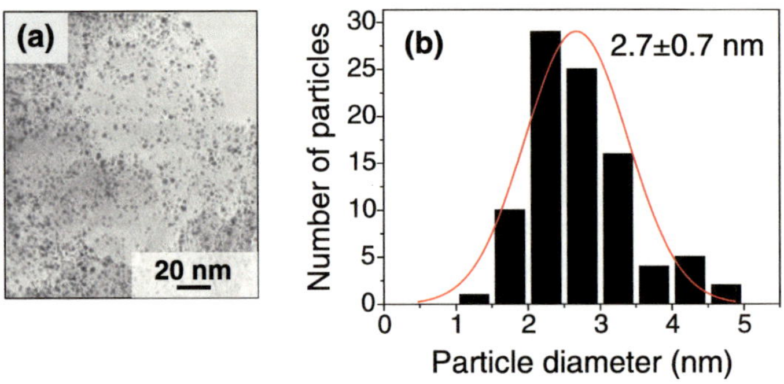

Figure 19. (a) TEM image of CNWs supporting TiO_2 nanoparticles after SCF-MOCFD for 30 min and (b) the obtained particle size distribution [41].

4.2. Surface chemical modification of carbon nanowalls for the wide range control of surface wettability

In view of the practical applications using CNWs in sensors and platforms for cell culturing, the effects of morphologies of CNWs and their surface properties should be considered systematically. We investigated the surface wettability of CNWs with emphasis on the chemisorption effect by postprocessing using plasma treatments [51,52]. Here, CNW samples grown from CH_4/H_2 mixtures on the Si substrate were chosen as primary forms.

The surface of as-grown CNWs from CH_4/H_2 mixture was terminated with H atoms. After the preparation of CNW film sample using RI-PECVD for 15 min, its surface was oxidized using Ar atmospheric pressure plasma for 1 to 30 s [51,52]. The distance between the CNW film sample and the atmospheric pressure plasma source was 5 mm. We expect that soft oxidation by oxygen radicals was realized, while the effect of ion bombardment on the surface was negligible during the exposure to atmospheric pressure plasma due to the very short mean free path of ions at the atmospheric pressure. For comparison, the surface of CNW sample was fluorinated for 5 s to add hydrophobic properties to the CNWs. For the fluorination treatment, the CNW sample was exposed to CF_4 plasma generated in the VHF-CCP region of RI-PECVD chamber without using H_2 SWP [51,52].

After the plasma surface treatments using the Ar atmospheric pressure plasma for oxidation and the CF_4 plasma for fluorination in short duration, no noticeable change was observed in the morphology of CNW samples. These results indicate that such short-duration plasma treatments would induce surface chemical functionalization without etching or deposition. Figure 20(a) shows the water contact angles (WCAs) on the CNWs before and after the Ar atmospheric pressure plasma treatment, as a function of plasma treatment duration, together with the WCA after the CF_4 plasma treatment for 5 s. The inset shows SEM image of CNW film sample after Ar atmospheric pressure plasma treatment for 5 s. CNW film samples examined in this experiment have all the same morphology. The side view photographic images of the water droplets on the CNWs before and after the plasma treatments are shown in Figures 20(b)-20(e). The WCAs in the case of diamond films are reported to be approximately 75° on the H-terminated surface and 65° on the O-terminated surface [26]. In contrast, the WCA on the surface of as-grown CNWs prepared employing CH_4/H_2 mixture was 51° [Figure 20(b)]. The surface of as-grown CNWs prepared with CH_4/H_2, of which edges and defects would be partially H-terminated, was rather hydrophilic. After the Ar atmospheric pressure plasma treatment for just 1 s, the WCA was reduced drastically to 11°. Then the WCAs decreased gradually with further increase of the Ar atmospheric pressure plasma treatment duration. As a result of Ar atmospheric pressure plasma treatment for 30 s, the WCA on the CNWs was 5°, indicating that the CNW surface was completely superhydrophilic [Figure 20(d)]. On the other hand, after the CNW sample was exposed to CF_4 plasma for 5 s, the WCA on the CNWs increased significantly to 147°, indicating that the surface of fluorinated CNWs was superhydrophobic [Figure 20(e)]. From these experiments, it was found that the surface wettability of CNW films could be controlled from superhydrophilic to superhydrophobic by the postplasma treatments without changing morphology.

X-ray photoelectron spectroscopy (XPS) measurements were carried out ex situ to analyze the CNW surface exposed to Ar atmospheric pressure plasma for oxidation. Figure 21 shows the composition ratio of O to C (O/C) at the surface of the CNWs as a function of plasma treatment duration. The composition ratio O/C was calculated from the peak intensity ratio of O 1s to C 1s corrected using the relative intensity factors. O content was detected even for the as-grown CNWs without plasma treatment, as shown in Figure 21. Because of ex situ XPS measurements, CNW surface was oxidized when exposed to the atmosphere. Due of the slight existence of oxygen at the surface of CNWs, the as-grown CNWs prepared from CH_4/H_2 would exhibit hydrophilic property as shown in Figures 20(a) and 20(b), in contrast to the hydrophobic surface of H-terminated diamond [53]. As the duration of Ar atmospheric pressure plasma treatment increased, the composition ratio O/C at the surface of CNWs increased rapidly at

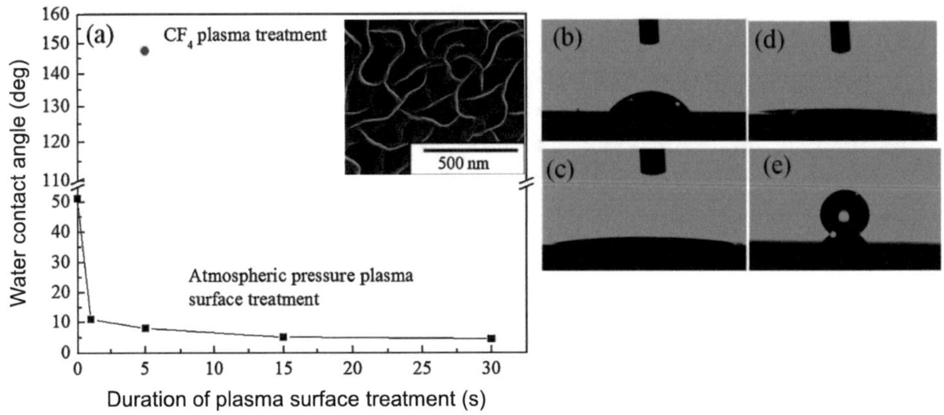

Figure 20. (a) WCAs on CNW films as a function of treatment duration using Ar atmospheric pressure plasma, together with WCA after CF_4 plasma treatment for 5 s. Inset shows SEM image of CNW sample after Ar atmospheric pressure plasma treatment for 5 s. Photos of water droplets on (b) as-grown CNWs, CNWs after Ar atmospheric pressure plasma treatment for (c) 5 s and (d) 30 s, and (e) CNWs after CF_4 plasma treatment for 5 s [52].

first, then very slowly from 5 s, and became almost constant after 15 s. C 1s photoelectron spectra after the plasma treatment were recorded (data not shown). There were various types of oxygen-related components in the CNWs after the Ar atmospheric pressure plasma surface treatment, although components related to the oxidized graphene were small. Therefore, the oxidation occurred only at the edges or surface defects, while the primary structure of CNWs has hardly been changed by the Ar atmospheric pressure plasma exposure.

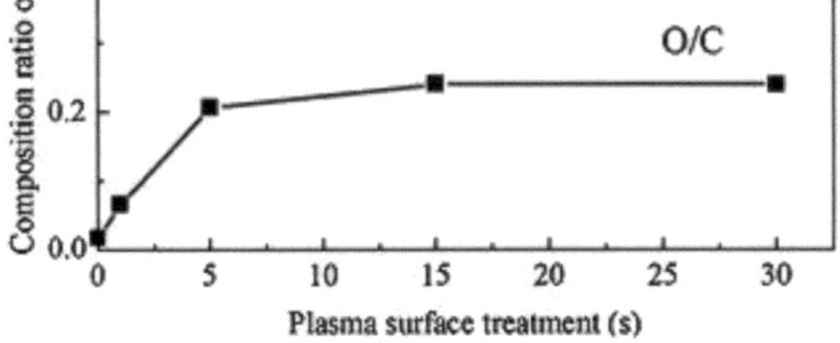

Figure 21. Composition ratio of O to C at the surface of CNWs evaluated from XPS results as a function of plasma treatment duration using the Ar atmospheric pressure plasma [52].

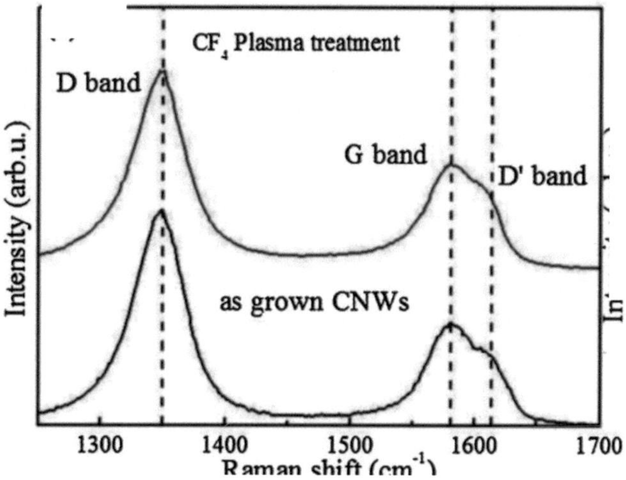

Figure 22. XPS C 1s spectrum of CNWs exposed to CF_4 plasma for 5 s [52].

As mentioned before, by exposure to CF_4 plasma for 5 s, the superhydrophobic surface of CNWs could be easily obtained from the as-grown H-terminated CNWs without changing the morphology of CNWs. The effect of CF_4 plasma treatment for surface fluorination was also investigated using XPS analysis. The composition ratio of F to C at the surface of the CNWs after CF_4 plasma treatment for 5 s was approximately 2.1. The composition ratio F/C was calculated from the ratio of the intensities of F 1s and C 1s peaks, corrected using the relative intensity factors. Figure 22 shows C 1s photoelectron spectrum of CNWs after the surface treatment using the CF_4 plasma for 5 s. The binding energy of 284.6 eV in the XPS spectrum of CNWs is attributed to the C—C (sp^2) bonds. The peaks at 289.1, 291.4, and 293.5 eV in the XPS spectrum shown in Figure 22 are assigned to the CF, CF_2, and CF_3 functional groups, respectively [54]. These three peaks in the XPS spectrum indicate that the F-terminated surface of CNWs was obtained by the CF_4 plasma surface treatment for 5 s, resulting in the realization of superhydrophobic surface.

4.3. Detection of protein using surface-modified carbon nanowalls as electrodes

Surface-oxidized CNW films were used as electrodes to detect bovine serum albumin (BSA) in phosphate-buffered solution (PBS). BSA, a serum albumin protein derived from cows, is often used as a protein concentration standard. CNWs were grown on SiO_2 substrates using RI-PECVD employing C_2F_6/H_2 mixture [5]. For the application of CNWs as an electrode of biosensor, the surface of CNW film was exposed to the Ar atmospheric pressure plasma for obtaining superhydrophilic surface. Electrochemical measurements were conducted using a standard three-electrode setup with an Ag/AgCl reference electrode and a Pt wire counter electrode. The cyclic voltammogram (CV) profiles of as-grown (bare) CNWs (500 nm in height), oxidized CNWs (500 nm), and oxidized CNWs of low height (350 nm) were recorded at scan rate of 100 mVs^{-1}. Figure 23 shows the CV profiles using these CNW electrodes in PBS containing BSA. In the CV profile measured using bare CNW electrode without the Ar atmospheric pressure plasma treatment, which had slightly hydrophilic surface, weak oxidation and reduction peaks were observed in anode peak potential at 0.2 V and cathodic peak potential at -0.3 V, respectively. In the CV profile using the typical oxidized CNW electrode, on the other hand, a broad oxidation and a high peak reduction currents were observed in anode peak potential of 0.2 V and cathodic peak potential at -0.75 V, respectively. The surface of as-grown CNW electrode could be easily modified into superhydrophilic one by the surface oxidation using the Ar atmospheric pressure plasma. In the case of oxidized CNW electrode with low height, the CV profile exhibited small peak currents due to the small surface area. The results in Figure 23 indicate that superhydrophilic surfaces of CNWs with large surface areas were useful as electrodes for biosensor.

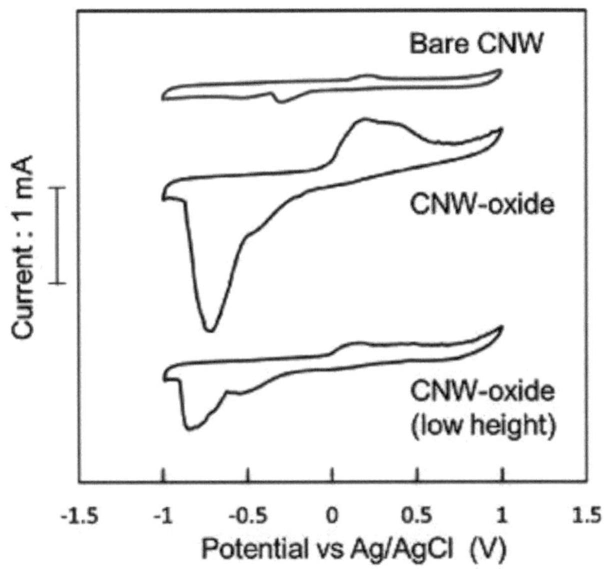

Figure 23. CV profiles of as-grown (bare) CNWs (500 nm in height), oxidized CNWs (500 nm), and oxidized CNWs of low height (350 nm) in PBS containing BSA. Scan rate: 100 mVs^{-1} [52].

4.4. Carbon nanowall scaffold to control culturing of cervical cancer cells

In recent years, cell culturing that uses carbon nanomaterials as scaffolds has been studied intensively [55-57]. The culturing rate is generally discussed with respect to the surface wettability of the scaffold. It has been reported that the cell-culturing rate would peak when the WCA on the scaffold surface is between 60° and 80° [58,59]. Moreover, many factors, including morphology, chemical termination, surface charge, scaffold surface stiffness, and the quantity of adsorbed protein are also essential for determining cell-culturing rates [60-63]. As mentioned in Section 4.2, the wide range control of surface wettability of CNWs was attained by postgrowth plasma treatments [52]. The unique features of CNWs and the variety of surface modification would give CNWs a high potential for scaffold application.

Here, the dependence of cell-culturing rate on the morphology and chemical termination of CNW scaffold was systematically investigated. Three types of CNW scaffolds with different densities (or different wall-to-wall distances) were prepared using RI-PECVD with CH_4/H_2 on quartz plates by changing the total pressure, CCP power, and growth period. The substrate temperature was 560°C. The flow rates of CH_4 and H_2 were fixed at 50 and 100 sccm, respectively. The total pressure was varied in the range of 1 to 5 Pa. The power applied to the SWP was 400 W and that to the CCP was changed in the range of 100 to 500 W. The growth period ranged from 8 to 80 min. Figures 24(a)-(c) show SEM images of the resulting CNW scaffolds with different densities. CNW scaffolds with average wall-to-wall distances of 95, 131, and 313 nm were obtained, which are denoted as "high-density [Figure 24(a)]", "medium-density [Figure 24(b)]", and "low-density [Figure 24(c)]" CNW scaffolds, respectively. The as-grown samples are denoted as H-terminated CNWs.

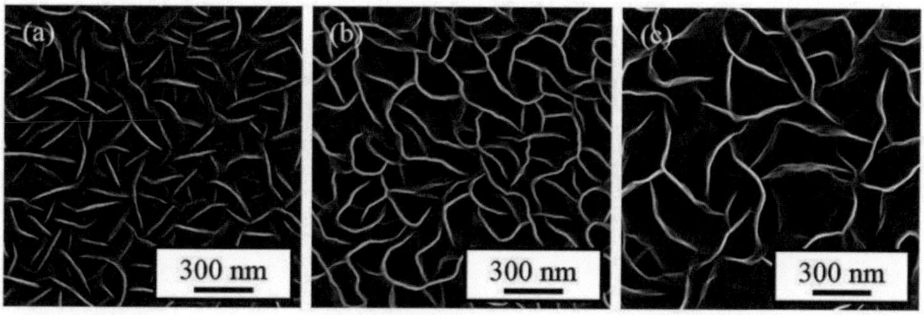

Figure 24. SEM images of (a) high-density CNW, (b) medium-density CNW, and (c) low-density CNW scaffolds prepared using RI-PECVD with CH_4/H_2 [25].

After the preparation of CNWs, some of them were subjected to various plasma treatments to realize the chemical termination of the edges and surfaces of CNWs. For oxygen termination, the CNW film was exposed to the atmospheric pressure plasma employing O_2 (50 sccm)/Ar (2000 sccm) at room temperature for 30 s (O-terminated CNWs). For fluorine termination, the CNW film was set in the VHF CCP employing CF_4 (36 sccm)/Ar (10 sccm) in the RI-PECVD system without SWP at room temperature for 5 s (F-terminated CNWs). The applied CCP power was 200 W, and the pressure was at 107 Pa during the plasma treatment. For nitrogen termination, the CNW film was set in the VHF-CCP region of the RI-PECVD system employing N_2 (12.5 sccm)/H_2 (37.5 sccm) at 560°C for 30 s (N-terminated CNWs). The applied SWP and CCP powers were 400 and 300 W, respectively. The pressure was 1 Pa. Prepared CNW scaffolds with different densities and terminations were put in multiwell cell-culturing plates. Cervical cancer cells (HeLa cells) at a density of 1.0×10^4 cell/cm^2 were seeded on each well. Incubation was conducted under a CO_2 (5%) atmosphere at 37°C for 96 h with 2 ml/well of the medium culture. The cells were maintained in a medium of minimum essential medium (MEM) Eagle, which consisted of 5 ml of L-glutamine (200 mM), 50 ml of fetal bovine serum (FBS), 5 ml of nonessential amino acids for MEM Eagle, and 5 ml of penicillin streptomycin. The HeLa cells cultured on the CNW scaffolds were picked up by using trypsin (0.5 w/v%, 5.3 mmol/l). After culturing for 24 h, the numbers of the cells with each of two different shapes were counted. One is the nonspreading cell with the circular shape and the other is the spreading cell with the noncircular shape [25].

Figure 25 shows the WCAs on the CNWs after various plasma treatments. The wettability of the CNWs depended on the types of chemical termination, not on the densities of the CNWs. The wettability of CNWs was controllable by plasma treatments in the range from superhydrophilic (WCA ≤ 10°) to near superhydrophobic (WCA ≥ 150°). The insets show the C 1s XPS profiles for the medium-density CNWs with (a) O-termination, (b) N-termination, (c) H-termination (as-grown), and (d) F-termination [52]. As shown in inset (a) of Figure 25, after the Ar/O_2 atmospheric pressure plasma treatment, a small broad peak related to C-O single bonding was observed at around 286.5 eV [64]. The composition ratio O/C was 0.21. After the N_2/H_2 plasma treatment, a broad peak tail was observed at around 285-287 eV, corresponding to nitrogen-related bondings [inset (b)] [65]. The composition ratio N/C was 0.08. In the case of as-grown CNWs, the composition ratio O/C was 0.05. A weak broad peak related to C-O single bonding was observed at around 286.5 eV [inset (c)]. Because of ex situ XPS measurements, CNW surface was oxidized when exposed to the atmosphere. Because of the slight existence of oxygen at the surface of CNWs, the as-grown CNWs exhibited slightly hydrophilic property. In contrast, after the CF_4/Ar plasma treatment, sharp peaks related to C-CF_X ($X \leq 3$) bonding structures were evident as shown in inset (d). The composition ratio F/C was estimated to be approximately 0.49.

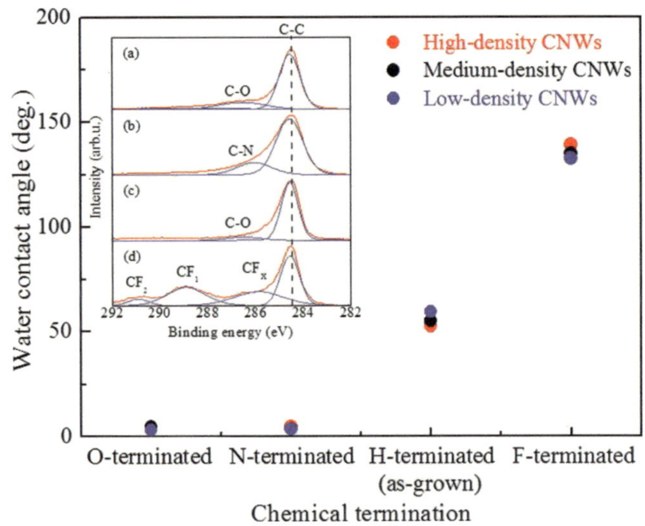

Figure 25. WCAs on the CNWs after various plasma treatments for different terminations. Inset shows C 1s XPS spectra of CNWs, in which the average wall-to-wall distance is 131 nm; (a) O-terminated, (b) N-terminated, (c) H-terminated (as-grown), and (d) F-terminated CNWs [25].

Optical microscope images of HeLa cells cultured on the CNW scaffolds with different terminations and wall densities after incubation for 24 h are shown in Figure 26 [25]. The numbers and morphologies of cultured HeLa cells somewhat depended on the chemical terminations as well as on the wall densities of CNW scaffolds. Figure 27 shows the numbers of HeLa cells cultured on the CNW scaffolds after incubation for 96 h as a function of WCA. For comparison, HeLa cells were also cultured on commercial glass plates, and the maximum number of cells was achieved at a WCA of 60°, as previously reported [66]. Similarly, on the medium density CNW scaffolds, the maximum number of cells was obtained at a WCA of 55°. In the case of low-density and high-density CNW scaffolds, however, the number of cells decreased and increased, respectively, with increasing WCAs. Hence, superhydrophilic surface is suitable for cell culturing on low-density CNW scaffolds, while superhydrophobic surface is suitable for high-density CNW scaffolds. As was shown in Figure 25, the surface wettability was nearly independent of the wall density. On the other hand, the cell-culturing rate was strongly dependent on the wall density of CNW scaffolds. These results suggest that the surface wettability is not dominant factor for determining the cell-culturing rate, while it is useful as approximate index of the expected cell-culturing rate. These experimental results indicate that the density of CNWs is the most essential factor for cell culturing rather than the surface wettability and types of chemical termination of CNW scaffolds. However, detailed mechanisms of cell/scaffold interactions in cell culturing on CNW scaffolds have not yet been clarified.

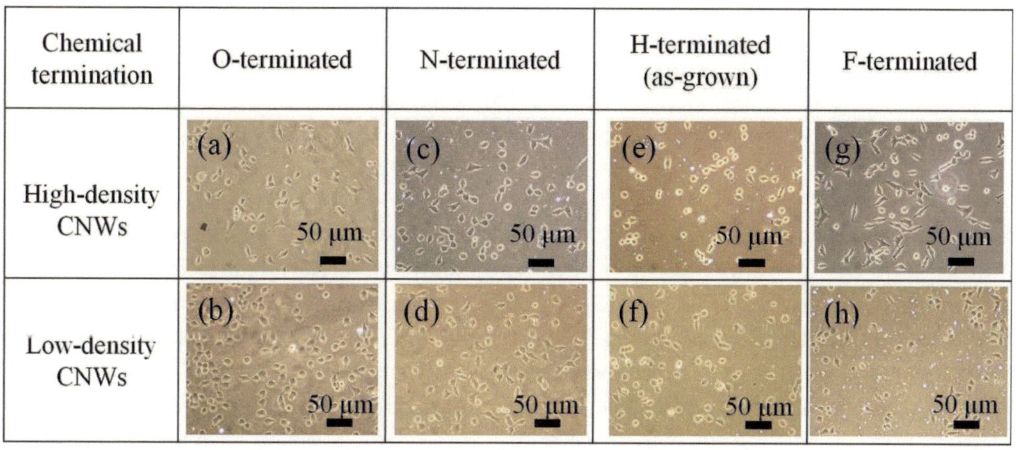

Figure 26. Optical microscope images of HeLa cells cultured on CNW scaffolds with different chemical terminations and wall densities after incubation for 24 h [25].

As a result of systematic investigation on the cell-culturing rates and morphological changes of HeLa cells on CNW scaffolds with respect to the wall densities and wettability of CNWs, it was found that the cell-culturing rates were significantly dependent on the CNW densities but seemed to be independent of the surface wettability of the CNW scaffolds. These results enable us to understand the detailed mechanisms of cell culturing on such scaffolds. Moreover, findings in the present study should also contribute to realize various nano-bioapplications using carbon nanomaterials.

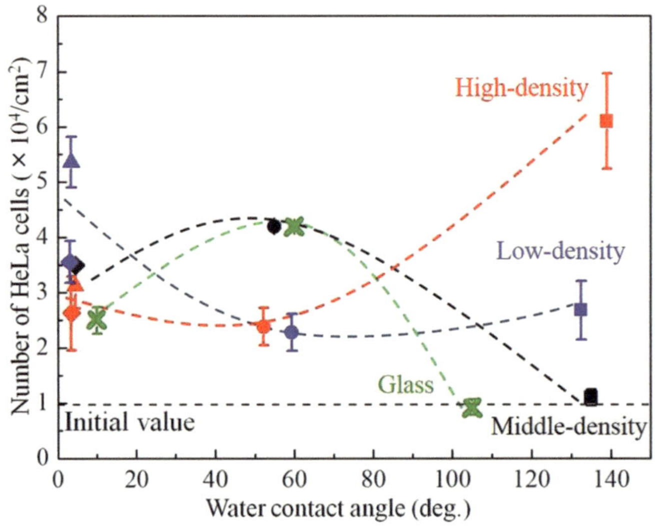

Figure 27. Number of HeLa cells cultured after incubation for 96 h as a function of the water contact angle. For comparison, data for a commercial glass substrate are also presented [25].

5. Conclusion

Carbon nanowalls and related nanocarbon structures composed of nanographene sheets standing vertically on a substrate have been studied intensively. The mazelike architecture of carbon nanowalls with large-surface-area graphene planes and a high density of graphene edges and domain boundaries could prove useful for a number of different applications. Fabrication techniques of carbon nanowalls and possible applications using carbon nanowalls as nanoplatform in the area of electrochemistry and tissue engineering have been described. A radical injection technique was successfully applied to fabricate straight and large-size monolithic carbon nanosheet. The morphology of carbon nanowalls was controlled by changing the total pressure and input power. In addition, the structure of carbon nanowalls was modified by O_2 plasma etching and H_2O_2 treatment.

Using carbon nanowalls as a platform would be the most promising and important application. Carbon nanowalls were used as electrode to detect several biomolecules. In addition, carbon nanowalls were oxidized by the surface treatment using the atmospheric pressure plasma, and proteins such as bovine serum albumin were immobilized on these surface. Moreover, carbon nanowalls were used as scaffold for cell culturing. The dependence of the cell-culturing rates and morphological changes of HeLa cells on carbon nanowall scaffolds with different densities and wettability were systematically investigated.

Nanoplatform based on vertical nanographene offers great promise for providing a new class of nanostructured electrodes for electrochemical sensing, biosensing, and energy conversion applications.

Author details

Mineo Hiramatsu[1*], Hiroki Kondo[2] and Masaru Hori[2]

*Address all correspondence to: mnhrmt@meijo-u.ac.jp

1 Meijo University, Japan

2 Nagoya University, Japan

References

[1] Berger, C.; Song, Z. M.; Li, X. B.; Wu, X. S.; Brown, N.; Naud, C.; Mayo, D.; Li, T. B.; Hass, J.; Marchenkov, A. N.; Conrad, E. H.; First, P. N. & de Heer, W.A. (2006). "Electronic confinement and coherence in patterned epitaxial graphene." *Science*, Vol. 312, No. 5777, pp. 1191-1196, DOI: 10.1126/science.1125925

[2] Yu, Q.; Lian, J.; Siriponglert, S.; Li, H.; Chen, Y. P. & Pei, S. S. (2008). "Graphene segregated on Ni surfaces and transferred to insulators." *Applied Physics Letters*, Vol. 93, Issue 11, 113103 (3 pages), DOI: 10.1063/1.2982585

[3] Li, X.; Cai, W.; An, J.; Kim, S.; Nah, J.; Yang, D.; Piner, R.; Velamakanni, A.; Jung, I.; Tutuc, E.; Banerjee, S.K.; Colombo, L. & Ruoff, R.S. (2009). "Large-area synthesis of high-quality and uniform graphene films on copper foils." *Science*, Vol. 324, No. 5932, pp. 1312-1314, DOI: 10.1126/science.1171245

[4] Wu, Y. H.; Qiao, P. W.; Chong, T. C. & Shen, Z. X. (2002). "Carbon nanowalls grown by microwave plasma enhanced chemical vapor deposition." *Advanced Materials*, Vol. 14, Issue 1, pp. 64-67, DOI: 10.1002/1521-4095(20020104)14:1<64::AID- ADMA64>3.0.CO;2-G

[5] Hiramatsu, M.; Shiji, K.; Amano, H. & Hori, M. (2004). "Fabrication of vertically aligned carbon nanowalls using capacitively coupled plasma-enhanced chemical vapor deposition assisted by hydrogen radical injection." *Applied Physics Letters*, Vol. 84, Issue 23, 4708 (3 pages), DOI: 10.1063/1.1762702

[6] Wang, J. J.; Zhu, M. Y.; Outlaw, R. A.; Zhao, X.; Manos, D. M.; Holloway, B. C. & Mammana, V. P. (2004). "Free-standing subnanometer graphite sheets." *Applied Physics Letters*, Vol. 85, Issue 7, 1265 (3 pages), DIO: 10.1063/1.1782253

[7] Zhao, X.; Outlaw, R. A.; Wang, J. J.; Zhu, M. Y.; Smith, G. D. & Holloway, B. C. (2006). "Thermal desorption of hydrogen from carbon nanosheets." *Journal of Chemical Physics,* Vol. 124, Issue 19, 194704 (6 pages), DOI: 10.1063/1.2187969

[8] Zhu, M.; Wang, J.; Holloway, B. C.; Outlaw, R. A.; Zhao, X.; Hou, K.; Shutthanandan, V. & Manos, D. M. (2007). "A mechanism for carbon nanosheet formation." *Carbon,* Vol. 45, Issue 11, pp. 2229-2234, DOI: 10.1016/j.carbon.2007.06.017

[9] Hiramatsu, M. & Hori, M. (2006). "Fabrication of carbon nanowalls using novel plasma processing." *Japanese Journal of Applied Physics,* Vol. 45, pp. 5522-5527, DOI: 10.1143/JJAP.45.5522

[10] Kondo, S.; Hori, M.; Yamakawa, K.; Den, S.; Kano, H. & Hiramatsu, M. (2008). "Highly reliable growth process of carbon nanowalls using radical injection plasma-enhanced chemical vapor deposition." *Journal of Vacuum Science & Technology* B, Vol. 26, Issue 4, pp. 1294-1300, DOI: 10.1116/1.2938397

[11] Takeuchi, W.; Ura, M.; Hiramatsu, M.; Tokuda, Y.; Kano, H. & Hori, M. (2008). "Electrical conduction control of carbon nanowalls." *Applied Physics Letters,* Vol. 92, 213103 (3 pages), DOI: 10.1063/1.2936850

[12] Mori, T.; Hiramatsu, M.; Yamakawa, K.; Takeda, K. & Hori, M. (2008). "Fabrication of carbon nanowalls using electron beam excited plasma-enhanced chemical vapor deposition." *Diamond and Related Materials,* Vol. 17, Issues 7-10, pp. 1513-1517, DOI: 10.1016/j.diamond.2008.01.070

[13] Hiramatsu, M.; Nihashi, Y.; Kondo, H. & Hori, M. (2013). "Nucleation control of carbon nanowalls using inductively coupled plasma-enhanced chemical vapor deposition," *Japanese Journal of Applied Physics,* Vol. 45, 01AK05 (6 pages), DOI: 10.7567/JJAP.52.01AK05

[14] Kondo, S.; Kawai, S.; Takeuchi, W.; Yamakawa, K.; Den, S.; Kano, H.; Hiramatsu, M. & Hori, M. (2009). "Initial growth process of carbon nanowalls synthesized by radical injection plasma-enhanced chemical vapor deposition." *Journal of Applied Physics,* Vol. 106, Issue 9, 094302 (6 pages), DIO: 10.1063/1.3253734

[15] Kawai, S.; Kondo, S.; Takeuchi, W.; Kondo, H.; Hiramatsu, M. & Hori, M. (2010). "Optical properties of evolutionary grown layers of carbon nanowalls analyzed by spectroscopic ellipsometry." *Japanese Journal of Applied Physics,* Vol. 49, 060220, DOI: 10.1143/JJAP.49.060220

[16] Seo, D. H.; Kumar, S. & Ostrikov, K. (2011). "Control of morphology and electrical properties of self-organized graphenes in a plasma." *Carbon,* Vol. 49, Issue 13, pp. 4331-4339, DOI: 10.1016/j.carbon.2011.06.004

[17] Krivchenko, V. A.; Dvorkin, V. V.; Dzbanovsky, N. N.; Timofeyev, M. A.; Stepanov, A.S.; Rakhimov, A. T.; Suetin, N. V.; Vilkov, O. Y. & Yashina, L. V. (2012). "Evolution of carbon film structure during its catalyst-free growth in the plasma of direct cur-

rent glow discharge." *Carbon,* Vol. 50, Issue 4, pp. 1477-1487, DOI: 10.1016/j.carbon.2011.11.018

[18] Yang, B.J.; Wu, Y.H.; Zong, B.Y. & Shen, Z.X. (2002). "Electrochemical synthesis and characterization of magnetic nanoparticles on carbon nanowall templates." *Nano Letters,* Vol. 2, No. 7, pp. 751-754, DOI: 10.1021/nl025572r

[19] Giorgi, L.; Makris, T.D.; Giorgi, R.; Lisi, N. & Salernitano, E. (2007). "Electrochemical properties of carbon nanowalls synthesized by HF-CVD." *Sensors and Actuators B: Chemical,* Vol. 126, Issue 1, pp. 144-152, DOI: 10.1016/j.snb.2006.11.018

[20] Shang, N. G.; Papakonstantinou, P.; McMullan, M.; Chu, M.; Stamboulis, A.; Potenza, A.; Dhesi, S. S. & Marchetto, H. (2008). "Catalyst-free efficient growth, orientation and biosensing properties of multilayer graphene nanoflake films with sharp edge planes." *Advanced Functional Materials,* Vo.18, Issue 21, pp. 3506-3514, DOI: 10.1002/adfm.200800951

[21] Luais, E.; Boujtia, M.; Gohier, A.; Tailleur, A.; Casimirius, S.; Djouadi, M.A.; Granier, A. & Tessier, P.Y. (2009). "Carbon nanowalls as material for electrochemical transducers." *Applied Physics Letters,* Vol. 95, Issue 1, 014104 (3 pages), DOI: 10.1063/1.3170033

[22] Tanaike, O.; Kitada, N.; Yoshimura, H.; Hatori, H.; Kojima, K. & Tachibana, M. (2009). "Lithium insertion behavior of carbon nanowalls by dc plasma CVD and its heat-treatment effect." *Solid State Ionics,* Vol. 180, Issues 4-5, pp. 381-385, DOI: 10.1016/j.ssi.2009.01.012

[23] Wang, Z.; Shoji, M. & Ogata, H. (2011). "Carbon nanosheets by microwave plasma enhanced chemical vapor deposition in CH_4-Ar system." *Applied Surface Science,* Vol. 257, Issue 21, pp. 9082-9085, DOI: 10.1016/j.apsusc.2011.05.104

[24] Mase, K.; Kondo, H.; Kondo, S.; Hori, M.; Hiramatsu, M. & Kano, H. (2011). "Formation and mechanism of ultrahigh density platinum nanoparticles on vertically grown graphene sheets by metal-organic chemical supercritical fluid deposition." *Applied Physics Letters,* Vol. 98, Issue 19, 193108 (3 pages), DOI: 10.1063/1.3583672

[25] Watanabe, H.; Kondo, H.; Okamoto, Y.; Hiramatsu, M.; Sekine, M.; Baba, Y. & Hori, M. (2014). "Carbon nanowall scaffold to control culturing of cervical cancer cells." *Applied Physics Letters,* Vol. 105, Issue 24, 244105 (4 pages), DOI: 10.1063/1.4902054

[26] Nemanich, R. J. & Solin, S. A. (1979). "First- and second-order Raman scattering from finite-size crystals of graphite." *Physical Review* B, Vol. 20, Issue 2, pp. 392-401, DOI: 10.1103/PhysRevB.20.392

[27] Kurita, S.; Yoshimura, A.; Kawamoto, H.; Uchida, T.; Kojima, K.; Tachibana, M.; Molina-Morales, P. & Nakai H. (2005). "Raman spectra of carbon nanowalls grown by plasma-enhanced chemical vapor deposition." *Journal of Applied Physics,* Vol. 97, Issue 10, 104320 (5 pages), DOI: 10.1063/1.1900297

[28] Ferrari, A. C.; Meyer, J. C.; Scardaci, V.; Casiraghi, C.; Lazzeri, M.; Mauri, F.; Piscanec, S.; Jiang, D.; Novoselov, K. S.; Roth, S. & Geim, A. K. (2006). "Raman spectrum of graphene and graphene layers." *Physical Review Letters,* Vol. 97, 187401 (4 pages), DOI: 10.1103/PhysRevLett.97.187401

[29] Ferrari, A. C. (2007). "Raman spectroscopy of graphene and graphite: disorder, electron-phonon coupling, doping and nonadiabatic effects." *Solid State Communications,* Vol. 143, Issues 1-2, pp. 47-57, DOI: 10.1016/j.ssc.2007.03.052

[30] Kondo, S.; Kondo, H.; Hiramatsu, M.; Sekine, M. & Hori, M. (2010). "Critical factors for nucleation and vertical growth of two dimensional nano-graphene sheets employing a novel Ar+ beam with hydrogen and fluorocarbon radical injection." *Applied Physics Express,* Vol. 3, 045102 (3 pages), DOI: 10.1143/APEX.3.045102

[31] Cho, H. J.; Kondo, H.; Ishikawa, K.; Sekine, M.; Hiramatsu, M. & Hori, M. (2014). "Density control of carbon nanowalls grown by CH_4/H_2 plasma and their electrical properties." *Carbon,* Vol. 68, pp. 380-388, DOI: 10.1016/j.carbon.2013.11.014

[32] Takeuchi, W.; Takeda, K.; Hiramatsu, M.; Tokuda, Y.; Kano, H.; Kimura, S.; Sakata, O.; Tajiri, H. & Hori, M. (2010). "Monolithic self-sustaining nanographeme sheet grown using plasma-enhanced chemical vapor deposition." *Physica Status Solidi (a),* Vol. 207, Issue 1, pp. 139-143, DOI: 10.1002/pssa.200925230

[33] Naito, S.; Ikeda, M.; Ito, N.; Hattori, S. & Goto, T. (1993). "Effect of rare gas dilution on CH_3 radical density in RF-discharge CH_4 plasma"', *Japanese Journal of Applied Physics,* Vol. 32, Issue 12R, pp. 5721-5725, DOI: 10.1143/JJAP.32.5721

[34] Tachibana, K. (1994). "Detection of H atoms in RF-discharge SiH_4, CH_4 and H_2 plasmas by two-photon absorption laser-induced fluorescence spectroscopy." *Japanese Journal of Applied Physics,* Vol. 33, No. 7B, pp. 4329-4334, DOI: 10.1143/JJAP.33.4329

[35] Takeuchi, W.; Sasaki, H.; Kato, S.; Takashima, S.; Hiramatsu, M. & Hori, M. (2009). "Development of measurement technique for carbon atoms employing vacuum ultraviolet absorption spectroscopy with a microdischarge hollow-cathode lamp and its application to diagnostics of nanographene sheet material formation plasmas." *Journal of Applied Physics,* Vol. 105, 113305 (6 pages), DOI: 10.1063/1.3091279

[36] Hori, M.; Kondo, H. & Hiramatsu, M. (2011). "Radical-controlled plasma processing for nanofabrication." Journal of Physics D: Applied Physics, Vol. 44, 174027 (15 pages), DOI: 10.1088/0022-3727/44/17/174027

[37] Kondo, S.; Kondo, H.; Miyawaki, Y.; Sasaki, H.; Kano, H.; Hiramatsu, M. & Hori, M. (2011). "Reactive ion etching of carbon nanowalls." Japanese Journal of Applied Physics, Vol. 50, 075101 (7 pages), DOI: 10.1143/JJAP.50.075101

[38] Shimoeda, H.; Kondo, H.; Ishikawa, K.; Hiramatsu, M.; Sekine, M. & Hori, M. (2013). "Atomic oxygen etching from the top edges of carbon nanowalls." Applied Physics Express, Vol. 6, 095201 (4 pages), DOI: 10.7567/APEX.6.095201

[39] Shimoeda, H.; Kondo, H.; Ishikawa, K.; Hiramatsu, M.; Sekine, M. & Hori, M. (2014). "Nanostructure modification to carbon nanowall surface employing hydrogen peroxide solution." Japanese Journal of Applied Physics, Vol. 53, 040305 (4 pages), DOI: 10.7567/JJAP.53.040305

[40] Simon, F.; Kukovecz, A. & and Kuzmany, H. (2003). "Controlled oxidation of single-wall carbon nanotubes: a Raman study," AIP Conference Proceedings Vol. 685, p. 185, DOI: 10.1063/1.1628014

[41] Horibe, T.; Kondo, H.; Ishikawa, K.; Kano, H.; Sekine, M.; Hiramatsu, M. & Hori, M. (2013). "Supercritical fluid deposition of high-density nanoparticles of photocatalytic TiO2 on carbon nanowalls." Applied Physics Express, Vol. 6, 045103 (3 pages), DOI: 10.7567/APEX.6.045103

[42] Zhou, M.; Zhai; Y. & Dong, S. (2009). "Electrochemical sensing and biosensing platform based on chemically reduced graphene oxide." Analytical Chemistry, Vol. 81, Issue 14, pp. 5603-5613, DOI: 10.1021/ac900136z

[43] Pumera, M.; Ambrosi, A.; Bonanni, A.; Chng, E. L. K. & Poh, H. L. (2010). "Graphene for electrochemical sensing and biosensing." Trends in Analytical Chemistry, Vol. 29, Issue 9, pp. 954-965, DOI: 10.1016/j.trac.2010.05.011

[44] Hill, E.W. (2011). "Graphene sensors." IEEE Sensors Journal, Vol. 11, Issue 12, pp. 3161-3170, DOI: 10.1109/JSEN.2011.2167608

[45] Hu, J.; Wisetsuwannaphum, S. & Foord, J. S. (2014). "Glutamate biosensors based on diamond and graphene platforms." Faraday Discussions, Vol. 172, pp. 457-472, DOI: 10.1039/c4fd00032c

[46] Kobayashi, K.; Tanimura, M.; Nakai, H.; Yoshimura, A.; Yoshimura, H.; Kojima, K. & Tachibana, M. (2007). "Nanographite domains in carbon nanowalls." Journal of Applied Physics, Vol. 101, Issue 9, 094306 (4 pages), DOI: 10.1063/1.2728781

[47] Machino, T.; Takeuchi, W.; Kano, H.; Hiramatsu, M. & Hori, M. (2009). "Synthesis of platinum nanoparticles on two-dimensional carbon nanostructures with an ultrahigh aspect ratio employing supercritical fluid chemical vapor deposition process." Applied Physics Express, Vol. 2, 025001 (3 pages), DOI: 10.1143/APEX.2.025001

[48] Hiramatsu, M. & Hori, M. (2010). "Preparation of dispersed platinum nanoparticles on a carbon nanostructured surface using supercritical fluid chemical deposition." Materials, Vol. 3, No. 3, pp. 1559-1572, DOI: 10.3390/ma3031559

[49] Hiramatsu, M.; Machino, T.; Mase, K.; Hori, M. & Kano, H. (2010). "Preparation of platinum nanoparticles on carbon nanostructures using metal-organic chemical fluid deposition employing supercritical carbon dioxide." Journal of Nanoscience and Nanotechnology, Vol. 10, No. 6, pp. 4023-4029, DOI: 10.1166/jnn.2010.1996

[50] Hiramatsu, M.; Mitsuguchi, S.; Horibe, T.; Kondo, H.; Hori, M. & Kano, H. (2013). "Fabrication of carbon nanowalls on carbon fiber paper for fuel cell application," Jap-

anese Journal of Applied Physics, Vol. 45, 01AK03 (5 pages), DOI: 10.7567/JJAP.52.01AK03

[51] Watanabe, H.; Kondo, H.; Sekine, M.; Hiramatsu, M. & Hori, M. (2012). "Control of super hydrophobic and super hydrophilic surfaces of carbon nanowalls using atmospheric pressure plasma treatments." Japanese Journal of Applied Physics, Vol. 51, 01AJ07 (4 pages), DOI: 10.1143/JJAP.51.01AJ07

[52] Watanabe, H.; Kondo, H.; Hiramatsu, M.; Sekine, M.; Kumar, S.; Ostrikov, K. & Hori, M. (2013). "Surface chemical modification of carbon nanowalls for wide-range control of surface wettability." Plasma Processes and Polymers, Vol. 10, pp. 582-592, DOI: 10.1002/ppap.201200141

[53] Kaibara, Y.; Sugata, K.; Tachiki, M.; Umezawa, H. & Kawarada, H. (2003). "Control wettability of the hydrogen-terminated diamond surface and the oxidized diamond surface using an atomic force microscope." Diamond and Related Materials, Vol. 12, Issues 3-7, pp. 560-564, DOI: 10.1016/S0925-9635(02)00373-4

[54] Park, H.; Kwon, K.; Lee, S.; Koak, B.; Nahm, S.; Lee, H.; Cho, K.; Kwon, O. & Kang, Y. (1994). "A study on modified silicon surface after CHF3/C2F6 reactive ion etching." ETRI Journal, Vol. 16, No. 1, pp. 45-57, DOI: 10.4218/etrij.94.0194.0014

[55] Cristina, T.; Marsi, O.; Santos, T. G.; Soares, C. P.; Corat, E. J.; Marciano, F. R. & Lobo, A. O. (2012). "Biomineralization of superhydrophilic vertically aligned carbon nanotubes." Langmuir, Vol. 28, Issue 9, pp. 4413-4424, DOI: 10.1021/la300111k

[56] Zhang, X.; Wang, X.; Q. Lu, Q. & Fu, C. (2008). "Influence of carbon nanotube scaffolds on human cervical carcinoma HeLa cell viability and focal adhesion kinase expression." Carbon, Vol. 46, Issue 3, pp. 453-460, DOI: 10.1016/j.carbon.2007.12.015

[57] Nayak, T. R.; Andersen, H.; Makam, V. S.; Khaw, C.; Bae, S.; Xu, X.; Ee, P. L.; Ahn, J. H.; Hong, B. H.; Pastorin, G. & Özyilmaz, B. (2011). "Graphene for controlled and accelerated osteogenic differentiation of human mesenchymal stem cells." ACS Nano, Vol. 5, Issue 6, pp. 4670-4678, DOI: 10.1021/nn200500h

[58] Grinnell, F.; Milam, M. & Srere, P. A. (1973). "Attachment of normal and transformed hamster kidney cells to substrata varying in chemical composition." Biochemical Medicine, Vol. 7, Issue 1, pp. 87-90, DOI: 10.1016/0006-2944(73)90102-6

[59] Maroudas, N. G. (1973). "Chemical and mechanical requirements for fibroblast adhesion." Nature, Vol. 244, Issue 1, pp. 353-354, DOI: 10.1038/244353a0

[60] Brammer, K. S.; Choi, C.; Frandes, C. J.; Oh, S. & Jin, S. (2011). "Hydrophobic nanopillars initiate mesenchymal stem cell aggregation and osteo-differentiation." Acta Biomaterialia, Vol. 7, Issue 2, pp. 683-690, DOI: 10.1016/j.actbio.2010.09.022

[61] Lee, J. H.; Lee, J. W.; Khangt, G. & Lee, H. B. (1997). "Interaction of cells on chargeable functional group gradient surfaces." Biomaterials, Vol. 18, Issue 4, pp. 351-358, DOI: 10.1016/S0142-9612(96)00128-7

[62] Engler, A. J.; Sen, S.; Sweeney, H. L. & Discher, D. E. (2006). "Matrix elasticity directs stem cell lineage specification." Cell, Vol. 126, Issue 4, pp. 677-689, DOI: 10.1016/j.cell. 2006.06.044

[63] Ozeki E. & Matsuda, T. (1990). "Development of an artificial extracellular matrix. Solution castable polymers with cell recognizable peptidyl side chain." ASAIO Transactions, Vol. 36, Issue 3, pp.M294-M296

[64] Arima, Y. & Iwata, H. (2007). "Effect of wettability and surface functional groups on protein adsorption and cell adhesion using well-defined mixed self-assembled monolayersiamond surface and the oxidized diamond surface using an atomic force microscope." Biomaterials, Vol. 28, Issue 20, pp. 3074-3082, DOI: 10.1016/j.biomaterials. 2007.03.013

[65] Ramanathan, T.; Fisher, F. T.; Ruoff, R. S. & Brinson, L. C. (2005). "Amino- functionalized carbon nanotubes for binding to polymers and biological systems." Chemistry of Materials, Vol. 17, pp. 1290-1295, DOI: 10.1021/cm048357f

[66] Luo, D.; Zhang, G.; Liu, J. & Sun, X. (2011). "Evaluation criteria for reduced graphene oxide." Journal of Physical Chemistry C, Vol. 115, Issue 23, pp. 11327-11335, DOI: 10.1021/jp110001y

Chapter 7

Planar Heterostructures Based on Graphene

Pavel V. Ratnikov and Andrei P. Silin

Additional information is available at the end of the chapter

Abstract

We investigate a planar heterostructures based on gapless graphene and its gap modifications such as a single single heterojunction, a quantum well, and a superlattice. We focus on electron properties of these heterostructures within the envelope wave functions. A new phenomena such as valley-polarized currents along the heterojunction and a pseudospin splitting of an energy spectrum in quantum wells are predicted.

Keywords: Graphene, planar heterostructures, pseudospin splitting, interface states, minibands.

1. Introduction

The creation of graphene, a monolayer of carbon atoms forming a regular hexagonal lattice [1–3], has stimulated extensive experimental and theoretical studies along various lines of research. Graphene's unique properties make it a promising material for a new generation of carbon-based nanoelectronic devices. In particular, carrier mobility in graphene amounts to $2 \cdot 10^5$ $\text{cm}^2/(\text{V s})$, and ballistic transport is possible on a submicrometer scale [4, 5].

Over a few past years, numerous theoretical and experimental results have been reported on electronic properties of nanometer-wide ribbons of graphene (nanoribbons). Among the first were studies of electronic states of graphene nanoribbons using the Dirac equation under appropriate boundary conditions [6, 7]. The electronic properties of a graphene nanoribbon strongly depend on its size and edge geometry [8]. In terms of transport properties, graphene nanoribbons are highly reminiscent of carbon nanotubes [9, 10] since free carrier motion inside them is also one-dimensional.

A field-effect transistor (FET) based on a 2 nm wide, 236 nm long graphene nanoribbon was fabricated in a recent study [11] (nanoribbons of widths between 10 and 60 nm were also studied). The graphene nanoribbon was made narrow enough to open a gap of width required for room-temperature transistor operation. However, it is less compact than the graphene quantum-dot transistor 30 nm in diameter discussed in [12].

The results of paper [13] confirm that graphene devices exhibit very high electronic mobility μ on a hexagonal boron nitride (h-BN) substrates, graphene devices on WS_2 substrates (G/WS_2) are equally promising for high quality electronic transport $\mu \sim 3.8 \times 10^4$ cm^2/(V s) at room temperature, followed by G/MoS_2 $\mu \sim 10^4$ cm^2/(V s) and G/GaSe $\mu \sim 2.2 \times 10^3$ cm^2/(V s). However, authors of [13] observed a significant asymmetry in electron and hole conduction in G/WS_2 and G/MoS_2 heterostructures, most likely due to the presence of sulphur vacancies in the substrate crystals.

Heterogeneous engineering of two-dimensional layered materials, including metallic graphene and semiconducting transition metal dichalcogenides, presents an exciting opportunity to produce highly tunable electronic and optoelectronic systems. In order to engineer pristine layers and their interfaces, epitaxial growth of such heterostructures is required. We report the direct growth of crystalline, monolayer tungsten diselenide (WSe_2) on epitaxial graphene (EG) grown from silicon carbide. Vertical transport measurements across the WSe_2/EG heterostructure provides evidence that an additional barrier to carrier transport beyond the expected WSe_2/EG band offset exists due to the interlayer gap [14].

The integration of graphene and other atomically flat, two-dimensional materials has attracted much interest and been materialized very recently. An in-depth understanding of transport mechanisms in such heterostructures is essential. The vertically stacked graphene-based heterostructure transistors were manufactured to elucidate the mechanism of electron injection at the interface [15]. In a careful analysis of current-voltage characteristics, an unusual decrease in the effective Schottky barrier height and increase in the ideality factor were observed with decreasing temperature. A model of thermionic emission with a Gaussian distribution of barriers was able to precisely interpret the conduction mechanism.

The intrinsic performance of vertical and lateral graphene-based heterostructure FETs have been theoretically investigated in [16]. Authors focused on three recently proposed graphene-based transistors, that in experiments have exhibited large current modulation. The analysis is based on device simulations including the self-consistent solution of the electrostatic and transport equations within the non-equilibrium Green's function formalism. It was shown that the lateral heterostructure transistor has the potential to outperform the graphene-based technology and to meet the requirements for the next generation of semiconductor integrated circuits. On the other hand, it was found that vertical heterostructure transistors miss these performance targets by several orders of magnitude, both in terms of switching frequency and delay time, due to large intrinsic capacitances, and unavoidable current/capacitance tradeoffs.

In this chapter, we focus on the theoretical aspects of the planar technology of the graphene-based heterostructures. We will consider three types of these heterostructures: the single heterojunction, the quantum well, and the superlattice. Our analysis of the electronic properties of such heterostructures is based on the envelope wave function approach. In the first instance, this rather simple method gives an information about an energy spectrum of charge carriers. Then we can obtain a knowledge about optical and transport properties.

Here, we suggest to widely use the gap modifications of graphene in the planar heterostructures. We believe that the planar heterostructures made of gapless and gapped graphene are as prospective building blocks in future carbon-based nanoelectronics. The use of only gapless graphene reduces the diverse opportunities offered by bandgap engineering in gapped graphene.

2. Theoretical basis

2.1. The parity operator

Let us consider the parity operator in (3+1)QED [17]

$$\widehat{P} = i\gamma_4 \widehat{\Lambda}_{\mathbf{n}}. \tag{1}$$

Here, $i\gamma_4$ is the inversion operator and

$$\widehat{\Lambda}_{\mathbf{n}} = e^{-i\frac{\pi}{2}\boldsymbol{\Sigma}\mathbf{n}} = -i\boldsymbol{\Sigma}\mathbf{n}$$

is the operator of rotation by π about an $\mathbf{n}$ axis perpendicular to the graphene plane. In standard representation,

$$\boldsymbol{\Sigma} = \begin{pmatrix} \boldsymbol{\sigma} & 0 \\ 0 & \boldsymbol{\sigma} \end{pmatrix},$$

where $\boldsymbol{\sigma}$ denotes Pauli matrices, and

$$\gamma_4 \equiv \beta = \begin{pmatrix} I & 0 \\ 0 & -I \end{pmatrix}$$

where I is the 2×2 unit matrix. It is clear that operator (1) is analogous to the parity operator $i\gamma_5\widehat{n}$ in QED, where $\widehat{n} = \boldsymbol{\gamma}\mathbf{n}$,

$$\boldsymbol{\gamma} = \begin{pmatrix} 0 & \boldsymbol{\sigma} \\ -\boldsymbol{\sigma} & 0 \end{pmatrix}, \quad \gamma_5 = \gamma_1\gamma_2\gamma_3\gamma_4 = i\begin{pmatrix} 0 & I \\ I & 0 \end{pmatrix}.$$

The eigen functions of this operator describe electron polarization states [18].

Charge carrier states in graphene can be described in terms of helicity defined as the eigenvalue of the operator $\widehat{h} = \boldsymbol{\sigma} \cdot \mathbf{p}/(2|\mathbf{p}|)$. The projection of pseudospin on the direction of quasimomentum p indicates the valley in the Brillouin zone where electrons or holes belong (K or K' point in Fig. 2). Positive helicity corresponds to electrons and holes with wavevectors near the K and K' points, respectively; negative helicity corresponds to electrons and holes with wavevectors near the K' and K points, respectively [19].

Massless states with opposite helicities are decoupled [20]. In addition, charge carriers have chiral symmetry (helicity is conserved), and parity can be defined for both massless and massive carriers[1]. In other words, a higher symmetry of massless charge carriers implies the existence of an additional quantum number: helicity. Whereas parity distinguishes only between the valleys where carrier states belong ($\lambda = +1$ and -1 for states close to the K and K' points, respectively), helicity differs between a particle (electron) and an antiparticle (hole). However, chiral symmetry is broken for massive charge carriers. So, helicity is not a good quantum number any longer. Carrier states in a planar heterostructure combining gapless and gapped graphene should be characterized by parity.

Recall that the Dirac equation describing massless carriers in graphene in terms of 4×4 matrices is derived by assuming that they are spinless and have two valley degrees of freedom [10]. When analysis is restricted to charge carriers in one valley, the Dirac equation can be reduced to a 2×2 matrix representation by Weyl's equation for a massless fermion analogous to neutrino in two Euclidean dimensions [22]. The carrier energy spectrum with a pseudospin splitting in a planar heterostructure combining gapless and gapped graphene cannot be correctly analyzed in the 2×2 representation. For similar reasons, the representation of the Dirac algebra in terms of 2×2 matrices is not sufficient for describing the chiral symmetry breaking in (2+1)QED [23].

Using the two-dimensional 4×4 Dirac equation to describe charge carriers in a graphene-based nanostructure, we can study pseudospin effects following an approach to narrow-gap semiconductor heterostructures based on the Dirac model [24]. This makes methods developed for solving problems in the spintronics of narrow-gap semiconductor heterostructures applicable to graphene-based ones [25–33].

2.2. The equation for the envelope wavefunction

To describe some phenomena in graphene-based heterostructures, an equation containing a mass term should be written for the envelope wavefunction. A bandgap opening in the energy spectrum of graphene results from the lack of symmetry between the two triangular sublattices of its hexagonal lattice. The corresponding tight-binding Hamiltonian taking into account nearest-neighbor hopping has the form [34]

$$\begin{aligned}\widehat{H} = & -t \sum_{\mathbf{B},i,\sigma} \left[a_\sigma^\dagger(\mathbf{B}+\mathbf{d}_i) b_\sigma(\mathbf{B}) + b_\sigma^\dagger(\mathbf{B}) a_\sigma(\mathbf{B}+\mathbf{d}_i) \right] \\ & + \Delta \sum_{\mathbf{B},\sigma} \left[a_\sigma^\dagger(\mathbf{B}+\mathbf{d}_1) a_\sigma(\mathbf{B}+\mathbf{d}_1) - b_\sigma^\dagger(\mathbf{B}) b_\sigma(\mathbf{B}) \right],\end{aligned} \tag{2}$$

where $t \approx 2.8$ eV is the nearest-neighbor hopping energy; the sum runs over the position vectors $\mathbf{B}$ of all B sublattice atoms; $\sigma = \uparrow, \downarrow$ is the (pseudo)spin index; a_σ ($a_\sigma^\dagger$) and b_σ ($b_\sigma^\dagger$) are the annihilation (creation) operators of A and B sublattice electrons, respectively (see Fig. 1); and the parameter Δ quantifies the on-site energy difference between the two sublattices (setting $\Delta = 0$ restores the symmetry between sublattices so that graphene becomes gapless, whereas nonzero Δ equals the half-gap width in gapped graphene as shown below).

[1] Massless states can be characterized by two quantum numbers: helicity and sign of energy or helicity and eigenvalue of the operator $i\gamma_5$ [21]. Parity is analogous to the eigenvalue of $i\gamma_5$.

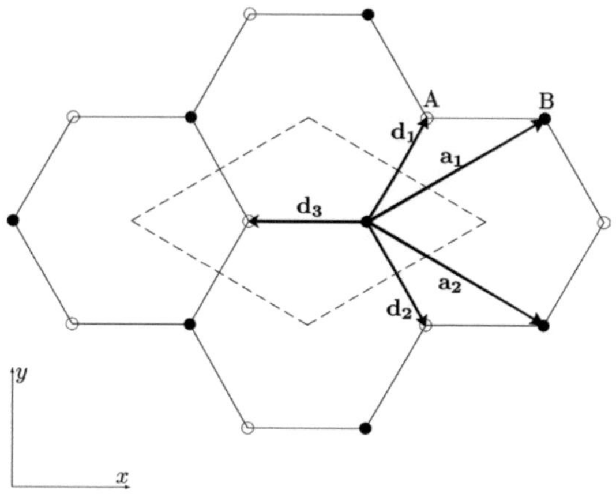

Figure 1. Part of hexagonal lattice, with highlighted vectors $\mathbf{d}_i$ from a B sublattice atom to the three nearest-neighbor A sublattice atoms.

Performing a Fourier transform, we change to the momentum representation

$$a_\sigma(\mathbf{A}) = \int\limits_{\Omega_B} \frac{d^2k}{(2\pi)^2} a_\sigma(\mathbf{k}) e^{i\mathbf{k}\cdot\mathbf{A}}, \; b_\sigma(\mathbf{B}) = \int\limits_{\Omega_B} \frac{d^2k}{(2\pi)^2} b_\sigma(\mathbf{k}) e^{i\mathbf{k}\cdot\mathbf{B}},$$

where Ω_B means integration over the first Brillouin zone (it is shown on Fig. 2).

Hamiltonian (2) is rewritten as

$$\widehat{H} = \sum_\sigma \int\limits_{\Omega_B} \frac{d^2k}{(2\pi)^2} \left(a_\sigma^\dagger(\mathbf{k}) \; b_\sigma^\dagger(\mathbf{k})\right) \begin{pmatrix} \Delta & -t\sum\limits_i e^{-i\mathbf{k}\cdot\mathbf{d}_i} \\ -t\sum\limits_i e^{i\mathbf{k}\cdot\mathbf{d}_i} & -\Delta \end{pmatrix} \begin{pmatrix} a_\sigma(\mathbf{k}) \\ b_\sigma(\mathbf{k}) \end{pmatrix}. \tag{3}$$

Conduction and valence band extrema lie at the corners of the first Brillouin zone. In the case of gapless graphene, they touch each other and there are the cone-like energy surfaces at the K and K' points (see Fig. 3). We use Hamiltonian (3) expanded around the K point with quasimomentum $\mathbf{q}_1 = \left(\frac{2\pi}{3a}, \frac{2\pi}{3\sqrt{3}a}\right)$ or around the K' point with $\mathbf{q}_2 = \left(\frac{2\pi}{3a}, -\frac{2\pi}{3\sqrt{3}a}\right)$

$$\widehat{H} = \sum_\sigma \int \frac{d^2k}{(2\pi)^2} \widehat{\Psi}_\sigma^\dagger(\mathbf{k}) \widehat{\mathcal{H}} \widehat{\Psi}_\sigma(\mathbf{k}).$$

where integration is performed over small neighbor-hoods of the K and K' points. Near the corners, the Hamiltonian reduces to

$$\widehat{\mathcal{H}} = \begin{pmatrix} v_F \boldsymbol{\sigma} \cdot \mathbf{k} + \Delta\sigma_z & 0 \\ 0 & v_F \boldsymbol{\sigma}^* \cdot \mathbf{k} + \Delta\sigma_z \end{pmatrix}, \tag{4}$$

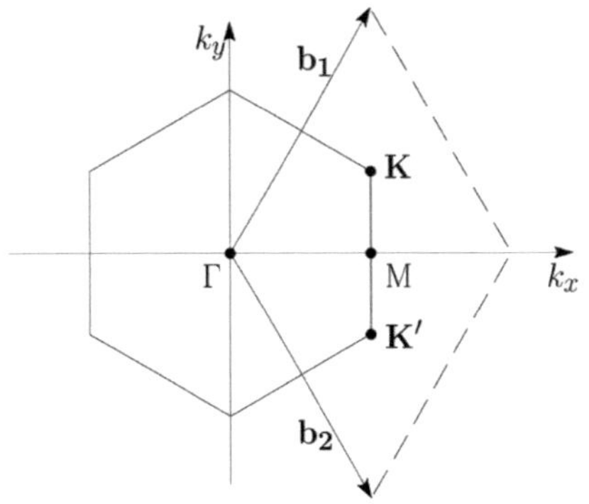

Figure 2. The first Brillouin zone of graphene, with linear energy spectrum at the corners (Dirac points). The reciprocal lattice vectors $\mathbf{b}_1 = (2\pi/3a, 2\pi/\sqrt{3}a)$ and $\mathbf{b}_2 = (2\pi/3a, -2\pi/\sqrt{3}a)$, where a=1.42 Å is the lattice spacing, combined with the dashed lines equivalently represent the first Brillouin zone as a rhombus.

where $v_F = \frac{3}{2}at$ is the carrier Fermi velocity, $\boldsymbol{\sigma} = (\sigma_x, \sigma_y)$ and $\boldsymbol{\sigma}^* = (\sigma_x, -\sigma_y)$ are Pauli matrices in the sublattice space, and $\widehat{\Psi}_\sigma(\mathbf{k})$ is the bispinor defined as

$$\widehat{\Psi}_\sigma(\mathbf{k}) = \begin{pmatrix} \widehat{\Psi}_\sigma^{(1)}(\mathbf{k}) \\ \widehat{\Psi}_\sigma^{(2)}(\mathbf{k}) \end{pmatrix},$$

in terms of

$$\widehat{\Psi}_\sigma^{(1,2)}(\mathbf{k}) = \exp\left(\frac{5\pi i}{12}\sigma_z\right)\sigma_z \begin{pmatrix} a_\sigma(\mathbf{q}_{1,2}+\mathbf{k}) \\ b_\sigma(\mathbf{q}_{1,2}+\mathbf{k}) \end{pmatrix}.$$

We write an equation for the envelope wavefunction in a planar heterostructure:

$$\left[v_{Fj}\left(\tau_0 \otimes \sigma_x \widehat{p}_x + \tau_z \otimes \sigma_y \widehat{p}_y\right) + \tau_0 \otimes \sigma_z \Delta_j + \tau_0 \otimes \sigma_0 \left(V_j - E\right)\right] \Psi(x,y) = 0. \tag{5}$$

Here, $\Delta_j = E_{gj}/2$ ($j = 1, 2, \ldots$) denotes half-width of bandgap; the respective work functions V_j depend on the mid-gap energies relative to the Dirac points for the corresponding materials; the 2×2 unit matrix σ_0 acts in the sublattice space; the 2×2 unit matrix τ_0 and the matrix τ_z defined similar to the Pauli matrix σ_z act in the valley space; $\otimes$ is the Kronecker product symbol; and $\widehat{p}_x = -i\partial_x$ and $\widehat{p}_y = -i\hbar\partial_y$ are the momentum operator components. Assuming that the carrier Fermi velocities may differ between the regions, we denote those for gapped jth regions by v_{Fj} for both gapless graphene and its gapped modifications.

Charge carriers move freely along the y axis:

$$\Psi(x,y) = \Psi(x)e^{ik_y y}.$$

The wavefunction $\Psi(x)$ is a bispinor:

$$\Psi(x) = \begin{pmatrix} \psi_K(x) \\ \psi_{K'}(x) \end{pmatrix},$$

where the spinors $\psi_K(x)$ and $\psi_{K'}(x)$ represent charge carriers in the K and K' valleys, respectively:

$$\psi_K(x) = \begin{pmatrix} \psi_{KA}(x) \\ \psi_{KB}(x) \end{pmatrix}, \quad \psi_{K'}(x) = \begin{pmatrix} \psi_{K'A}(x) \\ \psi_{K'B}(x) \end{pmatrix}.$$

In the present context, the parity operator is expressed as follows:

$$\widehat{P} = \tau_z \otimes \sigma_0. \tag{6}$$

Equation (5) is solved here in the parity basis. The eigenfunctions $\Psi_\lambda(x)$ of parity operator (6) are defined as follows:

$$\widehat{P}\Psi_\lambda(x) = \lambda\Psi_\lambda(x),$$

$$\begin{aligned} \Psi_{+1}(x) &= \begin{pmatrix} \psi_{+1,K}(x) \\ 0 \end{pmatrix}, \\ \Psi_{-1}(x) &= \begin{pmatrix} 0 \\ \psi_{-1,K'}(x) \end{pmatrix}. \end{aligned} \tag{7}$$

Rewriting Eq. (5) as the 2×2 matrix equations

$$\left(-iv_{Fj}\sigma_x\partial_x + v_{Fj}k_y\sigma_y + \lambda\Delta_j\sigma_z + V_j\right)\psi_{\lambda K}(x) = E_\lambda\psi_{\lambda K}(x), \tag{8}$$

$$\left(-iv_{Fj}\sigma_x\partial_x - v_{Fj}k_y\sigma_y - \lambda\Delta_j\sigma_z + V_j\right)\psi_{\lambda K'}(x) = E_\lambda\psi_{\lambda K'}(x). \tag{9}$$

We see that setting $\Delta_j = 0$ and $V_j = 0$ brings us back to the spinor wavefunctions describing chiral states near the K or K' point, where the operator $\widehat{h}$ can be defined. However, chiral symmetry is broken when $\Delta \neq 0$ (see previous section). Defining parity λ as the eigenvalue of operator (6), we find that it indicates the valley where charge carriers belong: by virtue of (7), $\lambda = +1$ for the states near the K point described by Eq. (8) and $\lambda = -1$ for the states near the K' point described by Eq. (9).

In both gapped and gapless graphene, the valleys transform into each other under time reversal. This is indicated by the opposite signs of the terms proportional to k_y in Eqs. (8) and (9), since $k_y \to -k_y$ under time reversal. It can be shown directly by using the time reversal operator $\mathcal{T}$ in explicit form that $\lambda \to -\lambda$ under $\mathcal{T}$. Indeed, if k_y is parallel to the line $K - M - K'$ (see Fig. 2) and its origin is set at M, then $K \to K'$ and $K' \to K$ under $\mathcal{T}$.

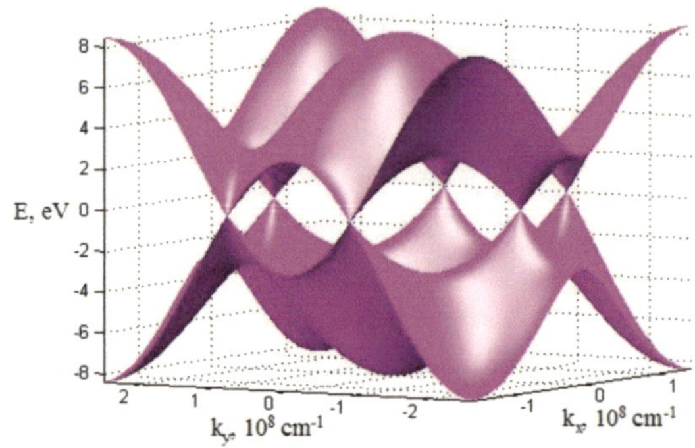

Figure 3. The energy surface for gapless graphene.

Equations (8) and (9) are equivalently rewritten as the 2×2 matrix equation

$$\left(-iv_{Fj}\sigma_x\partial_x + \lambda v_{Fj}k_y\sigma_y + \Delta_j\sigma_z + V_j\right)\psi_\lambda(x) = E_\lambda\psi_\lambda(x). \tag{10}$$

Hereinafter, valley indices K and K' are omitted as unnecessary since λ specifies the valley where charge carriers belong.

2.3. The boundary conditions

Now, let us discuss the boundary conditions at the interfaces between different graphene materials. At the outset, we note that they are easier to formulate than those used at the graphene–free-space interface in models of edge states [6, 35]. To derive boundary conditions in the present model, we must find a relation between $\psi_\lambda(l)$ and $\psi_\lambda(-l)$ as $l \to 0$ in the neighbor-hood of x = 0, where l goes down to an atomic scale (condition at x = d is derived similarly). Multiplying Eq. (10) by $\psi^\dagger(x)$ on the left, we integrate it over $[-l, l]$. Since a is small, we neglect all terms except those containing a derivative with respect to x to obtain[2]

$$\psi_\lambda^{(-)\dagger}(-l)v_F^{(-)}\psi_\lambda^{(-)}(-l) = \psi_\lambda^{(+)\dagger}(l)v_F^{(+)}\psi_\lambda^{(+)}(l),$$

where $\psi_\lambda^{(-)}$ and $\psi_\lambda^{(+)}$ are defined on the left- and right-hand sides of the boundary (at $x < 0$ and $x > 0$, respectively). Representing these functions as

$$\psi_\lambda^{(\pm)} = \left|\psi_\lambda^{(\pm)}\right| \exp\left(i\varphi^{(\pm)}\right),$$

we rewrite the equality above as

$$\sqrt{v_F^{(-)}}\left|\psi_\lambda^{(-)\dagger}(-l)\right| = \sqrt{v_F^{(+)}}\left|\psi_\lambda^{(+)\dagger}(l)\right|.$$

[2] From the given equality, we have the continuity of the current component normal to the interface in the heterostructure plane. It is necessary condition.

To formulate the boundary condition in final form, we assume that the difference between phases of $\psi_\lambda^{(-)}$ and $\psi_\lambda^{(+)}$ near the interface is a multiple of 2π:

$$\varphi^{(+)} = \varphi^{(-)} + 2\pi n, \quad n \in \mathbb{Z}.$$

As l goes to zero, we obtain the following wavefunction-matching condition [26, 27]

$$\sqrt{v_F^{(-)}}\psi_\lambda^{(-)} = \sqrt{v_F^{(+)}}\psi_\lambda^{(+)}, \tag{11}$$

where minus and plus signs refer to the materials on the left- and right-hand sides of the boundary, respectively.

3. Graphene heterojunctions

We consider a planar heterojunction composed of graphene and a gap modification of graphene [36]. When we say a gap modification of graphene we imply a graphene with an energy gap in the Dirac spectrum of charge carriers. There are several gap modifications of graphene.

First, the energy gap can open because graphene sheets are located not on SiO_2 substrate but on some other material, for example, h-BN, when two triangular sublattices of graphene become nonequivalent and a gap modification of graphene is formed with an energy gap of 53 meV [37]. Second, the energy gap opens in the epitaxially grown graphene on the SiC substrate [38], which is equal to 0.26 eV according to experimental results obtained by angular-resolved photoemission spectroscopy [39]. Third, recently another modification of graphene, i.e., graphane, was synthesized by hydrogenation [40], which has a direct energy gap of 5.4 eV at the Γ point according to the calculations [41]. In the first two cases, a graphene film deposited on inhomogeneous SiO_2–h-BN or SiO_2–SiC substrates can be used (Fig. 4a shows the case with h-BN). In the third case, an inhomogeneously hydrogenated graphene is used (a part of the graphene sample is left without hydrogenation, Fig. 4b).

We assume that the energy gap in the gap modifications of graphene opens at K and K' points of the first Brillouin zone, which correspond to the Dirac points of gapless graphene.

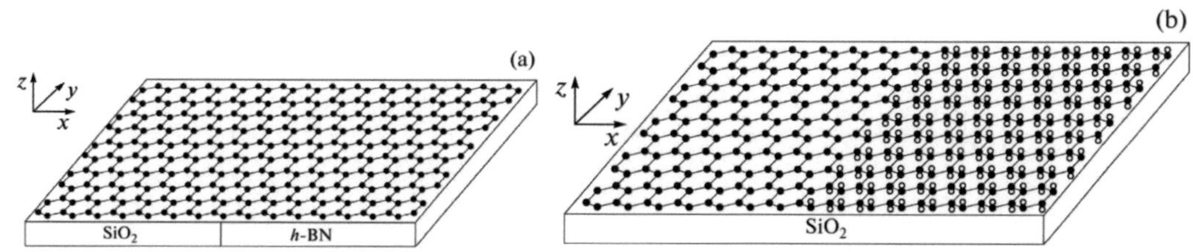

Figure 4. Two variants of the system in question: **(a)** graphene layer on the substrate composed of h-BN and SiO_2 and **(b)** nonuniformly hydrogenated graphene on the SiO_2 substrate. Open circles are hydrogen atoms, which are located so that they are bound to carbon atoms of one sublattice on one side of graphene sheet and to carbon atoms of the other sublattice on the other side.

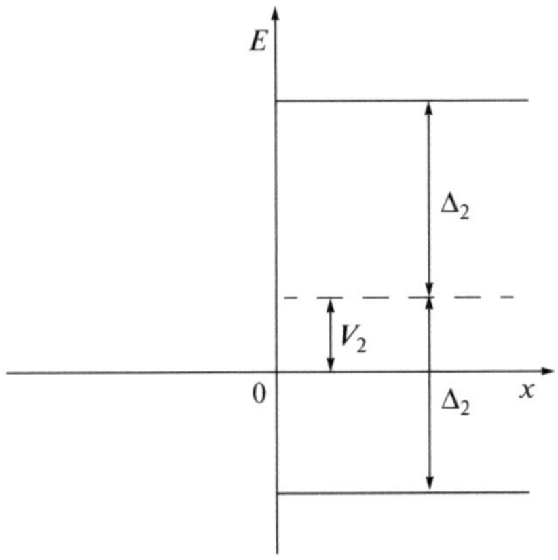

Figure 5. Graphene heterojunction under consideration.

Let us assume that the x axis is directed along the plane of the heterojunction perpendicular to the boundary between graphene and its gap modification and the y axis is directed along the boundary. The z axis is directed perpendicular to the plane of the heterojunction. The half-plane $x < 0$ is occupied by the gap-less graphene and the half-plane $x > 0$ belongs to the gap modification of graphene. So, the line $x = 0$ is the boundary under consideration.

In this case, the parameters in Eq. (5) with $j = 1$ are related to the gapless graphene and the parameters with $j = 2$ are related to the gap modification of the graphene: v_{F1} and v_{F2} are the Fermi velocities (in the general case, $v_{F2} \neq v_{F1}$, and $v_{F1} \approx 10^8$ cm/s); $\Delta_1 = 0$ and $\Delta_2 \neq 0$ are the half-widths of the energy gaps; V_1 and V_2 are the work functions (V_2 determines position of the middle of the energy gap for the gap modification of the graphene with respect to the Dirac points of the gapless graphene, and $V_1 = 0$ is chosen for the origin, see Fig. 5).

In order to avoid spontaneous generation of electron-hole pairs, we assume that the heterojunction in question is a junction of the first kind, i.e., the Dirac points of gapless graphene are located inside the energy gap of its gap modification. This limits value of the work function $|V_2| < \Delta_2$.

Equation (5) is solved within the class of wave eigenfunctions $\Psi_\lambda(x)$ of the parity operator (6) for two areas on both sides from the boundary.

Equation (5) can be easily represented as two 2×2 matrix equations

$$\left(-iv_{Fj}\sigma_x \frac{d}{dx} + v_{Fj}k_y\sigma_y + \lambda\Delta_j\sigma_z + V_j\right)\Psi_{\lambda K}(x) = E_\lambda \Psi_{\lambda K}(x), \tag{12}$$

$$\left(-iv_{Fj}\sigma_x \frac{d}{dx} - v_{Fj}k_y\sigma_y - \lambda\Delta_j\sigma_z + V_j\right)\Psi_{\lambda K'}(x) = E_\lambda \Psi_{\lambda K'}(x). \tag{13}$$

In this case, we have $\lambda = +1$ in Eq. (12) and $\lambda = -1$ in Eq. (13).

The solution to Eq. (12) for boundary states has the form

$$\Psi_{\lambda K}(x) = \begin{cases} C\binom{1}{a}\exp(\kappa_1 x), & x < 0, \\ C\binom{b}{qb}\exp(-\kappa_2 x), & x > 0, \end{cases} \tag{14}$$

where

$$a = i\frac{v_{F1}(k_y - \kappa_1)}{E_\lambda}, \quad q = i\frac{v_{F2}(k_y + \kappa_2)}{E_\lambda - V_2 + \lambda\Delta_2},$$

C is the normalization factor, $b = \sqrt{\frac{v_{F1}}{v_{F2}}}$ is the constant obtained when matching solutions for $x < 0$ and $x > 0$ at the line $x = 0$ under condition (11),

$$E_\lambda = \pm v_{F1}\sqrt{k_y^2 - \kappa_1^2}, \tag{15}$$

from which it follows that the necessary condition for the existence of the boundary states is given by inequality

$$\kappa_1 < |k_y|. \tag{16}$$

Equation (15) can be rewritten as

$$\kappa_1 = \sqrt{k_y^2 - E_\lambda^2/v_{F1}^2},$$

Therefore, the following inequality should also be valid

$$|E_\lambda| < v_{F1}|k_y|. \tag{17}$$

Expression for κ_2 is represented in the form

$$\kappa_2 = \frac{1}{v_{F2}}\sqrt{\Delta_2^2 - (E_\lambda - V_2)^2 + v_{F2}^2 k_y^2}.$$

Moreover, the matching leads to the inequality

$$\frac{v_{F1}(k_y - \kappa_1)}{E_\lambda} = \frac{v_{F2}(k_y + \kappa_2)}{E_\lambda - V_2 + \lambda\Delta_2}. \tag{18}$$

The solution to Eq. (13) is produced from Eq. (14) by the following substitutions in factors a and q: $k_y \to -k_y$ and $\lambda \to -\lambda$.

Let us discuss separately the case of zero mode $E_\lambda = 0$. Components of the envelope wave function in $x < 0$ region (gapless graphene) $\Psi_{\lambda K} = \binom{a_1}{a_2} \exp(\kappa_1 x)$ satisfy equations:

$$(\kappa_1 - k_y)a_1 = 0,$$
$$(\kappa_1 + k_y)a_2 = 0,$$

i.e., either $\kappa_1 = k_y$ ($k_y > 0$) and $a_2 = 0$, or $\kappa_1 = -k_y$ ($k_y < 0$) and $a_1 = 0$. Then it follows from the matching condition (11) that both components of the envelope wave function are zero in $x > 0$ region ($b = 0$); therefore, we have $a_1 = 0$ and $a_2 = 0$, i.e., $\Psi_{\lambda K}(x) \equiv 0$. Thus, there is no zero mode for the boundary states in question.

The following equations are easily obtained from Eq. (18):

$$\kappa_1 \kappa_2 = \frac{E_\lambda (E_\lambda - V_2)}{v_{F1} v_{F2}} - k_y^2, \tag{19}$$

$$\lambda \Delta_2 E_\lambda = v_{F1} v_{F2} k_y (\kappa_1 + \kappa_2). \tag{20}$$

The two latter equations are valid for either value of λ (for both valleys), because they are invariant in respect to simultaneous substitutions $k_y \to -k_y$ and $\lambda \to -\lambda$.

Since $\kappa_1 > 0$ and $\kappa_2 > 0$, right-hand side of Eq. (19) should be positive. Let us denote by $\varepsilon_0(k_y)$ such value of E_λ that the right-hand side of Eq. (19) turns zero,

$$\varepsilon_0(k_y) = \frac{V_2}{2} \pm \sqrt{\frac{V_2^2}{4} + v_{F1} v_{F2} k_y^2}, \tag{21}$$

where "+" corresponds to electrons and "−" to holes. Then, the condition $\kappa_1 \kappa_2 > 0$ is equivalent to the inequality

$$|E_\lambda| > |\varepsilon_0(k_y)|. \tag{22}$$

It follows from Eq. (20) that inequality $\lambda k_y > 0$ holds for electron boundary states ($E_\lambda > 0$), and $\lambda k_y < 0$ holds for hole boundary states ($E_\lambda < 0$). The boundary states are not degenerate in parity. That means that there is no Kramers degeneracy of energy spectrum for them. This is also true for boundary states in a planar quantum well based on graphene nanoribbon [42] and for boundary states localized on zigzag edges of gapless graphene [43]. Since parity determines charge carrier attribution to one of two valleys, the property mentioned above means also that there is a "valley polarization" of boundary states: electrons that move along the heterojunction boundary with $k_y > 0$ are located near K point and electrons with $k_y < 0$ are near K' point and vise versa in case of holes. Because of that, current that flows along the heterojunction boundary would be "valley-polarized" [36].

By squaring Eq. (20) we get a quadratic equation, solution of which produces dependence of energy on k_y:

$$E_\lambda(k_y) = \frac{v_{F1}v_{F-}k_y^2V_2 + \lambda v_{F1}k_y\Delta_2\sqrt{\Delta_2^2 + v_{F-}^2k_y^2 - V_2^2}}{\Delta_2^2 + v_{F-}^2k_y^2}, \tag{23}$$

where $v_{F-} = v_{F1} - v_{F2}$. Equation (23) takes into account that sign of λk_y determines type of charge carriers in the boundary states.

It is easy to verify that inequality (17) is always true if the energy is given by Eq. (23). Therefore, inequality (16) also holds.

Now, it is simple to analyze inequality (22). Let us introduce the following notation:

$$k_{y1} = \frac{|V_2|}{|v_{F-}|},$$

$$k_{y2,3} = \sqrt{\frac{v_{F2}V_2^2 + 2v_{F-}\Delta_2^2 \mp |V_2|\sqrt{v_{F2}^2V_2^2 + 4v_{F1}v_{F-}\Delta_2^2}}{2v_{F2}v_{F-}^2}}.$$

Under the condition

$$v_{F1} < v_{F2} < 2v_{F1}, \quad \frac{2}{v_{F2}}\sqrt{v_{F1}|v_{F-}|}\Delta_2 < |V_2| < \Delta_2, \tag{24}$$

the boundary states exist in the ranges[3]

$$0 < |k_y| < k_{y2}, \quad k_{y3} < |k_y| < k_{y1}$$

either for electrons, if $V_2 < 0$, or for holes, if $V_2 > 0$.

Under condition

$$v_{F1} < v_{F2} < 2v_{F1}, \ |V_2| < \frac{2}{v_{F2}}\sqrt{v_{F1}|v_{F-}|}\Delta_2 \tag{25}$$

the boundary states exist in the range

$$0 < |k_y| < k_{y1}$$

either for electrons, if $V_2 < 0$, or for holes, if $V_2 > 0$.

Under the condition

$$v_{F1} > v_{F2}, \ 0 < V_2 < \Delta_2 \tag{26}$$

[3] Here and below, we exclude the point $k_y = 0$, because it corresponds to $E_\lambda = 0$.

the electron boundary states exist in the range

$$k_{y3} < |k_y| < k_{y1},$$

and the hole boundary states exist in the range

$$0 < |k_y| < k_{y2}.$$

Under condition

$$v_{F1} > v_{F2}, \quad -\Delta_2 < V_2 < 0$$

the electron boundary states exist in the range

$$0 < |k_y| < k_{y2},$$

and the hole boundary states exist in the range

$$k_{y3} < |k_y| < k_{y1}.$$

Let us consider three special cases.

(1) Under condition $V_2 = 0$ and $v_{F-} \neq 0$, the boundary states exist for both electrons and holes in the following range if $v_{F1} > v_{F2}$

$$0 < |k_y| < \frac{\Delta_2}{\sqrt{v_{F1} v_{F-}}}. \tag{27}$$

(2) Under condition $v_{F1} = v_{F2}$, $0 < |V_2| < \Delta_2$ the boundary states exist in the range

$$0 < |k_y| < \frac{\Delta_2 \sqrt{\Delta_2^2 - V_2^2}}{v_{F2} |V_2|} \tag{28}$$

either for electrons, if $V_2 < 0$, or for holes, if $V_2 > 0$.

(3) Under condition $v_{F1} = v_{F2}$, $V_2 = 0$, the boundary states are absent both for electrons and holes, because $|E_\lambda(k_y)| = |\varepsilon_0(k_y)|$, which is in contradiction with inequality (16).

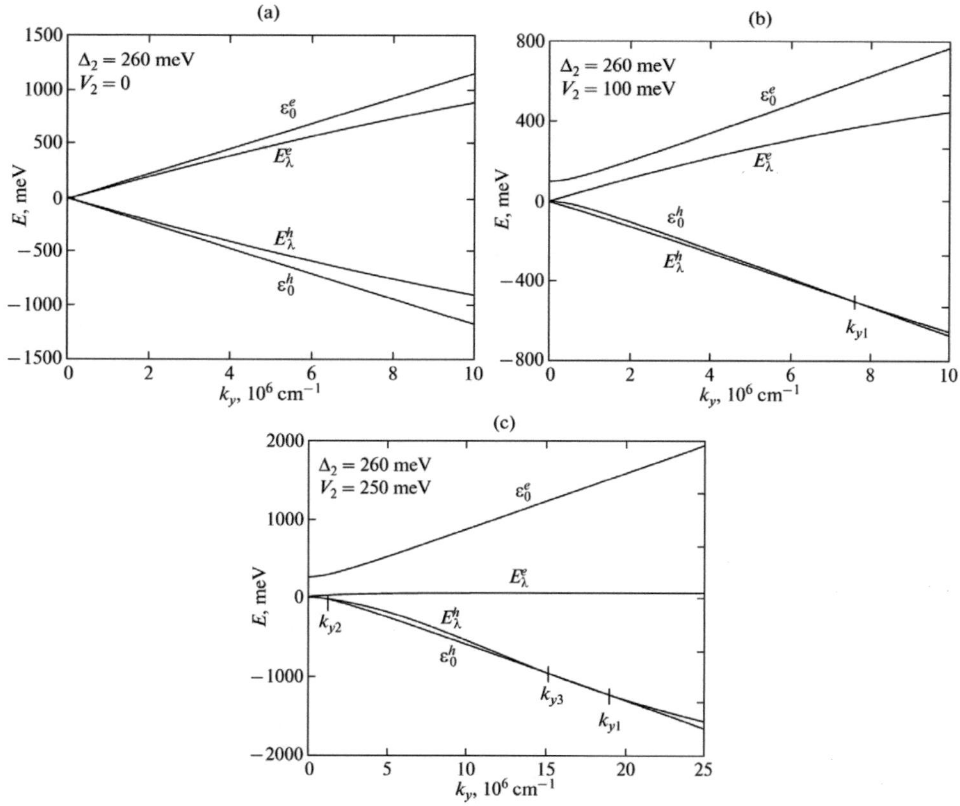

Figure 6. Dispersion curves $E_{\lambda}^{e,h}(k_y)$ and $\varepsilon_0^{e,h}(k_y)$: **(a)** there are no boundary states for electrons and holes at $V_2 = 0$, **(b)** there are only hole boundary states in the range $0 < |k_y| < k_{y1}$ at V_2 = 100 meV, and **(c)** there are only hole boundary states in the ranges $0 < |k_y| < k_{y2}$ and $k_{y3} < |k_y| < k_{y1}$ at V_2 = 250 meV.

Fig. 6 shows dispersion curves $E_{\lambda}^{e,h}(k_y)$ and $\varepsilon_0^{e,h}(k_y)$ for the electron and hole boundary states for three values of V_2 in the model of graphene-based heterojunction with Δ_2 = 260 meV and $v_{F2} = 1.2 \times 10^8$ cm/s for gap modification of graphene.

Our results remain in essence the same if instead of a sharp heterojunction we consider a smooth heterojunction. Indeed, let $v_F(x)$ and $\Delta(x)$ vary smoothly from their values for gapless graphene to their values in gap modification of graphene over a strip with the width $d \lesssim \kappa_{1,2}^{-1}$. Then change in energy of the boundary states is $|\delta E_{\lambda}(k_y)| \lesssim 1$ meV. Such insignificant variation in energy of the boundary states produces no noticeable qualitative changes. A similar result has been obtained for boundary states in heterojunctions of narrow-gap semiconductors with intercrossing dispersion curves in [44].

To conclude, we would like to point out that the new type of boundary states in graphene heterojunctions can be studied in experiment by tunnel spectroscopy of angular-resolved photoemission spectroscopy similar to how it have been done for boundary states in gapless graphene [45–47].

4. Graphene quantum well

4.1. General consideration

In this study, we examine a planar quantum well (QW) made of a graphene nanoribbon whose edges are in contact with gapped graphene sheets.

A bandgap opening in graphene can be induced by several methods. First, graphene can be deposited on h-BN substrate instead of a silicon-oxide one. This makes its two triangular sublattices nonequivalent, inducing in a bandgap of 53 meV [37]. Second, epitaxial graphene grown on a silicon-carbide substrate also has a nonzero bandgap [38]. According to angle-resolved photoemission data, a bandgap of 0.26 eV is produced by this method [39]. Third, a hydrogenated derivative of graphene synthesized recently, graphane [40], has been predicted to have a direct bandgap of 5.4 eV at the Γ point [41]. Fourth, *ab initio* calculations have shown that CrO_3 adsorption on graphene induces a gap of 0.12 eV [48]. In the first two methods, a heterogeneous substrate can be used, such as an h-BN — SiO_2 nanoribbon — h-BN or SiC — SiO_2 nanoribbon — SiC one (Fig. 7a depicts a substrate with h-BN). The last two methods produce a graphene sheet containing a nanoribbon without hydrogenation (as the nonhydrogenated one in Fig. 7b) or a graphene strip without adsorbed CrO_3 molecules, respectively. Furthermore, the bandgap can be varied by using partially hydrogenated graphene (where some carbon atoms are not bonded to hydrogen atoms). Combinations of these methods can also be employed. Extensive experimental studies of graphene on substrates made of various materials, including rare-earth metals, have been reported recently [49–51]. It may be possible to open a bandgap via adsorption of other molecules on graphene or by using other materials as substrates. The use of gapped graphene to create potential barriers opens up additional possibilities for bandgap engineering in carbon-based materials [52].

In the case of QW, in Eq. (10) $\Delta_j = E_{gj}/2$ $(j = 1,2,3)$ denotes half-width of bandgap ($\Delta_1 \neq 0$ and $\Delta_3 \neq 0$ in regions *1* and *3*, whereas $\Delta_2 = 0$ region 2); the respective work functions V_1 and V_3 of regions *1* and *3* depend on the mid-gap energies relative to the Dirac points for the corresponding materials (we set $V_2 = 0$ to be specific, see Fig. 8).

The solution to Eq. (10) is expressed as follows.

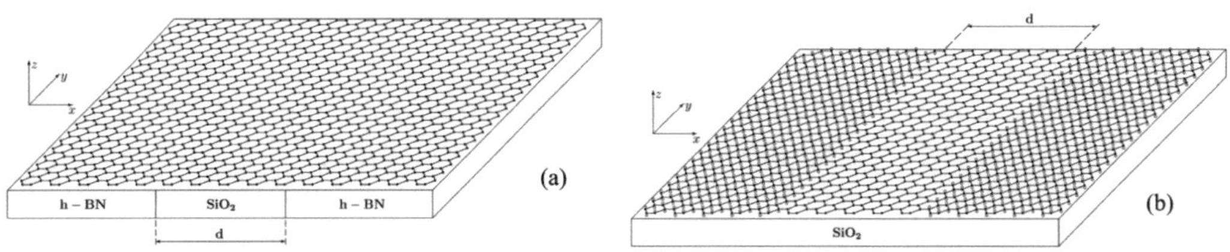

Figure 7. Two configurations of the system under study: **(a)** graphene sheet on a substrate consisting of a SiO_2 nanoribbon of width d inserted between h–BN nanoribbons; **(b)** a graphene sheet on a SiO_2 substrate containing a nonhydrogenated nanoribbon of width *d*, where open and closed circles are hydrogen atoms bonded to carbon atoms in different sublattices on opposite sides of the sheet, respectively.

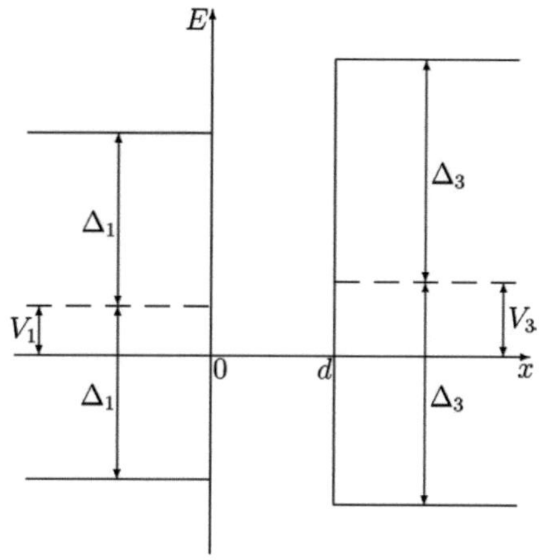

Figure 8. An energy scheme of QW under analysis.

1. At $x < 0$,

$$\psi_\lambda(x) = C \begin{pmatrix} 1 \\ q_1 \end{pmatrix} e^{k_1 x}, \tag{29}$$

where

$$q_1 = -i\frac{v_{F1}(k_1 - \lambda k_y)}{E_\lambda - V_1 + \Delta_1},$$

$$v_{F1}k_1 = \sqrt{\Delta_1^2 - (E_\lambda - V_1)^2 + v_{F1}^2 k_y^2}.$$

2. At $0 < x < d$,

$$\psi_\lambda(x) = C \begin{pmatrix} \kappa^* \\ q_2\kappa^* \end{pmatrix} e^{ik_2 x} + C \begin{pmatrix} \kappa \\ -q_2\kappa \end{pmatrix} e^{-ik_2 x}, \tag{30}$$

where

$$\kappa = \frac{1}{2}\sqrt{\frac{v_{F1}}{v_{F2}}}\left[1 + i\left(\frac{\lambda k_y}{k_2} + \frac{v_{F1}(k_1 - \lambda k_y)E_\lambda}{v_{F2}k_2(E_\lambda - V_1 + \Delta_1)}\right)\right],$$

$$q_2 = \frac{v_{F2}(k_2 + i\lambda k_y)}{E_\lambda}, \qquad E_\lambda = \pm v_{F2}\sqrt{k_2^2 + k_y^2},$$

with plus and minus corresponding to electrons and holes, respectively.

3. At $x > d$,

$$\psi_\lambda(x) = C \begin{pmatrix} \zeta \\ q_3\zeta \end{pmatrix} e^{-k_3(x-d)}, \tag{31}$$

where

$$\zeta = \sqrt{\frac{v_{F1}}{v_{F3}}}\left[\cos(k_2 d) + \left(\frac{\lambda k_y}{k_2} + \frac{v_{F1}(k_1 - \lambda k_y)E_\lambda}{v_{F2}k_2(E_\lambda - V_1 + \Delta_1)}\right)\sin(k_2 d)\right],$$

$$q_3 = i\frac{v_{F3}(k_3 + \lambda k_y)}{E_\lambda - V_3 + \Delta_3},$$

$$v_{F3}k_3 = \sqrt{\Delta_3^2 - (E_\lambda - V_3)^2 + v_{F2}^2 k_y^2}.$$

The constant C is found by using the normalization condition for wavefunctions (29)—(31),

$$\int_{-\infty}^{\infty} \Psi_\lambda^\dagger(x)\Psi_\lambda(x)dx = 1.$$

The carrier energy spectrum is determined by the dispersion relation

$$\tan(k_2 d) = v_{F2}k_2 f(\lambda k_y; k_1, k_3, E_\lambda), \tag{32}$$

where

$$\begin{aligned} f(\lambda k_y; k_1, k_3, E_\lambda) = &\left[v_{F1}(k_1 - \lambda k_y)(E_\lambda - V_3 + \Delta_3) + v_{F3}(k_3 + \lambda k_y)(E_\lambda - V_1 + \Delta_1)\right] \\ &\times \left[E_\lambda(E_\lambda - V_1 + \Delta_1)(E_\lambda - V_3 + \Delta_3) - v_{F2}v_{F3}\lambda k_y(k_3 + \lambda k_y)(E_\lambda - V_1 + \Delta_1)\right. \\ &\left.+ v_{F1}v_{F2}\lambda k_y(k_1 - \lambda k_y)(E_\lambda - V_3 + \Delta_3) - v_{F1}v_{F3}(k_1 - \lambda k_y)(k_3 + \lambda k_y)E_\lambda\right]^{-1} \end{aligned}$$

is a function of k_2 as well. Equation (32) must be solved for k_2, and then the energy E_λ is found.

In the case of an asymmetric QW, the dependence of (32) on λ gives rise to pseudospin splitting as the extrema of the dispersion curves shift away from Brillouin-zone corners. The dispersion relation predicts that $E_\lambda(k_y) \neq E_{-\lambda}(k_y)$, and an energy splitting appears near the conduction-band bottom at $k_y = k_{ye}^*$:

$$\delta E_s^e = |E_{-1}^e(k_{ye}^*) - E_{+1}^e(k_{ye}^*)|.$$

A similar energy splitting appears near the valence-band top at $k_y = k_{ye}^*$:

$$\delta E_s^h = |E_{-1}^h(k_{yh}^*) - E_{+1}^h(k_{yh}^*)|$$

Thus, a graphene nanoribbon becomes an indirect band-gap semiconductor analogous to silicon and germanium, where an electron-hole plasma can exist [53]. In the case of a symmetric QW ($\Delta_1 = \Delta_3$, $V_1 = V_3$, $v_{F1} = v_{F3}$) band structure is invariant under parity and there is no pseudospin splitting [42].

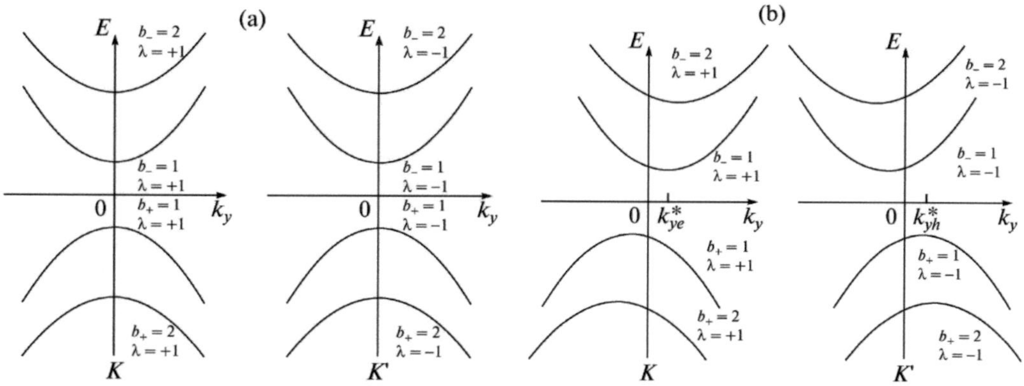

Figure 9. Energy spectra: **(a)** symmetric QW (no pseudospin splitting), with matching branches for $\lambda = +1$ and $\lambda = -1$ ($E_\lambda(k_y) = E_{-\lambda}(k_y) = E_\lambda(-k_y)$); **(b)** asymmetric QW, with pseudospin splitting manifested by the "spread-out" in quasimomentum between the extrema at k^*_{ye} for electrons, shown for $b_- = 1$, and at k^*_{yh} for holes, shown for $b_+ = 1$ ($E_\lambda(k_y) \neq E_{-\lambda}(k_y)$).

4.2. Size quantization

Solving Eq. (32), we determine the size-quantized energies

$$E_{\lambda b_\mp}(k_y) = \pm v_{F2}\sqrt{k^2_{2b_\mp}(\lambda k_y) + k_y^2},$$

where $b_\mp = 1, 2, \ldots$ labels electron $(-)$ and hole $(+)$ branches, respectively. The size-quantized energy spectra for symmetric and asymmetric QW are shown schematically in Fig. 9.

We now determine the carrier effective masses arising because of size quantization in the graphene nanoribbon in a planar heterostructure. Note that the effective masses are invariant under parity regardless of pseudospin splitting. Hereinafter, we omit indices $b_\mp$, restricting ourselves to a particular branch of the electron spectrum and a particular branch of the hole spectrum.

We write the dispersion law for electrons near an extremum at λk^*_{ye} as

$$\begin{aligned} E^e_\lambda &\approx E^e_0 + \frac{1}{2m^*_e}\left(k_y - \lambda k^*_{ye}\right)^2, \\ m^*_e &= \frac{1}{v_{F2}}\frac{\sqrt{k^2_{20e} + k^{*2}_{ye}}}{1 + k'^2_{20e} + k_{20e}k''_{20e}}, \end{aligned} \tag{33}$$

where the respective values k_{20e}, k'_{20e}, k''_{20e} of $k_{2e}(k_y)$ and its first and second derivatives at $k_y = \lambda k^*_{ye}$ are independent of λ; $E^e_0 = v_{F2}\sqrt{k^2_{20e} + k^{*2}_{ye}}$ is the energy at the extremum.

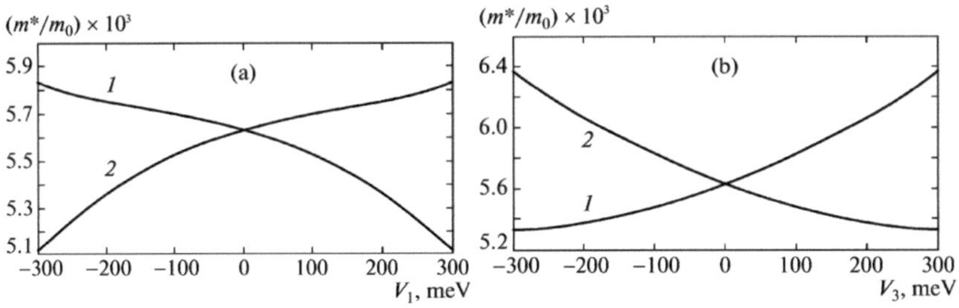

Figure 10. Electron (curves *1*) and hole (curves *2*) effective masses in the graphene nanoribbon (in units of free-electron mass m_0) as functions of V_1 for $V_3 = 0$ **(a)** and as functions of V_3 for $V_1 = 0$ **(b)**.

Analogous expressions are obtained for hole energies:

$$E_\lambda^h \approx E_0^h + \frac{1}{2m_h^*}\left(k_y - \lambda k_{yh}^*\right)^2,$$
$$m_h^* = \frac{1}{v_{F2}} \frac{\sqrt{k_{20h}^2 + k_{ye}^{*2}}}{1 + k_{20h}'^2 + k_{20e}k_{20h}''}, \tag{34}$$

where the respective values k_{20h}, k'_{20h}, k''_{20h} of $k_{2h}(k_y)$ and its first and second derivatives at $k_y = -\lambda k_{yh}^*$; $E_0^h = -v_{F2}\sqrt{k_{20h}^2 + k_{yh}^{*2}}$.

To estimate characteristic values, we consider the planar heterostructure combining a gapless nanoribbon with gapped graphene sheets with $\Delta_1 = 0.75$ eV, $v_{F1} = 1.1v_{F2}$, $\Delta_3 = 1$ eV, and $v_{F3} = 1.2v_{F2}$. The nanoribbon width is d = 2.46 nm (ten hexagonal cells). Since the unknown values of V_1 and V_3 can be found by comparing our results with experimental data, we seek the dependence of energy spectrum parameters on V_1 and V_3. Note that $|V_1| \leq \Delta_1$ and $|V_3| \leq \Delta_3$ to ensure that the heterostructure is type I.

Figures 10–13 show the results of numerical calculations of electron and hole effective masses in the graphe-ne nanoribbon, extremum energies, k_{xe}^* and k_{xh}^* values, and pseudospin splitting $\delta E_s^{e,h}$ plotted versus work function for one of the gapped graphene sheets given that the work function for the other is zero.

It is clear from Fig. 13 that the pseudospin splitting energy may amount to approximately 10 meV. To obtain a larger pseudospin splitting, QW must be more asymmetric. Both V_1 and V_3 can be varied by shifting the valley energies in gapped graphene under applied stress, with potential barriers playing the role of bandgaps in the gapped graphene sheets. An analogous effect is achieved by applying an electric field on the order of 10^6 V/cm perpendicular to the interfaces in the graphene plane [33].

As expected, the energy spectrum is symmetric under the change $E \to -E$ when V_1 and V_3 = 0; i.e., the electron and hole spectra have equal effective masses, extremum energies, extremum positions, and pseudospin splitting energies. The electron and hole effective masses in graphene are smaller than those in the gapped graphene sheets adjoining the gapless graphene nanoribbon ($m_1^* = \Delta_1/v_{F1}^2 \approx 0.11m_0$ and $m_3^* = \Delta_3/v_{F3}^2 \approx 0.15m_0$).

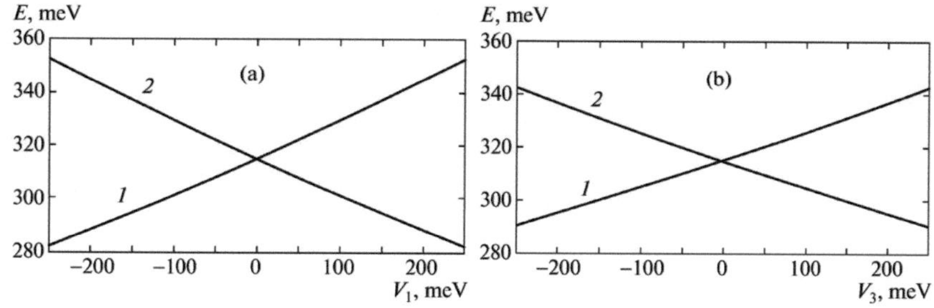

Figure 11. Electron (curves *1*) and hole (curves *2*) extremum energies and in the size-quantization spectra as functions of V_1 for $V_3 = 0$ **(a)** and as functions of V_3 for $V_1 = 0$ **(b)**. The effective bandgap $E_g^{eff} = E_0^e + |E_0^h| \approx 629$ meV varies insignificantly.

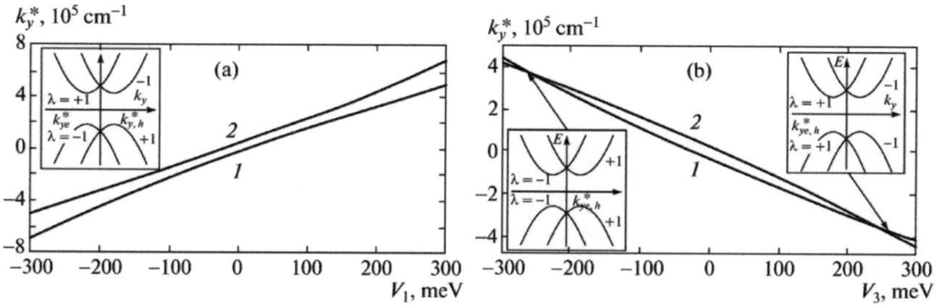

Figure 12. Extremum points of size-quantization branches for electrons ($k_y^* = k_{ye}^*$, curves *1*) and holes ($k_y^* = k_{yh}^*$, curves *2*) as functions V_1 for $V_3 = 0$ **(a)** and as functions of V_3 for $V_1 = 0$ **(b)**. Inserts show the relative positions of dispersion curves for $V_1 = V_3 = 0$ **(a)** and at $k_{ye}^* = k_{yh}^*$ **(b)**; K and K' points are set at the same position for simplicity.

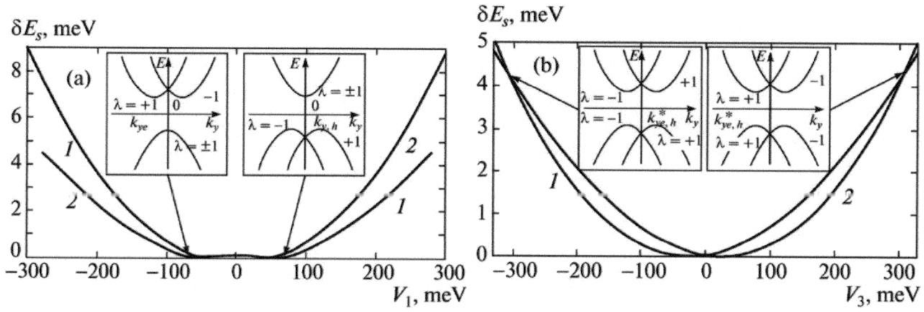

Figure 13. Pseudospin splitting in electron (curves *1*) and hole (curves *2*) spectra, δE_s^e and δE_s^h, as functions V_1 for $V_3 = 0$ **(a)** and as functions of V_3 for $V_1 = 0$ **(b)**. Vanishing $\delta E_s^{e,h}$ corresponds to vanishing $k_{ye,h}^*$ in Fig. 8, as shown in inserts to **(a)**. Inserts to **(b)** show positions of dispersion curves when k_{ye}^* and k_{yh}^* coincide.

4.3. Interface states

We consider interface states of a new type that arise in a narrow quasimomentum interval from the crossing of dispersion curves and are analogous to those in narrow-gap semiconductor heterostructures [44]. In the planar graphene-based heterostructure examined here, these states are localized near the heterojunction interfaces between the nanoribbon and the gapped graphene sheets. Interface states can exist not only in QWs but also in quantum barriers [10]. Note that interface states arise as well from the crossing of dispersion curves in a single heterojunction between different graphene materials [42].

The wave function describing an interface electronic state is expressed as follows.

1. At $x < 0$,

$$\widetilde{\psi}_\lambda(x) = \widetilde{C}\begin{pmatrix} 1 \\ \widetilde{q}_1 \end{pmatrix} e^{\kappa_1 x}, \tag{35}$$

where

$$\widetilde{q}_1 = -i\frac{u_1(\kappa_1 - \lambda k_y)}{E_\lambda - V_1 + \Delta_1},$$

$$v_{F1}\kappa_1 = \sqrt{\Delta_1^2 - (E_\lambda - V_1)^2 + v_{F1}^2 k_y^2}.$$

2. At $0 < x < d$,

$$\widetilde{\psi}_\lambda(x) = \widetilde{C}\begin{pmatrix} \widetilde{\varkappa}_- \\ \widetilde{q}_2\widetilde{\varkappa}_- \end{pmatrix} e^{-\kappa_2 x} + \widetilde{C}\begin{pmatrix} \widetilde{\varkappa}_+ \\ \widetilde{q}_2'\widetilde{\varkappa}_+ \end{pmatrix} e^{\kappa_2 x}, \tag{36}$$

where

$$\widetilde{\varkappa}_\pm = \frac{1}{2}\sqrt{\frac{v_{F1}}{v_{F2}}}\left[1 \pm \frac{\lambda k_y}{\kappa_2} \pm \frac{v_{F1}(\kappa_1 - \lambda k_y)E_\lambda}{v_{F2}\kappa_2(E_\lambda - V_1 + \Delta_1)}\right],$$

$$\widetilde{q}_2 = i\frac{v_{F2}(\kappa_2 + \lambda k_y)}{E_\lambda}, \; \widetilde{q}_2' = -i\frac{v_{F2}(\kappa_2 - \lambda k_y)}{E_\lambda}.$$

3. At $x > d$

$$\widetilde{\psi}_\lambda(x) = \widetilde{C}\begin{pmatrix} \widetilde{\zeta} \\ \widetilde{q}_3\widetilde{\zeta} \end{pmatrix} e^{-\kappa_3(x-d)}, \tag{37}$$

where

$$\widetilde{\zeta} = \sqrt{\frac{v_{F1}}{v_{F3}}}\left[ch(\kappa_2 d) + \left(\frac{\lambda k_y}{\kappa_2} + \frac{v_{F1}(\kappa_1 - \lambda k_y)E_\lambda}{v_{F2}\kappa_2(E_\lambda - V_1 + \Delta_1)}\right) sh(\kappa_2 d)\right]$$

$$\widetilde{q}_3 = i\frac{v_{F3}(\kappa_3 + \lambda k_y)}{E_\lambda - V_3 + \Delta_3},$$

$$v_{F3}k_3 = \sqrt{\Delta_3^2 - (E_\lambda - V_3)^2 + v_{F3}^2 k_y^2}.$$

The relation for energy of interface states is

$$E_\lambda = \pm v_{F2}\sqrt{k_y^2 - \kappa_2^2}, \tag{38}$$

with plus and minus corresponding to electrons and holes, respectively.

The expression for energy in (38) implies that an interface state exists only if[4]

$$|\kappa_2| < |k_y|.$$

We obtain the dispersion relation

$$\tanh(\kappa_2 d) = v_{F2}\kappa_2 f(\lambda k_y; \kappa_1, \kappa_3, E_\lambda). \tag{39}$$

which is similar to (32) up to the substitutions $k_1 \to \kappa_1$, $k_2 \to i\kappa_2$, and $k_3 \to \kappa_3$.

When $V_1 = 0$, the allowed quasimomenta for hole interface states ($\lambda = -1$) are similar to those for electron states, but the hole and electron energies have opposite signs; i.e., the spectrum is symmetric under the change $E_\lambda \to -E_\lambda$. When $V_1 = 100$ meV, the symmetry is broken and hole interface states exist only at negative quasimomenta.

4.4. Excitons

In gapless graphene, the carrier effective mass is zero and excitons do not exist. The existence of excitons in gapless graphene would lead to excitonic instability and excitonic insulator transition to a gapped state [54, 55].

The energy gap arising in a graphene nanoribbon due to the size quantization makes it possible to generate excitons by optical excitation or electron-hole injection. Excitons in QW strongly affect optical properties of the system considered here.

Excitons in similar quasi-one-dimensional carbon-based systems (semiconducting single- and multi-walled nanotubes) have been studied theoretically in [56]. The exciton spectrum is calculated here for a planar graphene quantum well by using the model applied to quantum wires in [57]. This model yields simple analytical expressions for exciton binding energy.

Since formulas (33) and (34) are obtained in the nonrelativistic limit, the two-particle exciton wave function depending on the electron and hole coordinates y_- and y_+ in a sufficiently narrow nanoribbon must obey the 1D Schrödinger equation with Coulomb potential:

$$\left(-\frac{1}{2m_e^*}\frac{\partial^2}{\partial y_-^2} - \frac{1}{2m_h^*}\frac{\partial^2}{\partial y_+^2} - \frac{\tilde{e}^2}{|y_- - y_+|}\right)\phi(y_-, y_+) = E'\phi(y_-, y_+), \tag{40}$$

where $E' = E - E_g^{eff}$ and $\tilde{e}^2 \equiv e^2/\kappa_{eff}$. The effective dielectric constant of graphene, $\kappa_{eff} = (\varepsilon + \varepsilon')/2$, may vary widely with the dielectric constants ε and ε' of the media in contact with graphene, such as free-space permittivity and substrate dielectric constant [58, 59].

[4] The zero mode corresponding to $|\kappa_2| = |k_y|$ (with $E_\lambda = 0$) is irrelevant here because $\widetilde{\psi}_\lambda(x) \equiv 0$.

The electron-hole Coulomb interaction in a 1D graphene nanoribbon is three-dimensional, but the problem can be reduced to one dimension (electron and hole y positions) for sufficiently narrow nanoribbons.

Rewriting Eq. (40) in terms of electron-hole separation $y = y_- - y_+$ and center-of-mass coordinate

$$Y = \frac{m_e^* y_- + m_h^* y_+}{m_e^* + m_h^*}$$

and introducing the function

$$\phi(y_-, y_+) = \psi_n(y) e^{iKY},$$

where K is the total exciton momentum, we obtain

$$\left(-\frac{1}{2\mu^*} \frac{\partial^2}{\partial y^2} - \frac{\tilde{e}^2}{|y|} \right) \psi_n(y) = E_n \psi_n(y), \tag{41}$$

where $\mu^* = m_e^* m_h^* / (m_e^* + m_h^*)$ is the reduced mass and E_n is the energy of the nth exciton level ($n = 0, 1, 2, \ldots$ is the principal quantum number). The total exciton energy E' is obtained by adding the total kinetic energy of the electron-hole pair to E_n:

$$E' = E_n + \frac{K^2}{2(m_e^* + m_h^*)}.$$

To find the solution at $y > 0$, we substitute $\psi_n(y)$ represented as

$$\psi_n(y) = B_n \exp\left(-y/a_n\right) F_n\left(\frac{2y}{a_n}\right).$$

into Eq. (41) and obtain the confluent hypergeometric differential equation

$$\xi F_n'' - \xi F_n' + \eta F_n = 0, \tag{42}$$

where $\xi = \frac{2y}{a_n}$ and $\eta = \mu^* \tilde{e}^2 a_n$. We also have

$$E_n = -\frac{1}{2\mu^* a_n^2}. \tag{43}$$

Equation (42) with $\eta = n$ is solved by the associated Laguerre polynomial

$$F_n(\xi) = \frac{1}{n!} \xi e^{\xi} \frac{d^n}{d\xi^n} \left(\xi^{n-1} e^{-\xi} \right) \equiv L_n^{-1}(\xi).$$

and the wavefunction is expressed as

$$\psi_n(y) = B_n \exp\left(-y/a_n\right) L_n^{-1}\left(\frac{2y}{a_n}\right).$$

Analogously, we find the solution to Eq. (41) at $y < 0$:

$$\psi_n(y) = \pm B_n \exp\left(y/a_n\right) L_n^{-1}\left(-\frac{2y}{a_n}\right),$$

where "+" and "−" are taken for $n = 0$ and $n \neq 0$, respectively, and the continuity of $\psi_n(y)$ and its first derivative $\psi_n'(y)$ are used as boundary conditions. Since $\psi_n(0) = 0$ and $\psi'(0) \neq 0$ for $n \neq 0$, the excited-state wavefunction $\psi_n(y)$ is odd (otherwise, it would be discontinuous at the origin), whereas the ground-state wavefunction is even.

The normalization condition

$$\int_{-\infty}^{\infty} |\psi_n(y)|^2 dy = 1$$

is used to determine the coefficient B_n in the expression for $\psi_n(y)$:

$$B_n = \left[a_n \int_0^{\infty} (L_n^{-1}(\xi))^2 e^{-\xi} d\xi\right]^{-1/2}$$

and $B_n = 1/\sqrt{2a_n}$ for $n = 1, 2, \ldots$ and $B_0 = 1/\sqrt{a_0}$ for $n = 0$. Here,

$$a_n = \frac{n}{\mu^* \tilde{e}^2} \tag{44}$$

($n = 1, 2 \ldots$) is the Bohr radius of an exciton in the nth excited state. Combining (43) with (44), we find the exciton energy spectrum:

$$E_n = -\frac{\mu^* \tilde{e}^4}{2n^2}. \tag{45}$$

The 1D ground-state ($n = 0$) Coulomb energy exhibits a logarithmic divergence at short distances [60]. Therefore, the lateral spread of the exciton wave function (along the x axis) due to the three-dimensional nature of Coulomb interaction should be taken into account by introducing a cutoff parameter $d_0 \lesssim d$. Averaging the kinetic energy operator

$$\widehat{T} = -\frac{1}{2\mu^*}\frac{\partial^2}{\partial y^2}$$

and the potential

$$V(y) = -\frac{\tilde{e}^2}{|y|}\theta\left(|y| - d_0\right)$$

over ground-state wave functions

$$\psi_0(y) = \frac{1}{\sqrt{a_0}}e^{-|y|/a_0}, \tag{46}$$

where the ground-state Bohr radius a_0 plays the role of a variational parameter, we express the ground-state exciton energy as [42]

$$E_0 = \frac{1}{2\mu^* a_0^2} - \frac{2\tilde{e}^2}{a_0}\ln\frac{a_0}{d}. \tag{47}$$

Minimizing (47) with respect to a_0, we obtain an equation for a_0:

$$a_0 = \frac{a_1}{2\left(\ln\frac{a_0}{d} - 1\right)}. \tag{48}$$

To logarithmic accuracy, when

$$\ln\frac{a_1}{d} \gg 1, \tag{49}$$

we find the relations

$$E_0 = 4E_1\ln^2\frac{a_1}{d}, \tag{50}$$

$$a_0 = \frac{a_1}{2\ln\frac{a_1}{d}}. \tag{51}$$

Using (48), we easily obtain the next-order correction to E_0:

$$\delta E_0^{(1)} = -8E_1\ln\frac{a_1}{d}\ln\left(2\ln\frac{a_1}{d}\right).$$

We now examine the applicability of the formulas derived here. The semiconducting state induced in a graphene nanoribbon is stable with respect to spontaneous electron-hole pair creation (excitonic insulator transition) only if the exciton binding energy $|E_0|$ is smaller than the effective bandgap in the graphene nanoribbon,

$$|E_0| < E_g^{eff}.$$

Furthermore, the quantum well width d must be much smaller than the exciton Bohr radius a_1,

$$d \ll a_1.$$

Logarithmically accurate formula (50) is correct only if condition (49) holds. However, the asymmetric QW analyzed here to examine pseudospin effects may not admit even a single size-quantization level if the graphene nanoribbon width d is too narrow. As d decreases, the effective bandgap increases, approaching $\Delta_+ + \Delta_-$, where $\Delta_\pm = \min\{\Delta_1 \pm V_1, \Delta_3 \pm V_3\}$ (with plus and minus corresponding to electrons and holes, respectively). When a certain d_c is reached, the size-quantization levels are pushed into the continuum. This imposes a lower limit on d:

$$d > d_c,$$

where d_c can be estimated as [42]

$$d_c \simeq \frac{\pi v_{F2}}{\Delta_+ + \Delta_-}.$$

As d increases, condition (49) is violated. In this case, a more accurate variational calculation should be performed using the modified three-dimensional Coulomb potential

$$\widetilde{V}(y) = -\frac{\widetilde{e}^2}{\sqrt{y^2 + d_0^2}},$$

where d_0 is a cutoff parameter. We average the Hamiltonian with potential $\widetilde{V}(y)$ over trial functions (46) to obtain

$$E_0 = \frac{1}{2\mu^* a_0^2} - \frac{2\widetilde{e}^2}{a_0} I(\rho), \tag{52}$$

where $I(\rho) = \frac{\pi}{2}\left[H_0(\rho) - Y_0(\rho)\right]$, $H_\nu(\rho)$ is a Struve function, $Y_\nu(\rho)$ is a Bessel function of the second kind, and $\rho = 2d_0/a_0$ (ν is 0 here and 1 below).

Minimizing (52) with respect to a_0, we obtain an equation for a_0:

$$\frac{2a_0}{a_1} I(\rho) + \frac{4d_0}{a_1} J(\rho) = 1, \tag{53}$$

where $J(\rho) = 1 - \frac{\pi}{2}\left[H_1(\rho) - Y_1(\rho)\right]$.

Figure 14 shows the numerical results obtained by using both methods to calculate $E_0(d)$ for the heterostructure, with $d_0 \propto d$ adjusted to match the curves at small d. Discrepancy at large d increases as $\ln(a_1/d)$ approaches unity.

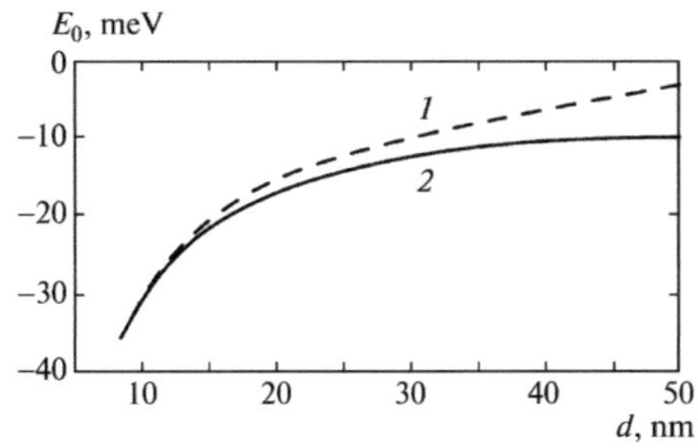

Figure 14. Exciton ground-state energy calculated by formula (50) (curve *1*) and by formula (52) after Eq. (53) is solved numerically for a_0 (curve 2), $d_0 = 0.22d$.

4.5. Electric field effect on excitons levels

Interaction between an exciton and an external electrostatic field $\mathcal{E}$ is described by the operator

$$\widehat{H}_i = -\mathbf{d}\mathcal{E} = |\widetilde{e}|(\mathcal{E}_x x + \mathcal{E}_y y)$$

where $x = |x_- - x_+|$ and $y = |y_- - y_+|$ are the electron-hole relative position vector components and **d** is dipole moment. The electric field is supposed to be weak enough to ensure that the energy level shift is not only smaller than the spacing between size-quantization levels but also smaller than the spacing between exciton levels. These conditions can be written as

$$d \ll a_1 \ll a_{\mathcal{E}},$$

where $a_{\mathcal{E}} = (\mu^* |\widetilde{e}| \mathcal{E})^{-1/3}$ is the electric length.

We consider two cases: (1) the electric field is applied parallel to the x axis and perpendicular to the nanoribbon edges in the graphene plane; (2) the electric field is applied along the y axis, parallel to the nanoribbon edges.

In the former case, the energy shift varies linearly with the difference between the average x components of the electron and hole position vectors:

$$E^{(1)}_{\perp\lambda\lambda'} = |\widetilde{e}|\mathcal{E}\left(\langle x_- \rangle_\lambda - \langle x_+ \rangle_{\lambda'}\right), \tag{54}$$

where average x components are calculated by using electron and hole single-particle wave functions, generally depending on the electron and hole eigenvalues λ and λ' of the operator $\widehat{P}$, respectively. Exciton energy shift (54) is independent of the principal quantum number n. It may vary with λ and λ', resulting in different exciton binding energies (more precisely, the binding energy of an electron–hole pair with $\lambda = \pm 1$ and $\lambda' = \pm 1$ may have four different values).

In the latter case, the first-order electric field-induced correction is zero[5],

[5] Note that $E^{(1)}_{\perp n} \equiv 0$ in the former case if the electron and hole spectra transform into each other under field inversion.

$$E_{\parallel n}^{(1)} = |\tilde{e}|\mathcal{E}\langle y\rangle_n \equiv 0, \tag{55}$$

because the integral of $y|\psi_n(y)|^2$ with respect to y vanishes. To evaluate the second-order electric field-induced correction, we make use of the Dalgarno–Lewis perturbation theory [61]. Defining a Hermitian operator such that

$$[\widehat{F},\, \widehat{H}_0]|n\rangle = \widehat{H}_i|n\rangle, \tag{56}$$

where

$$\widehat{H}_0 = -\frac{1}{2\mu^*}\frac{\partial^2}{\partial y^2} - \frac{\tilde{e}^2}{|y|}$$

is the zeroth-order Hamiltonian, $|n\rangle = \psi_n(y)$ is the zeroth-order wave function of the nth exciton level, and $\widehat{H}_i = |\tilde{e}|\mathcal{E}y$, we obtain

$$E_{\parallel n}^{(2)} = \langle n|\widehat{H}_i\widehat{F}|n\rangle - \langle n|\widehat{H}_i|n\rangle\langle n|\widehat{F}|n\rangle. \tag{57}$$

In the case in question, the second term in this formula vanishes by virtue of (55).

Rewriting Eq. (56) as

$$\psi_n\frac{\partial^2\widehat{F}}{\partial y^2} + 2\frac{\partial\psi_n}{\partial y}\frac{\partial\widehat{F}}{\partial y} = 2\mu^*\widehat{H}_i\psi_n, \tag{58}$$

we find

$$\widehat{F}(y) = 2\mu^*\int_{-\infty}^{y}\frac{dy'}{|\psi_n(y')|^2}\int_{-\infty}^{y'}dy''\psi_n^*(y'')\widehat{H}_i\psi_n(y''). \tag{59}$$

Combining (57) with (59), we have the exciton ground-state energy shift [42]

$$E_{\parallel 0}^{(2)} = -\frac{5}{128}\frac{a_1^3}{\ln^4\frac{a_1}{d}}\mathcal{E}^2, \tag{60}$$

which is very small compared to E_0 given by (50) because of the fourth power of a logarithm in the denominator and a small numerical factor.

For comparison, we write out the energy correction to the first excited exciton state [42]

$$E_{\parallel 1}^{(2)} = -\frac{3}{8}(31-6\gamma)a_1^3\mathcal{E}^2,$$

where $\gamma = 0.577\ldots$ is Euler's constant.

By analogy with layered heterostructures [62], the ionizing (exciton-breaking) field strength $\mathcal{E}_c$ is estimated as

$$\mathcal{E}_c = \frac{|E_0|}{8|\tilde{e}|\langle|y|\rangle_0}, \tag{61}$$

where $\langle|y|\rangle_0 = a_0/2$ is the average electron-hole separation for the ground-state exciton. To logarithmic accuracy, it follows that [42]

$$\mathcal{E}_c = \mu^{*2}|\tilde{e}|^5 \ln^3 \frac{a_1}{d}. \tag{62}$$

To get the order of magnitude of $\mathcal{E}_c$, consider QW discussed abode. Setting $m_e^* = m_h^* \approx 0.0056m_0$, the SiO_2 substrate dielectric constant $\kappa_{eff} \approx 5$, $d = 2.46$ nm, and $a_1 \approx 81$ nm, we use formula (62) to obtain $\mathcal{E}_c = 9$ kV/cm.

4.6. The effective theory

4.6.1. The effective Hamiltonian

According to the expressions (33) and (34) we have the approximations for dispersion curves of electrons and holes respectively

$$\begin{aligned} E_\lambda^e &\approx E_0^e + \frac{1}{2m_e^*}\left(k_y - \lambda k_{ye}^*\right)^2 = \frac{k_y^2}{2m_e^*} - \lambda\alpha_e k_y + \Delta_e \equiv \widetilde{E}_\lambda^e, \\ E_\lambda^h &\approx E_0^h - \frac{1}{2m_h^*}\left(k_y + \lambda k_{yh}^*\right)^2 = -\frac{k_y^2}{2m_h^*} - \lambda\alpha_h k_y + \Delta_h \equiv \widetilde{E}_\lambda^h, \end{aligned} \tag{63}$$

where the following notations are introduced

$$\alpha_{e,h} = \frac{k_{ye,h}^*}{m_{e,h}^*}, \quad \Delta_{e,h} = \pm\frac{k_{ye,h}^{*2}}{2m_{e,h}^*} + E_0^{e,h}$$

(in the latter formula plus corresponds to the case of electrons, minus for the case of holes).

We can write the effective Hamiltonians in the form including explicitly the parity λ

$$\begin{aligned} \widehat{H}_{eff}^{(\lambda)e} &= \frac{\widehat{p}_y^2}{2m_e^*} - \lambda\alpha_e\widehat{p}_y + \Delta_e, \\ \widehat{H}_{eff}^{(\lambda)h} &= \frac{\widehat{p}_y^2}{2m_h^*} - \lambda\alpha_h\widehat{p}_y + \Delta_h. \end{aligned} \tag{64}$$

We can also combine the Hamiltonians $\widehat{H}_{eff}^{(+1)e}$ and $\widehat{H}_{eff}^{(-1)e}$ into one 2×2 matrix Hamiltonian (analogously for $\widehat{H}_{eff}^{(+1)h}$ and $\widehat{H}_{eff}^{(-1)h}$)

$$\begin{aligned}\widehat{H}_{eff}^{e} &= \frac{\hat{p}_y^2}{2m_e^*} - \alpha_e \tau_z \hat{p}_y + \Delta_e,\\ \widehat{H}_{eff}^{h} &= \frac{\hat{p}_y^2}{2m_h^*} - \alpha_h \tau_z \hat{p}_y + \Delta_h.\end{aligned} \tag{65}$$

Here we emphasize by using the matrix τ_z that these Hamiltonians act in the valley space. The eigen wave functions of these Hamiltonians $\widetilde{\Psi}_{\lambda}^{e}$ and $\widetilde{\Psi}_{\lambda}^{h}$,

$$\widehat{H}_{eff}^{e}\widetilde{\Psi}_{\lambda}^{e} = \widetilde{E}_{\lambda}^{e}\widetilde{\Psi}_{\lambda}^{e} \text{ and } \widehat{H}_{eff}^{h}\widetilde{\Psi}_{\lambda}^{h} = \widetilde{E}_{\lambda}^{h}\widetilde{\Psi}_{\lambda}^{h},$$

are spinors in a class of eigen functions of the operator $\hat{\mathcal{P}} = \tau_z$ which can be considered as a "reduced" parity operator:

$$\begin{aligned}\widehat{\mathcal{P}}\widetilde{\Psi}_{\lambda}^{e,h} &= \lambda\widetilde{\Psi}_{\lambda}^{e,h},\\ \widetilde{\Psi}_{+1}^{e,h} = \begin{pmatrix}\widetilde{\psi}_{+1}^{e,h}\\ 0\end{pmatrix}&, \quad \widetilde{\Psi}_{-1}^{e,h} = \begin{pmatrix}0\\ \widetilde{\psi}_{-1}^{e,h}\end{pmatrix}.\end{aligned} \tag{66}$$

(Remember that the parity operator $\widehat{P} = \tau_z \otimes \sigma_0$ where the matrix σ_0 acts in the sublattice space.)

We can also see that the second term in (65) is an analogue of spin-orbit (SO) coupling in the Rashba form:

$$\widehat{H}_{SO}^{e,h} = \alpha_{e,h}\left[\boldsymbol{\tau}\widehat{\mathbf{p}}\right] \cdot \boldsymbol{\nu}, \tag{67}$$

where $\boldsymbol{\tau} = (\tau_x, \tau_y, \tau_z)$ is the matrix vector in the valley space ($\boldsymbol{\tau}$ is the pseudospin operator), $\widehat{\mathbf{p}} = (\widehat{p}_x, \widehat{p}_y, 0)$ is the vector operator of momentum in the xy-plane, and $\boldsymbol{\nu}$ is the unit vector of the normal to the heterojunction interface (in the coordinate system used here, $\boldsymbol{\nu} = \mathbf{e}_x$ is the unit coordinate vector of the x axis). The constants $\alpha_{e,h}$ can be named as the effective Rashba constants.

So, we have the effective theory with the non-relativistic Hamiltonian with the SO-like term describing the pseudospin splitting of the energy spectrum of charge carriers. An appearance of this term is due to an asymmetry of a QW potential profile. In an absence of the asymmetry in the case of the symmetrical QW, the effective Rashba constants $\alpha_{e,h}$ tend to zero for both electrons and holes and there is no the pseudospin splitting in its energy spectra.

4.6.2. The effective Hamiltonian in a presence of a magnetic field

Let us set a problem about a quantization of a charge carriers energy in the planar heterostructure based on graphene in a magnetic field applied perpendicular to its plane. To begin with, we make some remarks.

Firstly, to separate the size quantization and the magnetic field quantization, we assume that a magnetic field H is enough weak one. Its condition can be expressed as the following inequality

$$a_H \gg d, \tag{68}$$

where

$$a_H = \sqrt{\frac{c}{|e|H}}$$

is the magnetic length, i.e. a restriction of the wave function along the y axis due to a magnetic field is significantly smaller than its restriction along the x axis (perpendicular to the potential barriers). A complexity of the problem consists in an absence of the usual Landau quantization whereas we have it in layered heterostructures at an application of a magnetic field perpendicular to layers (in this case problems about the size quantization and the magnetic field quantization are automatically separated).

Secondly, a vector potential $\mathbf{A}$ must be chosen so that decreasing wave function $\widetilde{\Psi}_\lambda^{e,h}(x, y)$ in the direction of the y axis is taken into account in an explicit form. The equation for $\widetilde{\Psi}_\lambda^{e,h}(x, y)$ in a presence the of a magnetic field must include a y coordinate. From general considerations, it is clear that $\widetilde{\Psi}_\lambda^{e,h}(x, y)$ must decrease along the y axis on the scale of a_H (see the schematic picture on Fig. 15).

An appropriate choice of the vector potential is

$$\mathbf{A} = (-Hy, 0, 0)\,.$$

To have an opportunity to make the minimal substitution, the effective equation for $\widetilde{\Psi}_\lambda^{e,h}(x, y)$ must explicitly contain the momentum operator $\widehat{p}_x$. However, this operator was excluded from the effective Hamiltonian without a magnetic field (65). This circumstance makes us detach in the effective equation an artificial term corresponding to the "size-quantization energy" $E_0^{e,h}$ so that

$$\frac{\widehat{p}_x^2}{2m_{e,h}^*}\widetilde{\Psi}_\lambda^{e,h}(x, y) = E_0^{e,h}\widetilde{\Psi}_\lambda^{e,h}(x, y).$$

Moreover, we propose to use an approximate expression

$$E_0^{e,h} = \pm\frac{k_{xe,h}^{*2}}{2m_{e,h}^*},$$

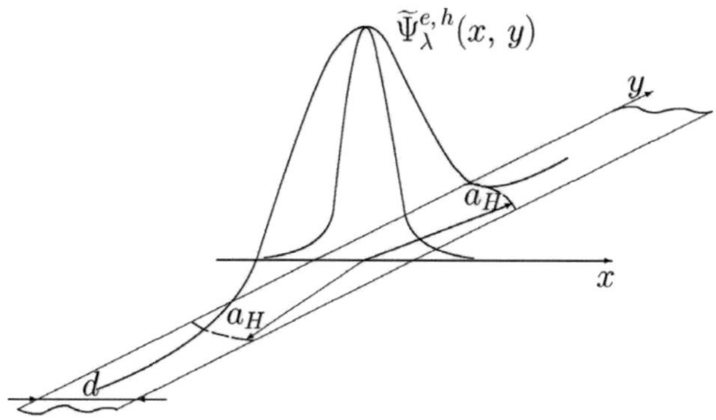

Figure 15. A behavior of the envelope wave function $\widetilde{\Psi}_{\lambda}^{e,h}(x, y)$ of charge carriers in the planar heterostructure in a presence of a magnetic field applied perpendicular to its plane.

where k_{xe}^* and k_{xh}^* are the size-quantized values of k_x as eigen values of the operator $\widehat{p}_x$ for electrons and holes near the extrema $k_y = \lambda k_{ye}^*$ and $k_y = -\lambda k_{yh}^*$, respectively.

So, we have in a presence of a magnetic field

$$\begin{aligned} \widehat{H}_{eff}^{\prime e} &= \frac{1}{2m_e^*}\left(k_{xe}^* + \frac{e}{c}Hy\right)^2 + \frac{1}{2m_e^*}\widehat{p}_y^2 + \widehat{H}_{SO}^e + \widetilde{\Delta}_e, \\ \widehat{H}_{eff}^{\prime h} &= \frac{1}{2m_h^*}\left(k_{xh}^* + \frac{e}{c}Hy\right)^2 + \frac{1}{2m_h^*}\widehat{p}_y^2 + \widehat{H}_{SO}^h + \widetilde{\Delta}_h, \end{aligned} \tag{69}$$

with

$$\widetilde{\Delta}_{e,h} = \pm\frac{k_{ye,h}^{*2}}{2m_{e,h}^*}.$$

We can introduce operators of a generalized momentum in a magnetic field

$$\begin{aligned} \widehat{P}_x^{e,h} &= k_{xe,h}^* - \frac{e}{c}\mathcal{A}_x, \\ \widehat{P}_y^{e,h} &= \widehat{p}_y - \frac{e}{c}\mathcal{A}_y. \end{aligned} \tag{70}$$

The operators $\widehat{P}_x^{e,h}$ are c-numbers, whereas the operators $\widehat{P}_y^{e,h}$ coincide with the differential operator $\widehat{p}_y = -i\partial_y$ in the determined above vector potential **A**. The commutation relation for these operators is

$$\left[\widehat{P}_x^{e,h}, \widehat{P}_y^{e,h}\right] = \frac{ie}{c}H. \tag{71}$$

We can also consider the combinations of these operators

$$\widehat{P}_{\pm}^{e,h} = \widehat{P}_{x}^{e,h} \pm i\widehat{P}_{y}^{e,h} \tag{72}$$

with the commutation relation

$$\left[\widehat{P}_{-}^{e,h}, \widehat{P}_{+}^{e,h}\right] = -\frac{2e}{c}H. \tag{73}$$

Following the Rashba's paper [63] we can introduce the annihilation and creation operators: for electrons as

$$\hat{a}_e = \sqrt{\frac{c}{2|e|H}}\widehat{P}_{-}^{e},\ \hat{a}_e^{\dagger} = \sqrt{\frac{c}{2|e|H}}\widehat{P}_{+}^{e},$$

for holes as

$$\hat{a}_h = \sqrt{\frac{c}{2|e|H}}\widehat{P}_{+}^{e},\ \hat{a}_h^{\dagger} = \sqrt{\frac{c}{2|e|H}}\widehat{P}_{-}^{h}$$

with the Bose commutation relation

$$\left[\hat{a}_{e,h}, \hat{a}_{e,h}^{\dagger}\right] = 1.$$

The effective Hamiltonians rewritten in the second-quantized representation are

$$\begin{aligned}\widehat{H}_{eff}^{\prime\prime e} &= \omega_c^{*e}\left(\hat{a}_e^{\dagger}\hat{a}_e + \frac{1}{2}\right) - \frac{i}{\sqrt{2}}\frac{\alpha_e}{a_H}\tau_z\left(\hat{a}_e - \hat{a}_e^{\dagger}\right) + \widetilde{\Delta}_e,\\ \widehat{H}_{eff}^{\prime\prime h} &= \omega_c^{*h}\left(\hat{a}_e^{\dagger}\hat{a}_e + \frac{1}{2}\right) + \frac{i}{\sqrt{2}}\frac{\alpha_h}{a_H}\tau_z\left(\hat{a}_h - \hat{a}_h^{\dagger}\right) + \widetilde{\Delta}_h,\end{aligned} \tag{74}$$

where $\omega_c^{*e,h} = \frac{|e|H}{m_{e,h}^{*}c}$ are the cyclotron frequencies.

Let us solve the equation for $\widetilde{\Psi}_{\lambda}^{e,h}$ with the Hamiltonians (74) in the class of eigen functions of the operator $\hat{\mathcal{P}}$. The matrix τ_z should be replaced by corresponding eigen value λ in the equations for the components of $\widetilde{\Psi}_{\lambda}^{e,h}$. After a nondimensionalization, these equations are written as

$$\begin{aligned}\left[\hat{a}_e^{\dagger}\hat{a}_e + \frac{1}{2} - i\lambda\delta_{SO}^{e}\left(\hat{a}_e - \hat{a}_e^{\dagger}\right)\right]\widetilde{\psi}_{\lambda}^{e} &= \epsilon_{\lambda}^{e}\widetilde{\psi}_{\lambda}^{e},\\ \left[\hat{a}_h^{\dagger}\hat{a}_h + \frac{1}{2} + i\lambda\delta_{SO}^{h}\left(\hat{a}_h - \hat{a}_h^{\dagger}\right)\right]\widetilde{\psi}_{\lambda}^{h} &= \epsilon_{\lambda}^{h}\widetilde{\psi}_{\lambda}^{h},\end{aligned} \tag{75}$$

where $\delta_{SO}^{e,h} = \frac{\alpha_{e,h}}{\sqrt{2}a_H\omega_c^{*e,h}} = \frac{1}{\sqrt{2}}a_H k_{ye,h}^*$ and $\pm\epsilon_\lambda^{e,h} = \frac{E_\lambda^{e,n}-\Delta_{e,h}}{\omega_c^{*e,h}}$, $\epsilon_\lambda^{e,h} > 0$ and plus corresponds to electrons (its energy $\widetilde{E}^e = \widetilde{\Delta}_e + \omega_c^{*e}\epsilon_\lambda^e$ has positive values), minus corresponds to holes (its energy $\widetilde{E}^h = \widetilde{\Delta}_h - \omega_c^{*h}\epsilon_\lambda^h$ has negative values).

Now we make an expansion of $\widetilde{\psi}_\lambda^{e,h}$ by the oscillator function basis ψ_n

$$\widetilde{\psi}_\lambda^{e,h} = \sum_{n=0}^{\infty} a_{\lambda n}^{e,h}\psi_n. \tag{76}$$

We have also the normalization condition in the form of the integral

$$\int_{-\infty}^{\infty} \left|\widetilde{\psi}_\lambda^{e,h}\right|^2 dy = 1$$

or in the form of the series

$$\sum_{n=0}^{\infty} \left|a_{\lambda n}^{e,h}\right|^2 = 1. \tag{77}$$

Taking into account the relations

$$\begin{aligned} \hat{a}_{e,h}\psi_n &= \sqrt{n}\psi_{n-1}, \\ \hat{a}_{e,h}^{\dagger}\psi_n &= \sqrt{n+1}\psi_{n+1}, \end{aligned}$$

we obtain two infinite systems of equations for coefficients $a_{\lambda n}^e$ and $a_{\lambda n}^h$

$$\begin{cases} \frac{1}{2}a_{\lambda 0}^e - i\lambda\delta_{SO}^e a_{\lambda 1}^e = \epsilon_\lambda^e a_{\lambda 0}^e, \\ \frac{3}{2}a_{\lambda 1}^e - i\lambda\sqrt{2}\delta_{SO}^e a_{\lambda 2}^e + i\lambda\delta_{SO}^e a_{\lambda 0}^e = \epsilon_\lambda^e a_{\lambda 1}^e, \\ \frac{5}{2}a_{\lambda 2}^e - i\lambda\sqrt{3}\delta_{SO}^e a_{\lambda 3}^e + i\lambda\sqrt{2}\delta_{SO}^e a_{\lambda 1}^e = \epsilon_\lambda^e a_{\lambda 2}^e, \\ \dots\dots\dots\dots\dots\dots\dots\dots\dots\dots \\ \left(n+\frac{1}{2}\right)a_{\lambda n}^e - i\lambda\sqrt{n+1}\delta_{SO}^e a_{\lambda n+1}^e + i\lambda\sqrt{n}\delta_{SO}^e a_{\lambda n-1}^e = \epsilon_\lambda^e a_{\lambda n}^e, \\ \dots\dots\dots\dots\dots\dots\dots\dots\dots\dots \end{cases} \tag{78}$$

$$\begin{cases} \frac{1}{2}a^h_{\lambda 0} + i\lambda\delta^h_{SO}a^h_{\lambda 1} = \epsilon^h_\lambda a^h_{\lambda 0}, \\ \frac{3}{2}a^h_{\lambda 1} + i\lambda\sqrt{2}\delta^h_{SO}a^h_{\lambda 2} - i\lambda\delta^h_{SO}a^h_{\lambda 0} = \epsilon^h_\lambda a^h_{\lambda 1}, \\ \frac{5}{2}a^e_{\lambda 2} + i\lambda\sqrt{3}\delta^h_{SO}a^h_{\lambda 3} - i\lambda\sqrt{2}\delta^h_{SO}a^h_{\lambda 1} = \epsilon^h_\lambda a^h_{\lambda 2}, \\ \cdots\cdots\cdots\cdots\cdots\cdots\cdots\cdots\cdots\cdots \\ \left(n+\frac{1}{2}\right)a^h_{\lambda n} + i\lambda\sqrt{n+1}\delta^h_{SO}a^h_{\lambda n+1} - i\lambda\sqrt{n}\delta^h_{SO}a^h_{\lambda n-1} = \epsilon^h_\lambda a^h_{\lambda n}, \\ \cdots\cdots\cdots\cdots\cdots\cdots\cdots\cdots\cdots\cdots \end{cases} \tag{79}$$

On the other hand, it is clear from the physical considerations that the energy spectrum of charge carriers should be independent on the quasimomentum shift. Here, this shift equals to the position of the dispersion curve extrema $k_y = \pm k^*_{ye}$ for electrons and $k_y = \pm k^*_{yh}$ for holes. The Landau levels spectrum in (quasi)one-dimensional system should also be the same as it in the case when we choose the dispersion curve extrema as the point of reference for quasimomentum:

$$\epsilon^{e,h}_\lambda = n + \frac{1}{2}. \tag{80}$$

The systems of equations for coefficients $a^e_{\lambda n}$ (78) and $a^h_{\lambda n}$ (79) are the recurrence relations for a calculation of these coefficients. So, the dependence on the parameters δ^e_{SO} and δ^h_{SO} in the problem appears only in the wave function $\widetilde{\psi}^{e,h}_\lambda$.

Lastly, our remark concerns a matter about a convergence of the series (77). Each term $|a^{e,h}_{\lambda n}|^2$ contains a summand $\sim 1/(\delta^{e,h}_{SO})^{2n}$, therefore it is necessary $\delta^{e,h}_{SO} > 1$. This inequality should be valid due to the condition (68) because $k^*_{ye,h} \simeq 1/d$ for QW with enough strong asymmetry.

4.7. Possible experiments on the heterostructure

Pseudospin splitting can be observed by means of Raman spectroscopy. The D' peak of interest for the present study (alternatively called $2D$ peak to emphasize that it is due to a two-phonon-assisted process) is located at 2700 cm^{-1} [66]. It arises from intervalley scattering involving phonons with wavenumbers $q > K$, where $K = 4\pi/3\sqrt{3}a \approx 1.7 \times 10^8$ cm^{-1} is the spacing between adjacent K and K' points. One process of this kind is indicated as $A \to B \to C \to D \to A$ in Fig. 16.

Pseudospin splitting enables intervalley scattering involving phonons with $q' \approx q \mp \Delta k$ (with plus for electrons and minus for holes), where $\Delta k = 2k^*_{ye}$ and $\Delta k = 2k^*_{yh}$ in electron and hole scattering, respectively. These processes contribute to a peak blueshifted from D' by $\Delta\omega^{(+)}_R$ and a peak redshifted from D' by $\Delta\omega^{(-)}_R$, giving rise to a doublet structure of the D' peak.

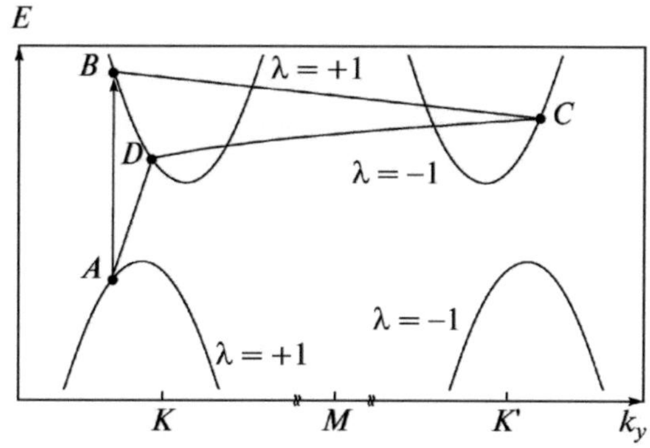

Figure 16. Possible double resonant Raman processes involving electron scattering between valleys. To simplify presentation, analogous processes involving hole scattering between valleys are not shown.

An estimate for $\Delta\omega_R$ can be obtained by using optical phonon dispersion $\omega_{ph}(q)$. The Raman shift is twice the optical phonon frequency:

$$\delta\omega_R(q) = 2\omega_{ph}(q).$$

The change in the Raman shift caused by pseudospin splitting is

$$\Delta\omega_R^{(\pm)} \approx |\delta\omega_R(\Delta K \mp \Delta k) - \delta\omega_R(\Delta K)|,$$

which amounts to $\Delta\omega_R \approx 24$ cm^{-1} for characteristic values of the heterostructure parameters. This value essentially exceeds the Raman spectral resolution of 1 cm^{-1} [67] and compares to the D' peak width for gapless graphene, $\Gamma_0 = 30$ cm^{-1} [68, 69].

Note that a blue shift of the D' peak has also been observed in the Raman spectrum of epitaxial graphene on a SiC substrate [67]. This effect is attributed to the strain induced by the substrate in quasi-free graphene since the SiC lattice constant exceeds substantially that of graphene.

Raman scattering contributions from gapped graphene sheets can be avoided either by using a laser beam whose width is smaller than that of the gapless graphene nanoribbon ($d \lesssim 10$ nm) or by pumping at a frequency ω such that the beam cannot be absorbed by gapped graphene materials,

$$E_g^{eff} + 2\omega_{ph} < \omega < min\{2\Delta_1, 2\Delta_3\}$$

The positions of the luminescence lines corresponding to exciton levels can be determined from optical experiments and compared to theoretical predictions. The splitting of exciton lines in an electric field applied in the plane of the heterostructure along the normal to its boundaries is evaluated by using formula (54).

Finally, interface states can manifest themselves in the I-V curve of the planar heterostructure carrying a current parallel to the gapless graphene nanoribbon. An increase in applied electric field may cause charge carriers to "drop" into interface states (preferable energy-wise), giving rise to a region of negative differential conductivity in the I-V curve.

4.8. Some intermediate conclusions

Before proceeding to the next section, let us recall some results obtained in this section. We think that it is important for understanding the next results.

We have analyzed the characteristics of planar graphene nanostructures. On the one hand, they retain the unique properties of infinite graphene sheets. On the other hand, bandgap opening makes them important building blocks in carbon-based nanoelectronics, which can be used to control electron motion. Parameters of graphene QWs can easily be manipulated by varying the gapless nanoribbon width or the potential barriers in the adjoining gapped graphene sheets.

We predict pseudospin splitting to occur in asymmetric graphene QWs and interface states to arise from the crossing of dispersion curves of gapless and gapped graphene materials. We have performed calculations of optical properties of planar graphene nanostructures and suggested possible experiments to study the effects in question.

Analysis of pseudospin (valley) characteristics in the heterostructure is simplified by using an effective Hamiltonian having a pseudospin-split energy spectrum. Note that an analogous spectrum was discussed in [63–65]. Therefore, the effective Hamiltonian must contain a Rashba-like spin-orbit coupling. We have developed the effective theory for describing graphene-based systems with the pseudospin splitting.

5. Planar graphene superlattices

5.1. Superlattice based on graphene on a strip substrate

5.1.1. Some remarks

Interest in graphene-based superlattices (SL) has increased in recent years. Calculations of graphene-based SL with periodic rows of vacancies were performed using the molecular dynamics method [70]. Calculations of single-atom-thick SL formed by lines of pairs of adsorbed hydrogen atoms on graphene were carried out with the density functional theory [71].

Rippled graphene that can be treated as a SL with the one-dimensional periodic potential of ripples was investigated in [72–74]. SL obtained when a periodic electrostatic potential [75–78] or periodically located magnetic barriers [79–82] were applied to graphene were analytically examined.

However, the investigation of the graphene-based SL with a periodic electrostatic potential disregarded the fact that the application of the electrostatic potential to a gapless semiconductor (graphene) results in the production of electron-hole pairs and the redistribution of charges: electrons move from the region where the top of the valence band lies above the Fermi level to the region where the bottom of the conduction band lies

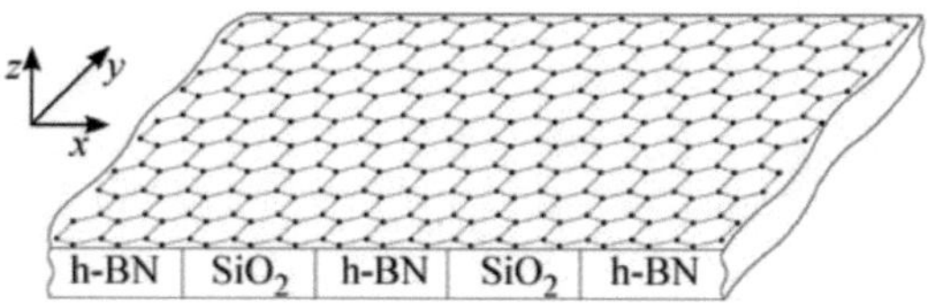

Figure 17. Graphene on the strip substrate consisting of alternating SiO_2 and h-BN strips.

below the Fermi level. The SL becomes a structure consisting of positively charged regions, where the electrostatic potential displacing the Dirac points upward in energy is applied, alternating with negatively charged regions. The strong electrostatic potential of induced charges appears and strongly distorts the initial step electrostatic potential and, therefore, the electronic structure of SL calculated disregarding the electrostatic potential of induced charges.

To avoid the production of electron-hole pairs, SL appearing due to the periodic modulation of the band gap is considered.

SL in the form of the periodic planar heterostructure of graphene nanoribbons between which nanoribbons of h-BN are inserted was previously proposed in [83]. The band structure of such SL was numerically calculated. However, it is very difficult to implement this SL even using the advances of modern lithography, because problems inevitably arise with the control of periodicity in the process of the etching of nanoribbons in a graphene sheet and the insertion of h-BN nanoribbons. Moreover, h-BN is an insulator with a band gap of 5.97 eV, which significantly hinders the tunneling of carriers between graphene nanoribbons. Such a heterostructure is most probably a set of QWs where the wavefunctions of carriers from neighboring QWs almost do not overlap.

Here, SL formed by a graphene sheet deposited on a strip substrate is proposed. The strip substrate is made of periodically alternating strips of SiO_2 (or any other material that does not affect the band structure of graphene) and h-BN, as shown in Fig. 17. The h-BN layers are located so that its hexagonal crystal lattice is under the hexagonal crystal lattice of graphene. Owing to this location, a band gap of 53 meV appears in the band structure of graphene in the graphene-sheet regions under the h-BN layers [37, 84].

It is assumed that all of the contacts between the regions of different band gaps are first-kind contacts (Dirac points of graphene are located in the band gaps of the gap modification of graphene). Such SL is a first-type SL (classification of SL can be found, e.g., in [85]).

The main advantage of the proposed SL is the simplicity of the manufacture and control of its periodicity. It is worth noting that some problems can arise in SL. The difference between the lattice constants of h-BN and graphene is about 2% [37]. If about 100 hexagonal graphene cells are packed into one period of the superlattice, the formation of the band gap in the gap modification of graphene in the graphene sheet regions above h-BN is violated owing to the inaccurate arrangement of carbon atoms above boron or nitrogen atoms. Since contacts between graphene and its gap modification are not heterocontacts (contacts between substances with different chemical compositions), the edges of quantum wells can be insufficiently sharp and QWs cannot be considered as square QWs. A transient

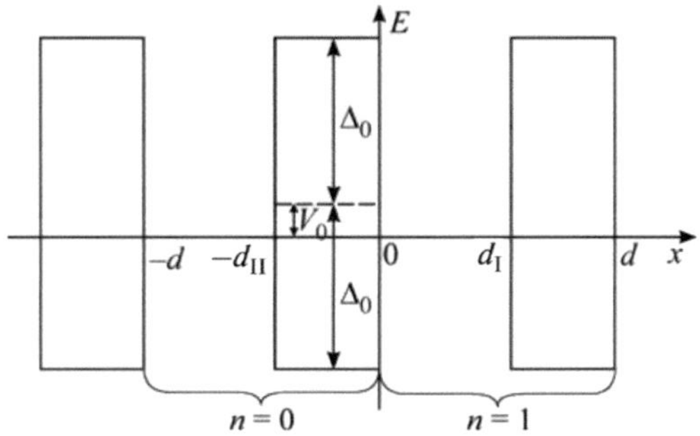

Figure 18. One-dimensional periodic Kronig-Penney potential of SL shown in Fig. 17: the periodically alternating gap modification of graphene on h-BN with a band gap of $2\Delta_0$ = 53 meV and gapless graphene on SiO_2.

layer with a spatially varying band gap can exist instead of the sharp edge. Finally, the substrate can be stressed. The appearing periodic stress field of the substrate can also affect the band structure of the proposed SL, but this effect is very small [86].

5.1.2. The model for describing SL

The x and y axes are perpendicular and parallel to the interfaces of h-BN and SiO_2 strips, respectively (see Fig. 16). SL is described by the Dirac equation

$$(v_F \boldsymbol{\sigma} \widehat{\mathbf{p}} + \Delta \sigma_z + V) \Psi(x y) = E \Psi(x, y), \tag{81}$$

where $v_F \approx 10^8$ cm/s is the Fermi velocity, $\boldsymbol{\sigma} = (\sigma_x, \sigma_y)$ and σ_z are the Pauli matrices, and $\widehat{\mathbf{p}} = -i\boldsymbol{\nabla}$ is the momentum operator (the system of units with $\hbar = 1$ is used). The half-width of the band gap is periodically modulated:

$$\Delta = \begin{cases} 0, & d(n-1) < x < -d_{II} + dn, \\ \Delta_0, & -d_{II} + dn < x < dn, \end{cases}$$

where n is an integer enumerating the supercells of the superlattice; d_I and d_{II} are the widths of the SiO_2 and h-BN strips, respectively; and $d = d_I + d_{II}$ are the period of the superlattice (the size of the supercell along the x axis). The periodic scalar potential V can appear due to the difference between the energy positions of the middle of the band gap of the gap modification of graphene and conic points of the Brillouin zone of gapless graphene (see Fig. 18):

$$V = \begin{cases} 0, & d(n-1) < x < -d_{II} + dn, \\ V_0, & -d_{II} + dn < x < dn. \end{cases}$$

In order for SL to be a first-type superlattice, the inequality $|V_0| \leq 0$ should be satisfied. The solution of Eq. (81) for the first supercell has the form

$$\Psi(x, y) = \psi_1(x)e^{ik_y y}, \qquad 0 < x < d.$$

For the nth supercell, in view of the periodicity of the superlattice,

$$\psi_n(x) = \psi_1(x + (n-1)d).$$

In the region of QW ($0 < x < d_I$), the solution of Eq. (81) is a plane wave

$$\psi_n^{(1)}(x) = N_{k_1}\begin{pmatrix} a_n^{(1)} \\ b_n^{(1)} \end{pmatrix} e^{ik_1 x} + N_{k_1}\begin{pmatrix} c_n^{(1)} \\ d_n^{(1)} \end{pmatrix} e^{-ik_1 x}, \tag{82}$$

where N_{k_1} is the normalization factor. The substitution of Eq. (82) into Eq. (81) provides the relation between the lower and upper spinor components

$$b_n^{(1)} = \lambda_+ a_n^{(1)}, \quad d_n^{(1)} = -\lambda_- c_n^{(1)}, \quad \lambda_\pm = \frac{v_F(k_1 \pm ik_y)}{E}.$$

The relation of E with k_1 and k_y has the form

$$E = \pm v_F\sqrt{k_1^2 + k_y^2}.$$

It is convenient to represent Eq. (82) in a more compact form [76]

$$\begin{aligned} \psi_n^{(1)}(x) &= \Omega_{k_1}(x)\begin{pmatrix} a_n^{(1)} \\ c_n^{(1)} \end{pmatrix}, \\ \Omega_{k_1}(x) &= N_{k_1}\begin{pmatrix} 1 & 1 \\ \lambda_+ & -\lambda_- \end{pmatrix} e^{ik_1 x\sigma_z}. \end{aligned} \tag{83}$$

When the inequality

$$\Delta_0^2 + v_F^2 k_y^2 - (E - V_0)^2 \geq 0 \tag{84}$$

is satisfied, the solution of Eq. (81) in the barrier region ($d_I < x < d$) has the form

$$\begin{aligned} \psi_n^{(2)}(x) &= \Omega_{k_2}(x)\begin{pmatrix} a_n^{(2)} \\ c_n^{(2)} \end{pmatrix}, \\ \Omega_{k_2}(x) &= N_{k_2}\begin{pmatrix} 1 & 1 \\ -\tilde{\lambda}_- & \tilde{\lambda}_+ \end{pmatrix} e^{k_2 x\sigma_z}, \end{aligned} \tag{85}$$

where

$$\widetilde{\lambda}_{\pm} = \frac{iv_F(k_2 \pm k_y)}{E + \Delta_0 - V_0}, \; k_2 = \frac{1}{v_F}\sqrt{\Delta_0^2 + v_F^2 k_y^2 - (E - V_0)^2}.$$

The solution of Eq. (81) in the barrier region under the condition

$$\Delta_0^2 + v_F^2 k_y^2 - (E - V_0)^2 < 0 \tag{86}$$

is given by Eq. (85) with the change $k_2 \to i\varkappa_2$, i.e., it is oscillating.

The possibility of existing *Tamm* minibands formed by localized states near the interface between graphene and its gap modification will be considered below. In this case, $k_1 \to i\varkappa_1$ k_2 is real. A necessary condition for existing Tamm states has the form

$$|k_y| \geq |\varkappa_1|;$$

under this condition, the energy $E = \pm v_F\sqrt{k_y^2 - \varkappa_1^2}$ is real.

5.1.3. The derivation of the dispersion relation

The dispersion relation is derived using the transfer matrix (T matrix) method. The T matrix relates the spinor components for the nth supercell to the spinor components of the solution of the same type for the (n + 1)th supercell. For example, for the solution in the quantum well region,

$$\begin{pmatrix} a_{n+1}^{(1)} \\ c_{n+1}^{(1)} \end{pmatrix} = T\begin{pmatrix} a_n^{(1)} \\ c_n^{(1)} \end{pmatrix}. \tag{87}$$

To determine the T matrix, the following conditions of the continuity of the solution of the Dirac equation describing the considered superlattice are used:

$$\psi_n^{(1)}(d_I - 0) = \psi_n^{(2)}(d_I + 0),$$
$$\psi_n^{(2)}(d - 0) = \psi_{n+1}^{(1)}(+0).$$

These conditions provide the equalities

$$\begin{pmatrix} a_n^{(2)} \\ c_n^{(2)} \end{pmatrix} = \Omega_{k_2}^{-1}(d_I)\Omega_{k_1}(d_I)\begin{pmatrix} a_n^{(1)} \\ c_n^{(1)} \end{pmatrix},$$
$$\begin{pmatrix} a_{n+1}^{(1)} \\ c_{n+1}^{(1)} \end{pmatrix} = \Omega_{k_1}^{-1}(0)\Omega_{k_2}(d)\begin{pmatrix} a_n^{(2)} \\ c_n^{(2)} \end{pmatrix}.$$

According to definition (87) of the T matrix and the last two equalities[6],

$$T = \Omega_{k_1}^{-1}(0)\Omega_{k_2}(d)\Omega_{k_2}^{-1}(d_I)\Omega_{k_1}(d_I). \tag{88}$$

The substitution of Eqs. (84) and (85) with the corresponding arguments into Eq. (88) yields the expressions

$$\begin{aligned} T_{11} &= \alpha e^{ik_1 d_I}\Big[(\lambda_- + \widetilde{\lambda}_+)(\lambda_+ + \widetilde{\lambda}_-)e^{-k_2 d_{II}} - \\ &\quad -(\lambda_- - \widetilde{\lambda}_-)(\lambda_+ - \widetilde{\lambda}_+)e^{k_2 d_{II}}\Big], \\ T_{12} &= 2\alpha e^{-ik_1 d_I}(\lambda_- + \widetilde{\lambda}_+)(\lambda_- - \widetilde{\lambda}_-)sh(k_2 d_{II}), \\ T_{21} &= T_{12}^*, \quad T_{22} = T_{11}^*, \end{aligned} \tag{89}$$

where

$$\alpha = \frac{1}{(\lambda_+ + \lambda_-)(\widetilde{\lambda}_+ + \widetilde{\lambda}_-)}.$$

The last two relations in Eqs. (89) are the general properties of the T matrix.

The derivation of the dispersion relation with the use of the T matrix is briefly as follows.

Let $N = L/d$ be the number of supercells in the entire SL, where L is the length of SL along the x axis, i.e., the direction of the application of the periodic potential. The Born-Karman cyclic boundary conditions for SL have the form

$$\psi_N^{(1,2)}(x) = \psi_1^{(1,2)}(x).$$

At the same time,

$$\psi_N^{(1,2)}(x) = T^N \psi_1^{(1,2)}(x),$$

from which, $T^N = \mathcal{I}$, where $\mathcal{I}$ is the 2×2 unit matrix.

It is convenient to diagonalize the T matrix by means of the transition matrix S:

$$T_d = STS^{-1} = \begin{pmatrix} \lambda_1 & 0 \\ 0 & \lambda_2 \end{pmatrix},$$

where $\lambda_{1,2}$ are the eigenvalues of the T matrix and have the property $\lambda_2 = \lambda_1^*$. According to $T_d^N = \mathcal{I}$

$$\lambda_1 = e^{2\pi i n/N}, \quad -N/2 < n \leq N/2.$$

[6] Note that the cyclic permutations of the factors of Ω matrices are possible in the definition of the T matrix; these permutation do not change dispersion relation (90). This can be verified by comparing Eq. (88) with Eq. (23) in [76].

In view of the property $TrT = TrT_d$ and in terms of the notation $k_x = 2\pi n/L$ $(-\pi/d < k_x \leq \pi/d)$, the dispersion relation is obtained in the form

$$TrT = 2\cos(k_x d). \tag{90}$$

Taking into account the last relation in Eqs. (89), Eq. (90) can also be written in the form

$$ReT_{11} = \cos(k_x d).$$

Dispersion relation (90) under condition (84) gives the equation [86]

$$\frac{v_F^2 k_2^2 - v_F^2 k_1^2 + V_0^2 - \Delta_0^2}{2v_F^2 k_1 k_2} \sinh(k_2 d_{II}) \sin(k_1 d_I) + \cosh(k_2 d_{II}) \cos(k_1 d_I) = \cos(k_x d). \tag{91}$$

According to this equation, the passage to the single-band limit is performed by two methods: first, $V_0 = \Delta_0$ (QW only for electrons) and, second, $V_0 = -\Delta_0$ (QW only for holes). The result of the passage coincides with the known nonrelativistic dispersion relation (see, e.g., [87]), although the expressions for k_1, k_2, and E are different.

If inequality (86) is satisfied, the change $k_2 \to i\varkappa_2$ should be made in Eq. (91)

$$\frac{-v_F^2 \varkappa_2^2 - v_F^2 k_1^2 + V_0^2 - \Delta_0^2}{2v_F^2 k_1 \varkappa_2} \sin(\varkappa_2 d_{II}) \sin(k_1 d_I) + \cos(\varkappa_2 d_{II}) \cos(k_1 d_I) = \cos(k_x d). \tag{92}$$

For Tamm minibands, the change $k_1 \to i\varkappa_1$ should be made in Eq. (91):

$$\frac{v_F^2 k_2^2 + v_F^2 \varkappa_1^2 + V_0^2 - \Delta_0^2}{2v_F^2 \varkappa_1 k_2} \sinh(k_2 d_{II}) \sinh(\varkappa_1 d_I) + \cosh(k_2 d_{II}) \cosh(\varkappa_1 d_I) = \cos(k_x d). \tag{93}$$

Equation (93) has the solution only under the condition

$$v_F^2 k_2^2 + v_F^2 \varkappa_1^2 + V_0^2 - \Delta_0^2 < 0.$$

This condition can be rewritten as

$$v_F^2 k_y^2 - E^2 < -EV_0. \tag{94}$$

At the same time, for Tamm minibands $E^2 = v_F^2 k_y^2 - v_F^2 \varkappa_1^2$, i.e. the left-side of (94) is positive. The allowed values of the energy should be negative if $V_0 > 0$ and vice versa. It is not difficult to show that the inequality (94) has the solutions when [88]

$$v_F^2 k_y^2 < \frac{\Delta_0^2(\Delta_0^2 - V_0^2)}{V_0^2}. \tag{95}$$

Formally, this condition coincides with the condition of the intersection of the dispersion curves for graphene and its gap modification [44].

5.1.4. The results of the numerical calculation

The numerical calculations of the dependence of the energy on k_x were performed for two values $k_y = 0$ and 0.1 nm^{-1} at $V_0 = 0$ (see Fig. 19). The energy of carriers is assumed to be low, $|E| \lesssim 1$ eV, because the Dirac dispersion relation for carriers and, correspondingly, Dirac equation (81) are invalid for high energies.

The electron minibands are separated from hole minibands by a band gap, which increases with $|k_y|$. For $d_I = d_{II}$ at $k_y = 0$, it is $E_g \simeq$ 10–30 meV when d=10–100 nm. In this case, the solution of Eq. (91) is transformed to the solution of Eq. (92). The band gap can increase strongly when d_{II} increases with respect to d_I: $E_g \gtrsim 100$ meV, i.e., is several times larger than $2\Delta_0$.

The width of the minibands decreases with an increase in the period of the superlattice d. The dependence of the width of the minibands on V_0 was also examined. The widths of the electron and hole minibands increase and decrease, respectively, at $V_0 > 0$ and vice versa at $V_0 < 0$.

5.1.5. The possible applications of SL

The described superlattice can be used as FET where the substrate serves as a gate. The ratio of the current through the superlattice to the current through the gate at a substrate thickness of about 10 nm can reach $\sim 10^6$ as for FET based on graphene nanoribbons [11]. The main advantage of the considered superlattice is the absence of the effect of the scattering of carriers on the edges of a nanoribbon on their mobility. The mobility of the carriers in gapless graphene reaches $\mu_0 = 2 \times 10^5$ $\text{cm}^2/(\text{V s})$ [4, 5]. However, the mobility of carriers in FET based on the graphene nanoribbon with a width of $w \sim 3$ nm is three orders of magnitude smaller than μ_0. The cause of such a strong decrease is possibly the scattering of carriers at the edges of graphene nanoribbons. The mean free path between two acts of the scattering of carriers at the edge of the graphene nanoribbon $\lambda_{edge} \propto w/P$, where P is the probability of backscattering [11]. For sufficiently good edges, $P \ll 1$. The problem of scattering on edges is absent for the proposed superlattice; for this reason, the mobility of the carriers in the superlattice is expected to be $\sim \mu_0$ in the absence of the problems with the periodicity of the potential. At the same time, a sufficiently large E_g value, which provides the operation of FET at room temperature, can be reached.

If an Au film is deposited on the lower side of the substrate and graphene is optically pumped, the superlattice can be used as a terahertz laser similar to a terahertz laser based

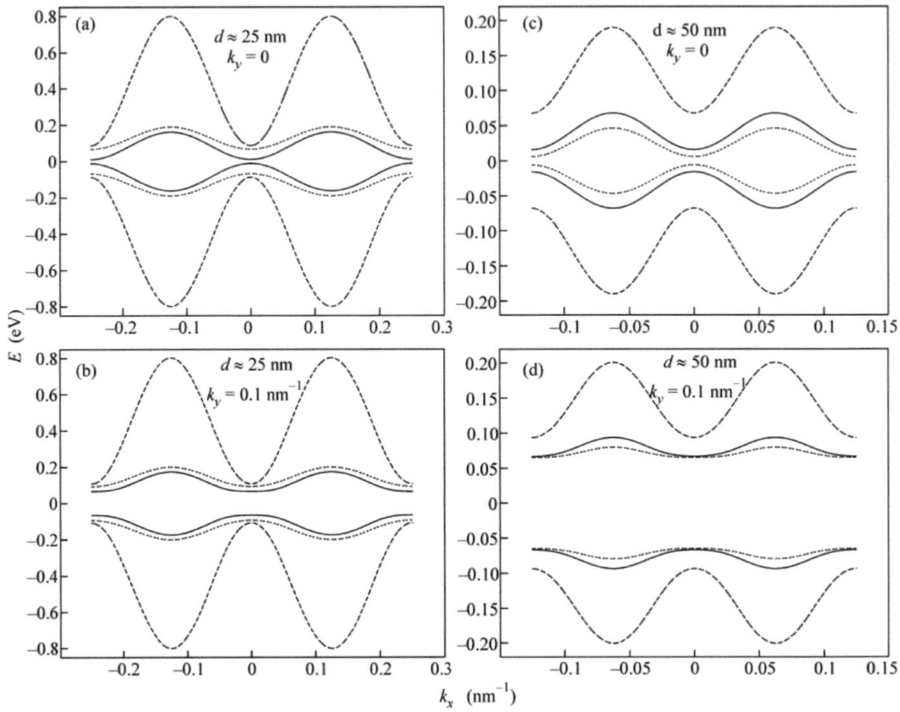

Figure 19. Numerically calculated dependence of the energy on k_x for two k_y values and two superlattice periods d. The dispersion curves for the superlattices with d_I = (solid lines) d_{II}, (dashed lines) $d_{II}/2$, and (dotted lines) $2d_{II}$.

on gapless graphene [89]. In this case, terahertz radiation will be emitted from the regions of the SiO_2 substrate.

5.2. Superlattice based on gapless graphene with the alternating Fermi velocity

5.2.1. Preliminary remarks

Now, we suggest to consider SL based on gapless graphene with alternating regions characterized by different values of the Fermi velocity [90]. In our case, the *Fermi velocity engineering* is based on the usage of the surrounding graphene materials, which have different values of permittivity [91]. It should be pointed out that the idea to control the Coulomb interaction between charge carriers in graphene by the choice of substrate materials with the necessary values of dc permittivity was first put forward in [92].

In such heterostructures, it is possible to achieve the energy quantization for charge carriers even in the absence of potential barriers (regions with wider band gaps) and QWs (regions with narrower band gaps), and even without any variations in the work function [25]. Note that the Tamm minibands are absent here since the straight dispersion lines do not intersect anywhere except for the Dirac point.

Such structure can be produced by the deposition of graphene on striped substrates where either the composition parameter x in an alloy of SiO_{2-x}, or the density of some

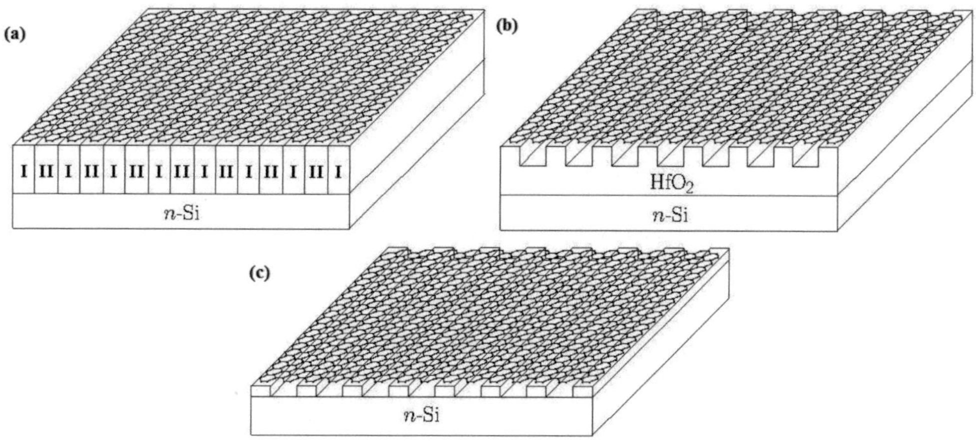

Figure 20. Three variants of SL under study: **(a)** graphene sheet placed on a striped substrate consisting of alternating layers of materials with substantially different values of the permittivity, e.g., SiO_2 with $\varepsilon = 3.9$ (I) and HfO_2 with $\varepsilon = 25$ (II); **(b)** graphene sheet placed on the HfO_2 substrate with periodically arranged grooves; and **(c)** graphene sheet deposited on a periodic array of parallel metallic strips. A plate of heavily doped silicon n-Si is used as a gate.

(nonmagnetic) impurities, or dc permittivity ε exhibit periodic variations. Here, we treat in detail the latter possibility.

According to the results of the theoretical [93–97] and experimental [91, 98–102] studies, the Fermi velocity becomes substantially renormalized. To estimate the renormalized Fermi velocity, we can use the relation [95]

$$\frac{v_F}{v_{F0}} = 1 - 3.28\alpha^* \left[1 + \frac{1}{4}\ln\left(1 + \frac{1}{4\alpha^*} - 1.45\right)\right],$$

where $\alpha^* = \tilde{e}^2/\hbar v_{F0}$ is the analog of the fine structure constant, v_{F0} is the initial unrenormalized Fermi velocity ($v_{F0} = 0.85 \times 10^8$ cm/s) [91, 101], $\tilde{e}^2 = e^2/\varepsilon_{eff}$, and $\varepsilon_{eff} = (\varepsilon_1 + \varepsilon_2)/2$ is the effective dc permittivity for the charge carriers in graphene depending on the values ε_1 and ε_2 of dc permittivity characterizing the materials surrounding graphene. Note that here the band gap is not open; this is confirmed in experiment with an accuracy of 0.1 meV [101].

Within the graphene region located over the strip with the lower value of ε, we have larger α^*. Hence, the corresponding renormalized Fermi velocity should be higher than that over the strip with the higher value of ε. This suggests the possibility of modulating v_F by varying the substrate permittivity. Note that such a system is a one-dimensional photonic crystal.

The first version of the suggested system is a graphene sheet placed on a striped substrate consisting of alternating layers of materials with substantially different values of the permittivity. A schematic image of such a system is shown in Fig. 20a.

It is also possible to use a substrate with periodically arranged grooves prepared by etching. The graphene sheet placed on such substrate should have the periodically alternating regions

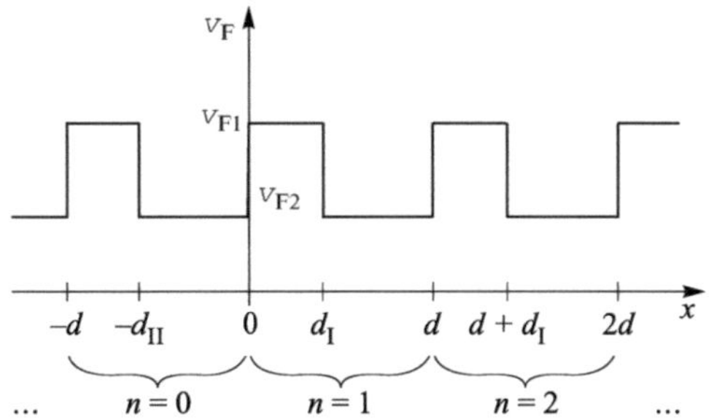

Figure 21. Fermi velocity profile in SL under study ($v_{F1} > v_{F2}$ case). The enumeration of supercells in SL and the sizes of its regions are indicated in the lower part of the figure: d_I is the width of the graphene strip with the Fermi velocity v_{F1}, d_{II} is the width of the graphene strip with the Fermi velocity v_{F2}, and $d = d_I + d_{II}$ is the SL period.

suspended over the grooves and those being in contact with the substrate material (see Fig. 20b). The renormalization of the Fermi velocity should be the most clearly pronounced just in the suspended graphene regions since here we have $\varepsilon_{eff} = 1$. According to the experimental data, the renormalized Fermi velocity in suspended graphene increases to 3×10^8 cm/s [101].

In the regions with graphene in contact with the narrow gap semiconducting material, where $\varepsilon_{eff} \gg 1$, the renormalized Fermi velocity differs only slightly from the unrenormalized one. In addition, the substrate itself is a diffraction grating. Therefore, the system should exhibit rather interesting optical characteristics, demanding a separate study.

There is another version of the system under study. It is possible to deposit graphene on a periodic array of parallel metallic strips (Fig. 20c). This is the limiting case: in the suspended graphene regions, we have $\varepsilon_{eff} = 1$ (the strongest renormalization of the Fermi velocity), whereas in the regions with graphene in contact with metallic strips, we have $\varepsilon_{eff} = \infty$ (vanishing renormalization of the Fermi velocity [91]).

We see that a whole class of such type of systems, which were not studied earlier, is possible. Without doubt, the studies of such systems should lead to important advances in the implementation of the technologies based on the controlled Fermi velocity.

5.2.2. The model

The model for the description of the suggested SL is similar to that used earlier to study SL on the striped substrate with the periodic variation in the band gap [86].

In our case, we assume that the band gap remains unchanged and is equal to zero (gapless graphene) and the work function is the same over all regions of SL (its value is chosen as the energy reference point). We have only a modulation of the Fermi velocity. In gapless graphene, a change in the work function leads to the electrical breakdown and to the creation of electron-hole pairs. We also assume that the near-border region corresponding to the gradual change in the Fermi velocity is much narrower than the SL period. Therefore, the v_F profile can be considered to be sharp enough (see Fig. 21).

We consider the charge carriers located close to the K point of the Brillouin zone (the results should be the same for the charge carriers located in the vicinity of the K' point). Let the x

axis be perpendicular to the strips as is shown in Fig. 21. The envelope of the wave-function $\Psi(x, y)$ for the charge carriers obeys the Dirac-Weyl equation with variable Fermi velocity[7]

$$v_F \boldsymbol{\sigma} \hat{\mathbf{p}} \Psi(x, y) = E\Psi(x, y), \tag{96}$$

$$v_F = \begin{cases} v_{F1}, & d(n-1) < x < -d_{II} + dn \\ v_{F2}, & -d_{II} + dn < x < dn. \end{cases} \tag{97}$$

Here, $\widehat{\mathbf{p}} = -i\boldsymbol{\nabla}$ is the momentum operator (here and further on, $\hbar = 1$). Integers n enumerate supercells (see Fig. 21). The Pauli matrices $\boldsymbol{\sigma} = (\sigma_x, \sigma_y)$ act in the space of two sublattices. The motion of charge carriers in SL along the y axis is free; hence, a solution to Eq. (96) has the form $\Psi(x, y) = \psi(x)e^{ik_y y}$.

Similarly to [86, 90], we find a solution of Eq. (96) with respect to $\psi(x)$ for the nth supercell

(i) at $0 < x < d_I$

$$\psi_n^{(1)}(x) = \Omega_{k_1}(x)\begin{pmatrix} a_n^{(1)} \\ c_n^{(1)} \end{pmatrix},$$

$$\Omega_{k_1}(x) = N_{k_1}\begin{pmatrix} 1 & 1 \\ \lambda_+^{(1)} & -\lambda_-^{(1)} \end{pmatrix} e^{ik_1 x\sigma_z},$$

$$\lambda_\pm^{(1)} = \frac{v_{F1}(k_1 \pm ik_y)}{E}, \; k_1 = \frac{\sqrt{E^2 - v_{F1}^2 k_y^2}}{v_{F1}},$$

(ii) at $d_I < x < d$

$$\psi_n^{(2)}(x) = \Omega_{k_2}(x)\begin{pmatrix} a_n^{(2)} \\ c_n^{(2)} \end{pmatrix},$$

$$\Omega_{k_2}(x) = N_{k_2}\begin{pmatrix} 1 & 1 \\ \lambda_+^{(2)} & -\lambda_-^{(2)} \end{pmatrix} e^{ik_2 x\sigma_z},$$

[7] In the general case, one should write the anticommutator of the Fermi velocity $v_F(x)$ with the term containing the momentum operator $\hat{p}_x$

$$\frac{1}{2}\{v_F(x), \boldsymbol{\sigma}\hat{\mathbf{p}}\}\Psi(x, y) = E\Psi(x, y).$$

Such symmetrization of the Hamiltonian is necessary for retaining its Hermitian form. Similar problems were considered in [103, 104]. In the case of the stepwise profile (97) of the Fermi velocity, we obtain the equation for $Psi(x, y)$ in form (96). This limitation is not significant since allowance for a smooth dependence $v_F(x)$ will complicate the calculations, but will insignificantly change the final results.

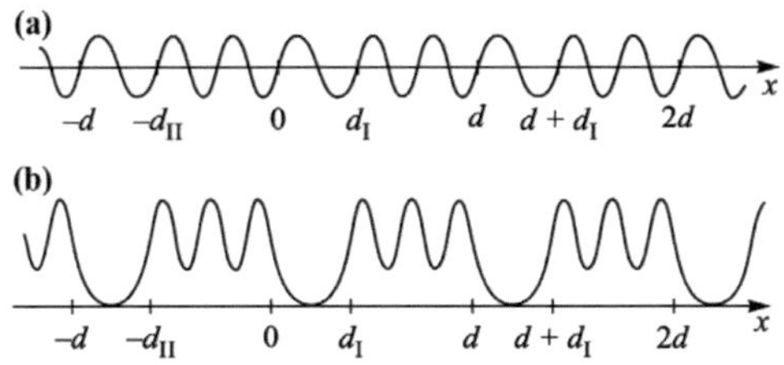

Figure 22. Schematic image illustrating the behavior of the envelope of the wavefunction of charge carriers in SL under study: **(a)** the oscillatory solution in all regions and **(b)** the solution being oscillatory in one region and exhibiting exponential decay deep into another region ($v_{F1} > v_{F2}$ case).

$$\lambda_{\pm}^{(2)} = \frac{v_{F2}(k_2 \pm ik_y)}{E}, \; k_2 = \frac{\sqrt{E^2 - v_{F2}^2 k_y^2}}{v_{F2}}.$$

Here, N_{k_1} and N_{k_2} are the normalization factors.

For the case $v_{F1} > v_{F2}$, the condition for the existence of the solution of Eq. (96), which oscillates in all regions of the SL (it is schematically illustrated in Fig. 22a), is reduced to the inequality

$$k_2^2 > \left(\frac{v_{F1}^2}{v_{F2}^2} - 1 \right) k_y^2. \tag{98}$$

The existence of a *solution of the mixed type* is also possible (see Fig. 22b). In this case, we have an oscillatory solution in some regions (effective QWs), whereas in the other regions, it exhibits exponential decay (effective potential barriers) deep into these regions. The condition for the existence of the mixed type solution is determined by the inequality inverse to (98) and it is met only for finite k_y values.

The effective quantum barrier of the new type is the region with the higher Fermi velocity because the energy of the charge carriers with the same momentum **k** in it is higher than that in the effective QW with the lower Fermi velocity [25]. In contrast to the usual QW, which is formed owing to the change in the width of the band gap, the height of the barrier in SL under study grows with k_y. At k_y=0, the barrier vanishes and our problem is reduced to the *empty lattice model* [105]. In the latter model, the potential is absent, but the periodicity is retained. As a result, energy bands corresponding to the symmetry of the problem arise, but we have zero band gaps.

5.2.3. The dispersion relation

To derive the dispersion relation, we use the transfer matrix (T-matrix) method in the way similar to that employed in [86, 90].

The transfer matrix determines the relation between the coefficients appearing in the expressions for the envelopes of the wavefunctions for the neighboring supercells

$$\begin{pmatrix} a_{n+1}^{(1)} \\ c_{n+1}^{(1)} \end{pmatrix} = T\begin{pmatrix} a_{n}^{(1)} \\ c_{n}^{(1)} \end{pmatrix}, \begin{pmatrix} a_{n+1}^{(2)} \\ c_{n+1}^{(2)} \end{pmatrix} = T\begin{pmatrix} a_{n}^{(2)} \\ c_{n}^{(2)} \end{pmatrix}.$$

We use the following boundary conditions for matching of the envelopes of the wavefunctions [27, 42]

$$\sqrt{v_{F1}}\psi_n^{(1)} = \sqrt{v_{F2}}\psi_n^{(2)}$$

and also the Bloch conditions in the form

$$\psi_n^{(1)}(x+d) = \psi_n^{(1)}(x)e^{ik_xd}$$

and

$$\psi_n^{(2)}(x+d) = \psi_n^{(2)}(x)e^{ik_xd}.$$

Then, the expression for the T-matrix has the form [86, 90]

$$T = \Omega_{k_1}^{-1}(0)\Omega_{k_2}(d)\Omega_{k_2}^{-1}(d_I)\Omega_{k_1}(d_I).$$

The dispersion relation is determined from Eq. (90), which for the oscillatory type solution, can be written as

$$\frac{v_{F1}v_{F2}k_y^2 - E^2}{v_{F1}v_{F2}k_1k_2}\sin(k_1d_I)\sin(k_2d_{II}) + \cos(k_1d_I)\cos(k_2d_{II}) = \cos(k_xd). \tag{99}$$

For the solution of the mixed type, the dispersion relation is found from (99) through the use of the formal substitution $k_1 \to i\kappa_1$, where $\kappa_1 = \frac{1}{v_{F1}}\sqrt{v_{F1}^2k_y^2 - E^2}$.

At $k_y = 0$, transcendental equation (99) has the form

$$\cos(k_1d_I + k_2d_{II}) = \cos(k_xd) \tag{100}$$

for which the exact solution can be found

$$E_l(k_x) = \pm v_F^*\left(k_x + \frac{2\pi l}{d}\right),\ l = 0, 1, 2, \ldots$$

Here, the effective Fermi velocity is introduced as

$$v_F^* = \frac{v_{F1}v_{F2}d}{v_{F1}d_{II} + v_{F2}d_I}. \tag{101}$$

For the lth miniband, the energy at the K point is equal to

$$E_l^0 = \pm\frac{2\pi l v_F^*}{d},\ l = 0, 1, 2, \ldots$$

We can see that the lower electron miniband ($l = 0$) touches the upper hole miniband at the K point and graphene remains gapless.

From Eq. (100), we find that, at the edge of the lth miniband, the energy at $k_x = \pm\pi/d$ is equal to

$$E_l\left(\pm\frac{\pi}{d}\right) = \pm\frac{\pi(2l+1)v_F^*}{d},\ l = 0, 1, 2, \ldots$$

The minibands are separated by the direct band gaps

$$E_G = E_{l+1}\left(\pm\frac{\pi}{d}\right) - E_l\left(\pm\frac{\pi}{d}\right) = \frac{2\pi v_F^*}{d}.$$

In the case of $k_y = 0$, indirect gaps are absent

$$E_l\left(\frac{\pi}{d}\right) = E_{l+1}\left(-\frac{\pi}{d}\right),$$

which corresponds to the empty lattice model [105].

5.2.4. The numerical calculation of the energy spectrum

Let us calculate the lower electron miniband for SL shown in Fig. 19c. According to [101], for it we have $v_{F1} = 3\times 10^6$ cm/s (suspended graphene) and $v_{F2} = 0.85\times 10^6$ cm/s (in the region with the contact of graphene with the metal, the Fermi velocity coincides with v_{F0}).

In the weak coupling model, the problem concerning the edge type at the interface turns out to be unimportant. Let us assume that we have a zigzag-type boundary at the interface (see Fig. 20) and, in each of two regions of the supercell, integer numbers N_I and N_{II} of graphene unit cells are packed up. Then, we have $d_I = 3N_I a$ and $d_{II} = 3N_{II} a$, where $a = 1.42$ Å is the lattice constant of graphene. For calculations, we assume that $N_I = N_{II} = 50$, i.e., $d_I = d_{II} = 21.3$ nm.

In the framework of the suggested model, it is necessary to introduce the upper limit on the wave vector component characterizing the free motion of charge carriers, $|k_y| \ll k_c$. Momentum k_c corresponds to the energy of the ultraviolet cutoff, $\Lambda \approx 3$ eV [101]. As a result, we find $k_c \approx 4.3$ nm^{-1}. This, in turn, imposes the limitation on the SL period, $d \gg a$.

The results of numerical calculations are represented in the form of two $E(k_x, k_y)$ plots for the lower electron miniband: (*i*) $E(k_x)$ at fixed values of k_y (Fig. 23a) and (*ii*) $E(k_y)$ at fixed values of k_x (Fig. 23b). In Fig. 23a, we can see, in particular, that $k_y = 0$ corresponds to the

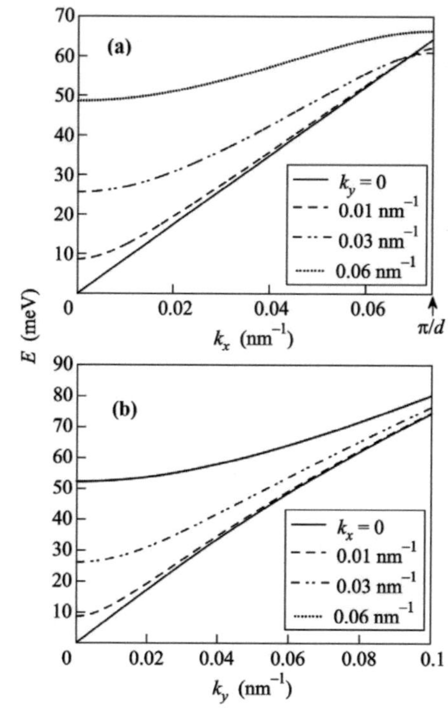

Figure 23. Numerical calculation of the dispersion curves for the lower electron miniband **(a)** versus k_x at fixed k_y values and **(b)** versus k_y at fixed k_x values.

linear dispersion law and the effective Fermi velocity is $v_F^* \approx 1.325 \times 10^8$ cm/s. The lower curve in Fig. 23b exhibits a nearly linear growth. This means that the $E(k_x, k_y)$ surface has the conical shape near the Dirac point.

Thus, we confirm by numerical calculations that at $k_y = 0$, the Fermi velocity of electrons (holes) has a constant value, does not vanish up to the boundaries of minibands, and is determined by Eq. (101) (this is true for all minibands). In this sense, the particles do not feel the boundaries of minibands. Note that, for $k_y \neq 0$, the velocity of particles always vanishes at the miniband boundaries.

5.2.5. *The qualitative analysis of the current-voltage characteristics*

Let us briefly discuss at the qualitative level the effect of the SL potential on the transport phenomena.

Having in mind the aforementioned qualitative difference between the $k_y = 0$ and $k_y \neq 0$ cases, we should expect that the current-voltage characteristics (I–V curves) of SL under study should be significantly different for these two cases.

At $k_y = 0$, the transport characteristics of SL under study should be the same as for effective gapless graphene with the average Fermi velocity v_F^* given by Eq. (101). In particular, at any arbitrarily low charge carrier density, we should observe nonzero minimum conductivity

σ_{min}. According to the experimental data, we have $\sigma_{min} = 4e^2/h$ [2], which coincides with the ballistic conductivity of graphene. The I–V curve should exhibit a linear growth similar to that characteristic of graphene samples with high enough mobility of charge carriers, $\mu \gtrsim 10^4$ cm^2/(V s) [106].

In the case of $k_y \neq 0$, the situation is more complicated. At a nonzero transverse field V_y and at a sufficiently small longitudinal field V_x, the I–V curve should be a growing one and the differential conductivity at small values of V_x is about or higher than the minimum conductivity

$$\sigma_{dif}(V_x \approx 0) \gtrsim \sigma_{min}.$$

Now, we calculate the velocity of electrons for the case of fixed longitudinal ($\mathcal{E}_x$) and nonzero transverse ($\mathcal{E}_y$) electric fields. For the corresponding implementation of such situation in experiment, it is possible to use the standard Hall device.

For simplicity, we assume that transport is ballistic; i.e., the mean free path λ is so large that an electron accelerated by the applied electric field can reach the miniband boundary without any scattering. To distinguish the spectrum related to the potential of the superlattice, the mean free path should be much larger than the period of SL [85]

$$\lambda_f \gg d. \tag{102}$$

For the sufficiently pure graphene samples, we have $\lambda_f \simeq 1\mu$m.

The direction of the electron motion is characterized by the polar angle $\phi = \arctan(k_y/k_x)$. Its value remains unchanged in the whole $-\pi/d \leq k_x \leq \pi/d$ range. The contribution to the conductivity related to the intraminiband transitions is determined by the electron velocity, which we seek:

$$v_\phi = \left.\frac{\partial E}{\partial k}\right|_{k_y = k_x \tan\phi}.$$

In Fig. 24, we illustrate the calculated dependence of the electron velocity on k_x for the same SL parameters as above for the polar angles $\phi = 5^o, 10^o$, and 15°. We can see that the velocity indeed vanishes at the miniband boundary and its abrupt decrease takes place within a quite narrow range near the miniband boundary. For low momenta, we have $v_\phi \approx v_F^*$.

An application of the superlattice at nonzero temperatures requires the existence of a quite clearly pronounced Fermi velocity profile; i.e., we should use rather large ϕ and $\delta v_F = |v_{F1} - v_{F2}|$ values:

$$\pi \frac{\delta v_F}{d} \sin\phi \gg T.$$

However, at large ϕ values close to $\pi/2$, the condition according to which charge carriers pass a large number of supercells at the mean free path can be violated. Then, condition (102) turns out to be unimportant: condition $\lambda_f \cos\phi \gg d$ should be met.

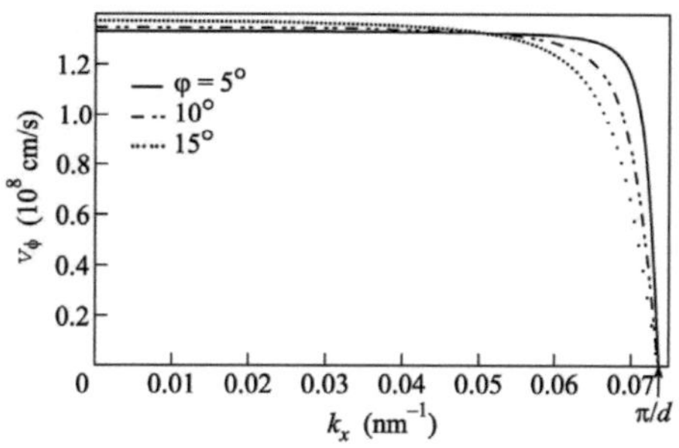

Figure 24. Numerical calculation of the electron velocity in the lower miniband along the direction specified by the fixed polar angle ϕ.

Similarly to the situation occurring in semiconductor SLs, the motion of charge carriers at sufficiently strong electric field $\mathcal{E}_x$ is finite. They oscillate with the Stark frequency [85]

$$\Omega = e\mathcal{E}_x d.$$

This stems from the nonlinearity of the I–V curve manifesting itself in the negative differential conductivity at a certain section of it. Charge carriers in the nonlinear regime undergo a large number of the Bloch oscillations during the mean free time τ:

$$\Omega\tau \gg 1. \tag{103}$$

We estimate the mean free time as $\tau \approx \lambda_f / v_F^*$ (the velocity of charge carriers is $v_\phi \approx v_F^*$ everywhere except for a narrow range near the miniband boundaries). Then, condition (103) can be rewritten as

$$\mathcal{E}_x \gg \frac{v_F^*}{ed\lambda_f}. \tag{104}$$

Condition (104) automatically gives an estimate for the minimum longitudinal voltage above which negative differential conductivity becomes possible

$$V_{x\,\min} \simeq \frac{v_F^*}{ed}\frac{L_x}{\lambda_f},$$

where L_x is the size of the system along the x axis. Assuming that $L_x \simeq \lambda_f$, we arrive at the estimate $V_{x\,min} \simeq 0.02$ V for SL with the same parameters as above.

In Fig. 25, we represent the qualitative behavior of the I–V curve for SL under study. At $k_y = 0$ (zero applied voltage in the transverse direction, $V_y = 0$), we observe its linear growth. At $k_y \neq 0$ (nonzero transverse voltage, $V_y \neq 0$), a section with negative differential conductivity arises in the curve. In this case, for higher V_y values, this section is more pronounced and more shifted toward lower V_x values. However, as is mentioned above, this section can arise only at a sufficiently high longitudinal voltage

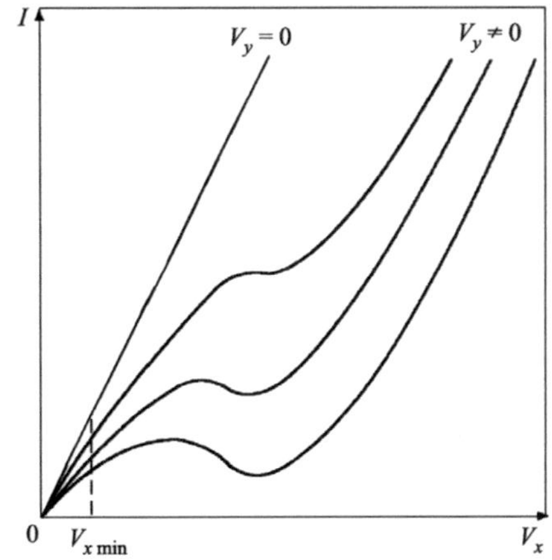

Figure 25. Qualitative behavior of the I–V curve for SL under study. Three $I(V_x)$ plots under the linear I–V curve correspond to the growth of the transverse voltage V_y (from top to bottom).

$$V_x \gg V_{x\,\mathrm{min}}.$$

Note finally that the characteristics of the system under study can depend on the gate voltage V_g (at different values of the charge carrier density n_{2D}) owing to the dependence of the renormalized Fermi velocity on n_{2D} [101, 102]. In this case, a controlling factor is the filling of minibands with electrons (holes). For the experimental observations, it is convenient to have partially filled either the lower electronic miniband or the upper hole one (in this case, the higher electronic or lower hole minibands are distinguishable). This takes place if $n_{2D} \ll n^*_{2D} = 4/d^2$. This condition can be rewritten in the form of a limitation imposed on the gate voltage [90]

$$|V_g| \ll 4\pi e n^*_{2D} L_g/\varepsilon^*_s,$$

where L_g is the gate thickness and ε^*_s is the effective dc permittivity of the substrate. For the layered substrate structure (see Fig. 19a), we have

$$\varepsilon^*_s = \frac{\varepsilon_{s1} d_I + \varepsilon_{s2} d_{II}}{d}.$$

6. Conclusions

We presented the method for the theoretical research of the electron properties of the planar heterostructures based on graphene, namely, the single heterojunctions, QWs, and SLs. The usage of the gap modifications of graphene in the planar heterostructures is a novel idea which can help to push the boundaries of science.

The valley-polarized currents must exist at the single heterojunctions. The novel phenomena of pseudospin splitting in the energy spectrum of the asymmetric QWs had been theoretically predicted. Some optical properties of the graphene-based QWs was considered (observation of the excitonic lines).

A model describing SL based on graphene on a strip substrate has been proposed. The dispersion relation has been derived, which is transferred to the known nonrelativistic dispersion relation in the passage to the single-band limit. The numerical calculations have been performed for a pair of the nearest electron and hole minibands using the derived dispersion relation. Possible applications of SL as a transistor or a terahertz laser have been pointed out.

We suggested a novel class of graphene-based systems, which are at the same time both photon crystals and graphene SLs with periodically varying Fermi velocity. Such a modulation appears to be possible owing to the renormalization of the Fermi velocity in the energy spectrum of graphene. New prospects become open for the implementation of the technologies based on controlled Fermi velocity. We point out some specific features of the transport phenomena in such systems, in particular, appearance of the sections with negative differential conductivity in the I–V curves. It is clear that, similarly to photon crystals, these systems should exhibit interesting optical characteristics.

Author details

Pavel V. Ratnikov* and Andrei P. Silin

*Address all correspondance to: ratnikov@lpi.ru

Lebedev Physical Institute, Russian Academy of Sciences, Moscow, Russia

References

[1] K.S. Novoselov, A.K. Geim et al., Science **306**, 666 (2004).

[2] K.S. Novoselov, A.K. Geim et al., Nature **438**, 197 (2005).

[3] Y. Zhang, Y.-W. Tan, H. L. Stormer, and P. Kim, Nature **438**, 201 (2005).

[4] S.V. Morozov, K.S. Novoselov et al., Phys. Rev. Lett. **100**, 016602 (2008).

[5] X. Du, I. Skachko, A. Barker, and E.Y. Andrei, Nat. Nanotech. **3**, 491 (2008).

[6] L. Brey and H.A. Fertig, Phys. Rev. B **73**, 235411 (2006).

[7] L. Brey and H.A. Fertig, Phys. Rev. B **75**, 125434 (2007).

[8] Y.-W. Son, M.L. Cohen, and S.G. Louie, Phys. Rev. Lett. **97**, 216803 (2006).

[9] R. Saito, G. Dresselhaus, and M.S. Dresselhaus, *Physical Properties of Carbon Nanotubes* (Imperial College Press, London, 1998).

[10] T. Ando, J. Phys. Soc. Jpn. **74**, 777 (2005).

[11] X. Wang, Y. Ouyang et al., Phys. Rev. Lett. **100**, 206803 (2008).

[12] L.A. Ponomarenko, F. Schedin et al., Science **320**, 356 (2008).

[13] J. Y. Tan, A. Avsar et al., Appl. Phys. Lett. **104**, 183504 (2014).

[14] Y.-C. Lin, C.-Y.S. Chang et al., Nano Lett. **14**, 6936 (2014).

[15] Y.F. Lin, W. Li et al., Nanoscale **6**, 795 (2014).

[16] D. Logoteta, G. Fiori, and G. Iannaccone, Scientific Rep. **4**, 6607 (2014).

[17] B.G. Idlis and M.Sh. Usmanov, Sov. Phys. Semicond. **26** (2), 186 (1992).

[18] A. I. Akhiezer and V. B. Berestetskii, *Quantum Electrodynamics* (Wiley, New York, 1965; Nauka, Moscow, 1969).

[19] A.H. Castro Neto, F. Guinea et al., Rev. Mod. Phys. **81**, 109 (2009).

[20] Yu.E. Lozovik, S.P. Merkulova, and A.A. Sokolik, Phys.–Usp. **51** (7), 727 (2008).

[21] S.S. Schweber, *Introduction to Relativistic Quantum Field Theory* (Halper and Row, New York, 1961).

[22] A.M. Tsvelik, *Quantum Field Theory in Condensed Matter Physics* (Cambridge University Press, 1998).

[23] T.W. Appelquist, M. Bowick, D. Karabali, and L.C.R. Wijewardhana, Phys. Rev. D **33**, 3704 (1986).

[24] B.A. Volkov, B.G. Idlis, and M.Sh. Usmanov, Phys.–Usp. 38 (7), 761 (1995).

[25] A.V. Kolesnikov and A.P. Silin, JETP **82**, 1145 (1996).

[26] A.V. Kolesnikov and A.P. Silin, J. Phys.: Condens. Matter **9**, 10929 (1997).

[27] A.P. Silin and S.V. Shubenkov, Phys. Solid State **40** (7), 1223 (1998).

[28] E.A. Andryushin, Sh.U. Nutsalov, and A.P. Silin, Phys. Low-Dim. Struct. **7/8**, 85 (1999).

[29] E.A. Andryushin, S.A. Vereshchagin, and A.P. Silin, Kratk. Soobshch. Fiz., No. 6, 21 (1999).

[30] E.A. Andryushin, A.P. Silin, and S.A. Vereshchagin, Phys. Low-Dim. Struct. **3/4**, 85 (2000).

[31] A.P. Silin and S.A. Vereshchagin, Phys. Low-Dim. Struct. **9/10**, 115 (2001).

[32] E.A. Andryushin, Sh.U. Nutsalov, and A.P. Silin, Kratk. Soobshch. Fiz., No. 3, 3 (2001).

[33] P.V. Ratnikov and A.P. Silin, Kratk. Soobshch. Fiz., No. 11, 22 (2005).

[34] G.W. Semenoff, Phys. Rev. Lett. **53**, 2449 (1984).

[35] D.A. Abanin, P.A. Lee, and L.S. Levitov, Phys. Rev. Lett. **96**, 176803 (2006).

[36] P.V. Ratnikov and A.P. Silin, Phys. Solid State **52** (8), 1763 (2010).

[37] G. Giovannetti, P.A. Khomyakov et al., Phys. Rev. B **76**, 073103 (2007).

[38] A. Mattausch and O. Pankratov. Phys. Rev. Lett. **99**, 076802 (2007).

[39] S.Y. Zhou, G.-H. Gweon et al., Nature Mater. **6**, 770 (2007).

[40] D.C. Elias, R.R. Nair et al., Science **323**, 610 (2009).

[41] S. Lebègue, M. Klintenberg, O. Eriksson, M.I. Katsnelson. Phys. Rev. B **79**, 245117 (2009).

[42] P.V. Ratnikov and A.P. Silin, JETP **114**, 511 (2012).

[43] G. Tkachov. Phys. Rev. B **79**, 045429 (2009).

[44] A.V. Kolesnikov, R. Lipperheide, A.P. Silin, U. Wille. Europhys. Lett. **43**, 331 (1998).

[45] Y. Kobayashi, K.I. Fukui et al., Phys. Rev. B **71**, 193406 (2005).

[46] Y. Niimi, T. Matsui, et al., Phys. Rev. B **73**, 085421 (2006).

[47] S.Y. Zhou, G.-H. Gweon et al., Nature Phys. **2**, 595 (2006).

[48] I. Zanella, S. Guerini et al., Phys. Rev. B **77**, 073404 (2008).

[49] S. Marchini, S. Günther, and J. Witterlin, Phys. Rev. B **76**, 075429 (2007).

[50] D. Martoccia, P.R. Willmon et al., Phys. Rev. Lett. **101**, 126102 (2008).

[51] I. Pletikosić, M. Kralj et al., Phys. Rev. Lett. **102**, 056808 (2009).

[52] M.Y. Han, B. Özyilmaz, Y. Zhang, and P. Kim, Phys. Rev. Lett. **98**, 206805 (2007).

[53] T.M. Rice, J.C. Hensel, T. Phillips, and G.A. Thomas, *The Electron-Hole Liquid in Semiconductors: Theoretical Aspects* (Academic, New York, 1977; Mir, Moscow, 1980).

[54] A.A. Abrikosov, J. Low Temp. Phys. **2**, 37 (1970).

[55] S.A. Brazovskii, Sov. Phys. JETP **35** , 433 (1972).

[56] T. Ando, J. Phys. Soc. Jpn. **66**, 1066 (1997).

[57] V.S. Babichenko, L.V. Keldysh, and A. P. Silin, Sov. Phys. Solid State **22** (4), 723 (1980).

[58] Yu.E. Lozovik and V.I. Yudson, Phys. Lett. **56A**, 393 (1976).

[59] L.V. Keldysh, JETP Lett. **29** (11), 658 (1979).

[60] R. Loudon, Am. J. Phys. **27**, 649 (1959).

[61] A. Dalgarno and J.T. Lewis, Proc. R. Soc. A **233**, 70 (1955).

[62] D.A.B. Miller, D.C. Chemla et al., Phys. Rev. Lett. **53**, 2173 (1984).

[63] E.I. Rashba, Sov. Phys. — Solid State **2**, 1109 (1960).

[64] Yu.A. Bychkov and E.I. Rashba, JETP Lett. **39** (2), 78 (1984).

[65] Yu.A. Bychkov and E.I. Rashba, J. Phys. C: Solid State Phys. **17**, 6039 (1984).

[66] D. Graf, F. Molitor et al., Nano Lett. **7**, 238 (2007).

[67] Z.H. Ni, W. Chen et al., Phys. Rev. B **77**, 115416 (2008).

[68] A.C. Ferrari, J.C. Meyer et al., Phys. Rev. Lett. **97**, 187401 (2006).

[69] I. Calizo, A.A. Balandin et al., Nano Lett. **7**, 2645 (2007).

[70] L.A. Chernozatonskii, P.B. Sorokin et al., JETP Lett. **84**, 115 (2006).

[71] L. A. Chernozatonskii, P. B. Sorokin et al., JETP Lett. **85**, 77 (2007).

[72] A. Isacsson, L.M. Jonsson, J.M. Kinaret, and M. Jonson, Phys. Rev. B **77**, 035423 (2008).

[73] F. Guinea, M.I. Katsnelson, and M.A.H. Vozmediano, Phys. Rev. B **77**, 075422 (2008).

[74] T.O. Wehling, A.V. Balatsky, M.I. Katsnelson, and A.I. Lichtenstein, Europhys. Lett. **84**, 17003 (2008).

[75] C. Bai and X. Zhang, Phys. Rev. B **76**, 075430 (2007).

[76] M. Barbier, F.M. Peeters, P. Vasilopoulos, and J.M. Pereira, Phys. Rev. B **77**, 115446 (2008).

[77] C.-H. Park, L. Yang et al., Phys. Rev. Lett. **101**, 126804 (2008).

[78] C.-H. Park, Y.-W. Son et al., Nano Lett. **8**, 2920 (2008).

[79] M.R. Masir, P. Vasilopoulos, A. Matulis, and F.M. Peeters, Phys. Rev. B **77**, 235443 (2008).

[80] M.R. Masir, P. Vasilopoulos, and F.M. Peeters, Phys. Rev. B **79**, 035409 (2009).

[81] L. Dell'Anna and A. De Martino, Phys. Rev. B **79**, 045420 (2009).

[82] S. Ghosh and M. Sharma, J. Phys. Condens. Matter **21**, 292204 (2009).

[83] H. Sevinçli, M. Topsakal, and S. Ciraci, Phys. Rev. B **78**, 245402 (2008).

[84] S.Y. Zhou, G.-H. Gweon et al., Nature Mater **6**, 770 (2007).

[85] A. P. Silin, Sov. Phys. Usp. **28**, 972 (1985).

[86] P.V. Ratnikov, JETP Lett. **90**, 469 (2009).

[87] M.A. Herman, *Semiconductor Superlattices* (Academy, Berlin, 1986; Mir, Moscow, 1989).

[88] G.M. Maksimova, E.S. Azarova, A.V. Telezhnikov, and V.A. Burdov, Phys. Rev. B **86**, 205422 (2012).

[89] V.Ya. Aleshkin, A.A. Dubinov, and V. Ryzhii, JETP Lett. **89**, 63 (2009).

[90] P.V. Ratnikov and A.P. Silin, JETP Lett. **100**, 311 (2014).

[91] C. Hwang, D.A. Siegel et al., Sci. Rep. **2**, 590 (2012).

[92] P.V. Ratnikov, JETP Lett. **87**, 292 (2008).

[93] J. Gonzalez, F. Guinea, and M.A.H. Vozmediano, Nucl. Phys. B **424**, 595 (1994).

[94] J. Gonzalez, F. Guinea, and M.A.H. Vozmediano, Phys. Rev. B **59**, 2474 (1999).

[95] S. Das Sarma, E.H. Hwang, and W.-K. Tse, Phys. Rev. B **75**, 121406(R) (2007).

[96] M.S. Foster and I.L. Aleiner, Phys. Rev. B **77**, 195413 (2008).

[97] F. de Juan, A.G. Grushin, and M.A.H. Vozmediano,Phys. Rev. B **82**, 125409 (2010).

[98] A. Bostwick, T. Ohta et al., Solid State Commun. **143**, 63 (2007).

[99] Z.Q. Li, E.A. Henriksen et al., Nature Phys. **4**, 532 (2008).

[100] G. Li, A. Luican, and E.Y. Andrei, Phys. Rev. Lett. **102**, 176804 (2009).

[101] D.C. Elias, R.V. Gorbachev et al., Nature Phys. **7**, 701 (2011).

[102] J. Chae, S. Jung et al., Phys. Rev. Lett. **109**, 116802 (2012).

[103] M.R. Geller and W. Kohn, Phys. Rev. Lett. **20**, 3103 (1993).

[104] A.V. Kolesnikov and A.P. Silin, Phys. Rev. B **95**, 7596 (1999).

[105] J. Callaway, *Energy Band Theory* (Academic, New York, 1964; Mir, Moscow, 1969).

[106] N. Vandecasteele, A. Barreiro et al., Phys. Rev. B **82**, 045416 (2010).